Edward Charles Spitzka

Insanity

Its Classification, Diagnosis and Treatment

Edward Charles Spitzka

Insanity

Its Classification, Diagnosis and Treatment

ISBN/EAN: 9783337778866

Printed in Europe, USA, Canada, Australia, Japan

Cover: Foto ©berggeist007 / pixelio.de

More available books at **www.hansebooks.com**

INSANITY

ITS CLASSIFICATION, DIAGNOSIS AND TREATMENT

A MANUAL FOR

STUDENTS AND PRACTITIONERS OF MEDICINE

By E. C. SPITZKA, M.D.

PRESIDENT OF THE NEW YORK NEUROLOGICAL SOCIETY; FORMERLY PHYSICIAN TO THE DEPARTMENT OF NERVOUS DISEASES OF THE METROPOLITAN THROAT HOSPITAL; CONSULTING NEUROLOGIST OF THE NORTH EASTERN DISPENSARY; NEUROLOGIST TO THE GERMAN POLIKLINIK; W. & S. TUKE, PRIZE ESSAYIST, ETC.

NEW YORK:
E. B. TREAT, 771 BROADWAY.
1887.   Price, $2.75.

PREFACE TO THE SECOND EDITION.

In presenting this edition to the medical public, its author has nothing to add to the preface of the preceding one, beyond an acknowledgment of the gratifying encouragement and valuable suggestions which he has received at the hands of distinguished American and English alienists. If not all of these suggestions have been utilized in the present volume, it is because the limits of a manual intended for students and general practitioners did not permit it. In a larger treatise on a kindred subject in preparation, they will be found incorporated and acknowledged. As far as alterations have been made in the present edition, the author is under special obligations to R. H. Gundry of the Maryland Asylum, and S. V. Clevenger of the Cook County Asylum, Illinois, and Frank H. Hamilton as well as N. E. Brill of New York, the latter of whom assumed the task of seeing the work through the press during the author's disability.

The first edition was liberally patronized by the legal profession, and frequently used as a book of reference in medico-legal cases. The temptation to lay more stress on the medico-legal aspect of insanity was consequently great. But any attempt to reconcile the medical with the legal view of mental disorder, under the existing system, with its varying and inherently inconsistent definitions, is doomed to failure. As the author has always held the province of the physician in medico-legal cases to be that of an adviser and not of an advocate, so he believes that a treatise on mental disease ceases to be a medical work as soon as it enters the domain of jurisprudence in its present attitude.

Regarding those views expressed in the first edition which were to some extent innovations in English psychological literature, or the subject of dispute, it may be proper to define the author's present position. The discrimination between "systemized" and "non-systemized delusions" has been generally adopted and is therefore retained. "Monomania," repudiated by controversialists, who took advantage of the unfortunate character of that term—but never abandoned by the great body of alienists, has received substantial endorsement at the hands of Burr of Pontiac, Clouston of Morningside, Granger of Buffalo, and the International Committee on Classification. The alternative designation of Paranoia has been adopted in this volume as a preferable term, for reasons similar to those which actuated Mendel in his recent proposition to substitute it for the German vernacular term "Verrücktheit."

The statements regarding the existence of "moral imbecility" have received such potent support by the recent contributions of D. Hack Tuke, Workman, and Hughes, not to mention a host of continental authors, that it ceased to become necessary to longer regard this disorder as a subject of controversy. With regard to the once equally doubted psychosis designated as transitory frenzy, confirmatory evidence has come from a number of sources, one of them being the Utica Asylum, where Brush reported a typical case. Indeed some of the very strongest evidence in favor of the scientific reality of Paranoia, moral imbecility, and the Hereditary Transmission of Insanity, has been recently furnished by the physicians of that asylum, which was with some reason regarded as occupying the isolated position of antagonism to this recognition, at the time the first edition of this work was offered to the profession. If the mite which the author contributed with the object of facilitating the spread of views which are shared by the leaders of psychiatric thought on both sides of the Atlantic, had any part in effecting this change, his expectations have been realized in a peculiarly gratifying manner.

NEW YORK, 712 Lexington Ave.,
September 18th, 1886.

CONTENTS.

PART FIRST.

THE GENERAL CHARACTERS AND THE CLASSIFICATION OF INSANITY.

CHAPTER I.
                                                              PAGE
The Definition of Insanity..................................... 17

CHAPTER II.
The Delusions of the Insane.................................... 23

CHAPTER III.
Imperative Conceptions and Morbid Propensities................. 35

CHAPTER IV.
Hallucinations and Illusions................................... 43

CHAPTER V.
The Emotional Disturbances..................................... 54

CHAPTER VI.
The Memory and Consciousness in Insanity....................... 57

CHAPTER VII.
The Will in Insanity... 64

CHAPTER VIII.
The Physical Indications of Acquired Insanity.................. 65

CHAPTER IX.

Somatic Signs of Insanity Indicating the Existence of a Constitutional Taint of, or a Predisposition to Insanity.............. 81

CHAPTER X.

The Morbid Anatomy of Insanity..................... 92

CHAPTER XI.

The Classification of Insanity.................................. 113

PART SECOND.

THE SPECIAL FORMS OF INSANITY.

CHAPTER I.
Mania... 131

CHAPTER II.
Melancholia... 140

CHAPTER III.
Katatonia... 149

CHAPTER IV.
Transitory Frenzy.................................. 154

CHAPTER V.
Stuporous Insanity.................................. 158

CHAPTER VI.
Primary Confusional Insanity........................ 161

CHAPTER VII.
Primary Mental Deterioration........................ 163

CHAPTER VIII.
The Secondary and Terminal Deteriorations........... 166

CONTENTS. 13

CHAPTER IX.
                                                                PAGE
Senile Dementia... 171

CHAPTER X.
Insanity of Pubescence... 175

CHAPTER XI.
Paretic Dementia—Preliminary Considerations.................... 178

CHAPTER XII.
Paretic Dementia—Course and Symptoms........................... 184

CHAPTER XIII.
Paretic Dementia—Morbid Anatomy and Theory of the Disease..... 218

CHAPTER XIV.
Syphilitic Dementia... 243

CHAPTER XV.
Delirium Grave.. 247

CHAPTER XVI.
Alcoholic Insanity.. 251

CHAPTER XVII.
Hysterical Insanity... 256

CHAPTER XVIII.
Epileptic Insanity.. 258

CHAPTER XIX.
Periodical Insanity... 267

CHAPTER XX.
The States of Arrested Development............................ 275

CHAPTER XXI.
Paranoia (Monomania) Preliminary Considerations............... 286

CHAPTER XXII.
Paranoia—Its Course and Varieties............................. 301

PART THIRD.

INSANITY IN ITS PRACTICAL RELATIONS.

CHAPTER I.
PAGE
How to Examine the Insane.... 320

CHAPTER II.
The Differential Diagnosis of the Forms of Insanity 330

CHAPTER III.
The Recognition of Simulation 352

CHAPTER IV.
The Physical Causes of Insanity................................. 369

CHAPTER V.
The Psychical Causes of Insanity................................ 381

CHAPTER VI.
The Medicinal and Dietetic Treatment of Insanity 384

CHAPTER VII.
The Psychical Treatment and Management of the Insane........... 397

APPENDIX.

INSANITY,

Its Classification, Diagnosis and Treatment.

PART I.

THE GENERAL CHARACTERS AND THE CLASSIFICATION OF INSANITY.

CHAPTER I.

THE DEFINITION OF INSANITY.

Insanity is a term applied to certain results of brain disease and brain defect which invalidate mental integrity. It is inaccurate to state that insanity is itself a disease. It is, strictly speaking, merely a symptom which may be due to many different morbid conditions, having this one feature in common: that they involve the organ of the mind.

From this point of view the term insanity is used in a sense analogous to that in which "aphasia" is employed. The aphasic symptom-group comprises derangements of the function of articulate speech, which may be manifestations of a cerebral defect, of gross cerebral disease, and of disturbance in the cerebral circulation. These conditions differing so widely amongst themselves in their pathological character must, in order to produce aphasia, agree in one essential respect: their regional distribution—that is, they must affect the central speech organ, the speech centres, and the speech tracts in a similar way. While the point at, and the extent to which the cerebral mechanism is disturbed by a lesion, determines the existence of this prominent *symptom*, it is not the location but the intrinsic nature of the lesion be it a hemorrhage, a tumor, or a nutritive interference, that determines the name of the patient's *disease*.

Theoretically it should be the same with insanity. The disease manifesting mental derangement may be an inflammation of the brain and its membranes, or a brain injury—a brain wasting, or an original brain-defect—a mal-nutrition, or a reflex disturbance of that organ; and in such conditions must we seek for the groundwork of a scientific nomenclature and classification of the morbid states underlying mental derangement. Unfortunately, sufficient positive observations on which to base a thorough pathology of these diseased states are, as yet, desiderata. The scalpel, the test-tube, the microscope, the sphygmograph, the ophthalmoscope, and delicate electrical apparatus have been diligently employed by able and earnest investigators; but these appliances and the most elaborate methods of the laboratory have more frequently failed than succeeded in revealing the organic conditions responsible for mental disturbance.

It is for this reason that while the ideal aim of the progressive alienist will continue to be the solution of the great problem of the physical foundation of insanity, he is commonly limited when he makes a diagnosis to a recognition of disease manifestations, and not of disease; when he classifies, to a classification of symptom groups; and when he treats insanity, to the treatment of symptomatic states. The conceptions and processes of mental science are hence traditional and empirical to a great extent. In some degree this is also due to the fact that the conventional discrimination between insanity and other symptoms of disease of the brain, however practically necessary, is from a scientific point of view both arbitrary and artificial. This becomes most evident when we attempt to define insanity.

It may be safely asserted that, in the present state of our knowledge, it is impossible to frame a definition of insanity which, while it meets the practical every-day requirements, is constructed on *scientific* principles. The failure of the best authorities to furnish such a one proves that until the material elements of mental derangement become more accessible to observation than they now are, scientific definitions must in large part rest on hypotheses. The *practical* need, however, is for a definition which shall include neither ambiguous nor theoretical terms. That the brain is the organ of the mind is an axiom of physiology, that insanity is a manifestation of a brain disorder is a resulting dogma of medical psychology; but even if we could establish the existence of a brain-lesion in every case of brain disturbance, we would not be able to formulate the topographical and patho-histological conditions which determine the falling of its manifestations within the boundaries of insanity in one case, and without them in another. And neither the axioms of physiology nor the dogmas of medical psychology are regarded with sufficient respect in our courts of law—where the problem of an accurate definition of insanity is apt to be most emphatically presented to the medical mind—to render their use in filling this gap in our knowledge either satisfactory or profitable.*

The following appears to the writer to comply with the chief requirements of a practical definition, although it labors under the disadvantage of length:

Insanity is either the inability of the individual to correctly register and reproduce impressions (and conceptions based on these) in sufficient number and intensity to serve as guides to actions in harmony with the individual's age, circumstances, and surroundings, and to limit himself to the registration as subjective realities of impressions transmitted by the peripheral organs of sensation; or the failure to properly co-ordinate such impressions, and to thereon frame logical conclusions and actions: these inabilities and failures being in every instance considered as excluding the ordinary influence of sleep, trance, somnambulism, the common manifestations of the general neuroses, such as epilepsy, hysteria and chorea, of febrile delirium, coma, acute intoxications, intense mental preoccupation, and the ordinary immediate effects of nervous shock and injury.

The first condition of a definition is that it shall be descriptive of the subject to be defined. The above depicts the fundamental error at the root of every form of those

---

* It is significant in this connection, that none of the most recent German writers on insanity attempt to give a definition of "insanity." The chief discussion as to the possibility of concocting such a definition has taken place in the Anglo-Saxon countries, and this for reasons it is not necessary to dilate on, indicates that the chief need for the definition is a medico-legal one. If this want, were not the main motor, it is doubtful whether an English author on Lunacy (Shepard) would have offered such a definition as "insanity is a disease of the neurine batteries of the brain," with such an object as his avowed one of puzzling the lawyers! That a clearly formulated definition of insanity is not indispensable to the scientific psychiatrist is illustrated by the incontestable fact that mental pathology has made more rapid progress in Germany, Italy and France, where little stress is laid on such definitions, than in England or America.

acquired and congenital mental deficiencies and perversions which pass under the designation insanity. The "inability to correctly register experiences and impressions in sufficient number" to serve as guides to rational conduct is the essential feature of idiocy and imbecility, whether moral or intellectual. The inability to limit the registration of impressions to such as are "received by the peripheral organs of sensation," is the foundation of the insane hallucination and illusion. The failure to co-ordinate impressions as the basis of "logical conclusions," is the characteristic feature of those insanities manifesting themselves by the formation of delusive opinion, insane projects, and imperative conceptions. Finally, the impaired reproduction of those impressions and experiences which are to serve as guides to rational conduct, constitutes the main feature of those acute insanities and their secondary sequelæ which are not covered by the other terms of the definition. The latter, therefore, complies also with the second condition of a definition, in that it covers the whole ground included under the term which it is intended to define.

The qualifying clauses are added for the purpose of filling the third requirement: that nothing outside the limits of the field to be defined shall be included in the definition. The words "in harmony with the individual's age, circumstances, and surroundings," exclude the elements of immaturity, mal-development of the sense-organs and members, defective education, and the influence of great popular movements on the human mind. They predicate as the standard of mental health that of the age which the individual has reached, of the period of time in which he lives, and of the class of society to which he belongs. In other words, the definition accommodates itself to the fact that the standard of mental health is a variable one. A lack of understanding, of reasoning power and of responsiveness, which would be a grave indication of unsoundness of mind in a person of previously high intellectual power and culture, might not have so grave a meaning in the case of a member of the lower orders of society or of an inferior race. The normal intellectual status of a Tasmanian is quantitatively inferior to that of some Caucasian imbeciles; and indeed authors have compared certain manifestations of insanity to anthropological degenerations in an atavistic sense. Again, the conduct of a Hindoo fakir, of a religious fanatic, or a revivalist, while it oversteps the boundaries of reason in more than one direction, is not insanity; because, however strange and suggestive of insanity such conduct may be, it is in consonance with the individual's surroundings, and the result of educational influences and mental contagion occurring within physiological limits.

Similar qualifications apply to the more positive signs of insanity. Beliefs which in the earlier periods of history were creeds with the majority of mankind would to-day in members of a civilized race rank with the insane delusion. The behavior of the adolescent, if marking the conduct of a person of mature years, has a sinister signification which it does not have at the period of life to which such conduct naturally appertains. The fantastic tendencies of the child, the prodigality and romancing of youth, the conservatism, lethargy, and miserly inclinations of old age, are phases in the normal evolution and involution of mind; but phantastic tendencies in a person of middle life, prodigality and romancing in aged persons, or habitual suspicion, miserliness, and lethargy in a child or youth, constitute wide departures from the standard of mental health of those respective periods of life, and suggest, if they do not prove, the existence of insanity.

In characterizing the insane hallucination and illusion, that is, the registration of impressions which either are not received at all or imperfectly transmitted from the peripheral organs of sensation, it is necessary to add the words "subjectively real," because hallucinations not accepted as realities sometimes occur in the sane (as when Goethe met his own figure on the road from Strasburg to Sesenheim), and herein differ from the insane hallucination, whose reality the patient is convinced of for the time being. It is also necessary to specify that a registration does not enter the field of mental unsoundness unless it is unjustified by a peripheral impression. We thus exclude from the domain of insanity the deceptive impressions produced by diplopia, scotomata, photopsia, diseases of the peripheral nerves, and the entotic sounds as well as those arising from imperfections of the sense organs.

Here the definition might terminate from a strictly scientific point of view; it is thus far merely a paraphrase of the dictum that insanity is a deficiency or perversion of the mental faculties, not provoked by any external cause, but arising and developing within the *ego*. There are, however, many conditions of temporary deficiency or perversion of the mind, not provoked by any external cause, which do not fall within the strict conception of insanity, and must be excluded in its definition. The phenomena of hallucinatory insanity and stupor are closely simulated by febrile delirium and coma; Indian hemp produces a condition very much like a mild form of expansive mania; finally, alcoholic intoxication imitates in its various degrees almost every phase of paretic dementia.

In excluding the phenomena of intoxication it is necessary to employ the qualifying adjective "acute," because the *chronic* intoxications producing those results detailed in the definition, are actual insanities; alcoholic and opium insanity being examples.

The bearing of the remaining qualifying clauses is almost self-evident. Several of them would be unnecessary but for the desirability of meeting even the most finical and hypercritical objections in a definition whose chief uses are of a medico-legal nature. The ordinary phenomena of epilepsy, hysteria, and the immediate results of nervous injury and shock are not insanity; but the *extraordinary* complications of the ordinary nervous diseases and the remote consequences of nervous shock and injury, if these lead to the development of the signs included in the first clauses of the definition, respectively become insanity of the epileptic, hysterical, choreic, and traumatic variety.

Mental concentration and preoccupation lead to some of the results which might be included under the clause, "inability to correctly register and reproduce impressions in sufficient number and intensity to serve as guides to action in consonance with the individual's age and surroundings." The abstracted man stuffing his pipe with his neighbor's finger, the Scotch professor, who, reminded that he had omitted to apologize to a lady against whom he had stumbled, politely excused himself to the cow who next collided with him, Archimedes oblivious that the soldier rushing into his apartment, where he was engaged over a geometrical problem, was about to kill him, the soldier who, in the exaltation of battle, rushes on after receiving a fatal wound,— are all instances of the action of physiological laws, whose results have a superficial, but only a very superficial, resemblance to the abstraction of certain demented patients, and which require to be excluded by the physician when defining insanity to those who are, or affect to be, misled by surface resemblances,

On some occasion the question of defining what is called "legal insanity" may be presented to the reader of these lines. When that question is asked, he may safely challenge the questioner to show him a broken leg or a case of small-pox in a hospital ward which is not a broken leg or a case of small-pox in law; to show him a tumor or a softening of the brain, which is meningitis or sclerosis in law, or to define the conditions under which any disease-symptom becomes an indication of health. When these conditions are complied with, and not till then, may the physician attempt to define "insanity in law" as distinguished from insanity in science. In the mean time he may rest contented with the dictum of one of the best legal authorities that that cannot be sanity in law which is insanity in science, just as nothing can be a fact in science and a fiction in law at one and the same time.

CHAPTER II.

THE DELUSIONS OF THE INSANE.

The characteristic evidences of insanity, although they are all manifestations of disordered action of the central nervous axis, may for practical purposes be divided into two groups: the mental symptoms proper, and the somatic, or the physical indications. While the former are mainly objects of study for general diagnostic and medico-legal purposes, and furnish important hints for moral treatment, the latter constitute our guides for physical treatment, and enlighten us, as far as signs observed during life can, as to the fundamental nature of the disorder. The significance of these physical evidences, however profound it appears to the experienced mental pathologist, is appreciable and interesting to the student only when taken in connection with what even the untutored eye regards as the essential and characteristic feature of insanity: its mental results. It is therefore best to begin the consideration of the subject with the latter.

To probably no other class of symptoms of mental derangement does so much interest and interest of so manifold a character attach as to the DELUSIONS of the insane. These perversions of the conceptional sphere have indeed had the high medico-legal position assigned to them—it is scarcely necessary to add, erroneously—of constituting the criteria of insanity, and from the days of Willis, Haslam, and Esquirol down, practical alienists have based many important indications for prognosis and treatment of mental disorders, on the special character of the delusions accompanying them.

There is no evidence of insanity which constitutes so proper a starting-point for study as the insane delusion. Though even on first sight the most complex and mysterious of the symptoms of mental disorder, it is yet that manifestation which strikes the mind of the novice with the greatest force. It is the symptom to which the readiest expression is given by the patient himself; the one which can be most readily laid bare before a class in the course of clinical demonstration; and the one which offers to the beginner in psychiatry that obvious contrast with sanity which is the most satisfactory because the most tangible to his mind. For the very reason that the insane delusion is considered to be the criterion of insanity by the laity and the legal profession, the common presence and patency of this symptom in the insane, it should constitute the introduction to the study of the subject.

The lay conception of a lung disorder associates it with cough and expectoration. Now, while cough and expectoration do not constitute the most essential signs of lung affections, yet the clinical teacher who will analyze these phenomena before the new-comer, and point out their true meaning before proceeding to the physical signs—whose recognition and interpretation require experience and acumen—does that new-comer a far greater service than he who endeavors to override the untutored mind by ignoring all which the latter has hitherto been cognizant of, and by presenting abstractions which the beginner is altogether unfitted to comprehend.

Insane delusions, that is, *faulty ideas growing out of a perversion or weakening of the logical apparatus*, are primarily divided by authorities according to their expansive or depressive character; and further subdivisions are based on the contents of the delusion, as to whether it relates for example to sexual, political, religious, or bodily matters. Thus expansive delusions include: ambitious, erotic, and religious ones; the depressive group: hypochondriacal delusions and those of persecution.

All these terms are admissable designations of what are frequently accidental characters of the delusion; but any system of classification which bases fundamental distinctions on them, is necessarily faulty (see Paranoia). A paretic dement may entertain the delusion that he is a king, so may a monomaniac and so may an imbecile or a dement; but nowhere does an old German saying, which translated reads: when two do the same thing it is not therefore the same—apply so well as here. In the cases hinted at, although the external dress of the delusion is formally the same, yet its logical foundation and structure is different in each. To study that difference is to analyze the essential nature of the insanity of which the delusion is but an evidence. As a result of such a study it is found that the external dress of the delusion is of very subsidiary importance. Its essential features are the method in which it has been developed and the manner in which it is defended by the patient.

Delusions may be divided into the *genuine* and the *spurious*. The former group consists of those delusions which have been mainly created by the patient himself; the latter, of those which have been altogether adopted from others. The former alone are of intrinsic importance to the alienist; the latter have only a relative bearing to such extent as it may be necessary to consider the possibility of their existence as a factor in differential diagnosis. They are found in weak-minded patients as evidences of an imitative tendency.

The genuine delusions of the insane, when classified according to their synthesis, naturally fall into two great groups. We find that certain delusions have a complex logical organization—THE SYSTEMATIZED DELUSIONS; while others are devoid of such an organization, are not as plausibly based, as elaborately expressed, and as skilfully defended —THE UNSYSTEMATIZED DELUSIONS.

The various forms of delusion ordinarily admitted fall under both of these heads. That is, we may have either delusions of grandeur or depressive delusions of the systematized as well as of the unsystematized type, and the same applies to their sub-varieties. In order to fully characterize a given delusion, it hence becomes necessary to incorporate all these elements. If, for example, a medical student, as in a reported case, believes that he is suffering from spinal disease, and bases this delusion on alleged symptoms which might justify the belief, if they were not due to illusional misinterpretations and hallucinatory visions of his spinal cord, he is suffering from a *systematized hypochondriacal delusion*. But a person who is unable to give connected reasons for such beliefs as that his body is decaying, that his heart is turned to ice, and that his intestines are stopped up, is afflicted with an *unsystematized hypochondriacal delusion*. In like manner the patient who claims that he is pursued by enemies, because he imagines that people are looking at him in a peculiar way, dogging his footsteps, putting poison in his food, calling out his name at night; and that they do all this to prevent him from making good his claims to a throne, an authorship, or an invention, is a sufferer from *systematized delusions of persecution*. While he who alleges that he is pursued by voices and by persons, although he does this on the strength of similar hallucinations, and can assign no other ground than a subjective feeling of worthlessness or criminality, exhibits the *unsystematized delusion of persecution*.

To answer the question as to whether a given delusion is systematized or unsystematized, is of vastly greater importance than to determine its more superficial features. Take a persecutory delusion, for example! If it is systematized, it may be assumed that we have to deal with that chronic primary insanity, the *Verruecktheit* of the Germans, the *Monomanie* of the French, the *Monomania* of some English authors. If, on the other hand, it is unsystematized, we know with equal certainty that the patient's disorder is a melancholia, unless other evidences point to the existence of senile and secondary insanity, or of paretic dementia in its first stage. (See Diagnosis.)

When we proceed, a little later on, to analyze the mechanism of the principal varieties of systematized expansive delusions, namely, the *simple*, or those relating to social and political ambitions; the *erotic*, or those involving sexual relations, and the *religious*, we will find, that while they all indicate a certain degree of logical enfeeblement, that this enfeeblement is more pronounced in the case of the two latter varieties than of the former. The highest general mental activity is found with those lunatics who cherish systematized delusions of social ambition; the patients who are the political reformers, claimants to thrones, inventors of flying machines, or the *perpetuum mobile*, and panaceas for all earthly ills, the poets, the military and the diplomatic geniuses of asylums. The patient here acts consistently with his assumed character, and the continued existence of a certain amount of mental ability and energy is shown by the formation of projects which, whatever their ultimate feasibility—and they sometimes are feasible—are undertaken with some attention to the patient's actual circumstances and to detail. Sometimes, especially with patients of high previous culture, the systematized expansive delusion is not of such a character as to lead to an error in the patient's sense of identity, but limited to his self-esteem in the abstract. He writes doggerel or mediocre verse, for example, and imagines himself as great a poet as Byron; or he invents some unimportant mechanical contrivance, and lays claim to the gratitude of a nation or a king. Not infrequently he commits plagiarism in a *quasi*-unconscious manner.

Systematized delusions of an erotic character are found as the leading symptoms of the so-called "Erotomania." This perversion is not necessarily accompanied by animal sexual desire, and the adjective erotic is here used in its classical sense.* The patient, noted in his adolescence for his romantic tendencies, construes an ideal of the other sex in his day-dreams, and subsequently discovers the incorporation of this ideal in some actual or imaginary personage, usually in a more exalted social circle than his own. He then spins out a perfect romance with the adored personage as its subject, and according as the external circumstances appear to him momentarily favorable or unfavorable, expansive or depressive delusions are added to and incorporated with the erotic ones. As a rule the affection for the adored object remains as chaste and pure as it begins; a sort of distant, romantic worship, insane for the reason that unimportant occurrences, accidental resemblances, facts which have no natural connection with the individual or his or her real or imaginary contemplated partner, and hallucinations are woven into the delusive conception, which consequently assumes such a predominating position in the patient's mental horizon as to entirely overshadow it.

Systematized delusions of a religious character are usually rooted in an early developing devotional tendency, and

---

* In the treatise of Bucknill and Tuke erotomania, and the so-called "nymphomania," which is an entirely different condition and not necessarily accompanied by delusions, nor ordinarily by amatory conceptions of the higher kind, are thrown together.

brought to full bloom by incidental circumstances, either actual or in the shape of hallucinations. It is not uncommon to find such patients designated as cases of religious melancholia, because, supposing themselves assailed by inimical or diabolical forces, or commanded to fast for a long period, or forever, they become sad and anxious or refuse food. But to call a patient who, aside from these actions, which are consistently regulated by the false ideas, believes himself or herself to be God, the Christ, a saint, the Messiah, a religious reformer, or the Virgin Mary, and who, perhaps, the very next day passes into visionary or ecstatic states, all the time systematizing his or her acts and notions —a melancholiac, is to involve one's self in a profound contradiction with the established use of that term.

Among the systematized delusions of a depressive character, the antitheses of all the preceding forms are found. This is well exemplified in cases where the subjects develop delusions of persecution overwhelming the ambitious notions previously dominating the mental sphere. How very unessential the line of demarkation between the depressive and expansive delusion really is, is made apparent by the fact that an expansive religious, erotic, or socially ambitious delusion may within a few days become depressive through the development of persecutory ideas; just as the reverse is occasionally observed. In fact, such expansive and depressive delusions often are entertained by the patient at one and the same time.

When an individual, without any manifest disturbance of his emotional or affective state, in full possession of the memories accumulated in the receptive sphere, and able to carry out most or all of the duties of his particular position in life, as some of the sufferers are, is found to cherish such a gross error as an insane delusion, firmly believing in the reality of that which from his education and surroundings he would be expected to recognize as absurd, the observer is naturally puzzled to account for the phenomenon. The component elements of the systematized delusion are individually the same as those which are combined to form healthy conceptions; they differ in that they are faultily united. In this respect the insane delusion does not differ from the faulty conclusions which constitute the *error* of the sane. But while the error of the sane person is based on faulty and not on perverted perceptions, on imperfect or vicious training, and on lack of experience, and disappears before the corrected evidence of the senses, improved educational methods, and broader experience; the sufferer from the insane delusion is unable, for the time being, to correct that delusion by a similar process. His logical apparatus stands powerless before the controlling conception. There is a fundamental weakening of those logical inhibitions or checks which the collateral elements of the *ego* constitute for the day-dreams and speculations of a sane person. What is a wish or fear of the sane mind becomes an article of faith with the delusional lunatic; what is a mere possibility in reality becomes a fixed fact in support of that article of faith with the insane; and finally the perceptions themselves are perverted and enslaved to add new building material to the faulty mental structure. It is because the physiological *ego* is weakened that the morbid *ego* is permitted to arise out of the small beginnings which a healthy *ego* could have held in check, and for this reason the following legal definition of the insane delusion—that it is *a faulty belief, out of which the subject cannot be reasoned by adequate methods for the time being*—is a sound one.*

Notwithstanding the fact that the delusions of the systematized variety are correlated with the patient's surroundings, in contradistinction to the unsystematized delusions, which are not correlated at all, they are *faultily associated.* While there is a chain of reasoning connecting the items of the systematized delusion into an organized whole, which is absent in the unsystematized delusion, yet after all that reasoning is only *pseudo-logical.* The systematized delusion is more similar to a sane conception than the unsystematized delusion; yet this similarity does not pass beyond the degree of an analogy.

As illustrating the readiness with which systematic delusionists utilize casual occurrences in the construction and defence of their delusions, it is but necessary to refer to the common case where such subjects detect a connection between their delusive hopes or fears and an advertisement or a bill-poster containing their initials; that others sustain the allegation of a royal descent by a fancied resemblance to some member of a reigning family; and that still others lay claim to an important office on the strength

---

* This definition is somewhat modified from that given by Ray and others, which fails to provide for the possibility of the lunatic's correcting some of his delusions during his insanity, and of correcting all of them if he convalesce.

of a friendly interview with some little great man of the day. The case of a lady whom the writer treated several years ago illustrates this readiness with which trifling matters are framed into delusions, very aptly. From the facts that the irides of her child changed color (which was an occurrence corroborated by the grandmother and others), and that it had greatly altered in appearance owing to an exhausting illness, she concluded that it was not her own child. Her nursery maid and one of a neighboring family had been together a great deal, and she claimed, what was in itself not impossible, that her child had been exchanged. Although the observant relatives were able to prove the identity of the child by a number of circumstances which would have been held satisfactory in any court of law, and which would have convinced any sane person, she only interpreted the remonstrances made as attempts to make the best of the case; and because her husband affected to treat her suspicions with indifference she reasoned that he was becoming neglectful of his family. A few ordinary civilities exchanged by him with several ladies belonging to the same church convinced her that from being indifferent to his child he was becoming unfaithful to his wife. One night a large negro looked over the garden wall, and the watchdog did not, as was his wont, bark at him. Several robberies occurring in the neighborhood about the same time, she inferred, ready as she was to believe anything that was bad of her husband, that the negro, who must have been one of the robbers, had not been barked at by the dog, because the latter recognized in him one of his master's confederates. Those who defended her husband or attempted to explain his actions in the face of "such evidence" could only do so in the hypothesis that they had joined the conspiracy against her. Her cousins were members of the conspiracy because a package of chemises put up by them, which ought to have contained a dozen contained a lesser number; and when removed from her native place to New York for treatment she found that the custom-house officers had joined the ranks of her enemies, because after their examination of the contents of her trunk other articles were missing or ruined.

The absurdity of a delusion is not so much a test of the absolute mental rank and the form of insanity of a patient, as is its organization. A very absurd delusive conclusion may be reached by an elaborate and plausibly delivered ratiocination, and a less absurd delusion may be formulated on a very crude process of reasoning.

While the most important factors determining the nature of a delusion are the form of insanity of which it is a symptom and the manner in which it affects the cerebral mechanism, other elements must be admitted to have some influence in tinging it. The general disposition of the patient, whether it be sanguine or suspicious, will often determine the expansive or persecutory character of a given delusion. If the physical state is poor and the visceral functions are disordered, hypochondriacal delusions are more apt to arise than with a robust or fair state of health. The age in which the patient lives and his social circumstances have no inconsiderable influence in the moulding of his morbid ideas. With the development of republics and constitutional monarchies, the growth of the sciences, the arts, the press, and the emancipation of mankind from superstitious creeds; the kings, emperors, prophets, ambassadors from the planets, Holy Virgins and Gods have become less common in the asylum corridor than they once were. They have given way to insane inventors and communistic, journalistic, educational, musical, and scientific delusional project-makers.

The UNSYSTEMATIZED DELUSIONS are found with the acute insanities and the chronic deteriorations. They may be ranged in two great classes according as they are due to the subjective misinterpretations arising from an emotional disturbance, or the result of a destruction of the logical associating force. The delusions of true melancholia are instances of the first type, those of paretic dementia of the second.

When the emotional state of the melancholiac has overwhelmed the mental apparatus, and the logical faculty is thrown in the background by the predominant painful depression, the patient may, in his endeavor to account to himself for his painful feelings, in a vague way conclude that he is a bad person. Since he is a bad person, it must be because he has committed the unpardonable sin; but he cannot tell when nor why nor how he has committed it, nor very often what the unpardonable sin is. Or, again, such a patient feels that he is despised, that he is despised because he is hated, and hallucinatory whispers from all sides drive him to seek relief from a danger which was never clear in his own mind, in suicide. The great distinction between the systematized delusion of persecution and the unsystematized delusion of a depressive nature is, that while the former is distinct and fixed, the latter is vague and changeable; while the former incorporates every present circumstance in a pseudo-logical chain, the latter jumps over the gap; and while such logical power as the patient ever had, is utilized in the assertion and defence of the systematized, it is in abeyance, in part or in whole, with reference to the announcement and defence of the unsystematized delusions.

The unsystematized delusion of grandeur differs in a similar way from the systematized expansive delusion. In the former case—for example, in paretic dementia—the patient may assert that he is a king or president, or that he has a million dollars; because it is a desirable thing for the sanguine patient to be a king, a president, and to have money. But he will make no attempt to explain how he can be a king and yet be named Dennis Maginnis; how he can be Ulysses S. Grant and Samuel Silberstein at one and the same time; and why it is that he has a hundred thousand acres of land to-day, when he only had ten houses and lots yesterday. Such a delusion is never as consistent nor expressed with that firm conviction that characterizes the systematized variety. A systematic delusionist would, if challenged to explain why he answers to the name of a private citizen, when he claims to be a king, say that the name in question is that of the menial child with which he was " exchanged while in the cradle," or that it is the designation which he has adopted to put his persecutors, "the agents of the usurper," on a false track. Again, a systematized delusional lunatic, if claiming great personal attractions, which is rare, fortifies his claim by letters received, containing a "hidden meaning" and by poems and advertisements " referring to him." In paretic dementia, with which affection this delusion is common, the patient will content himself with the vague announcement of the assertion that he is handsome, boast that the women are enraptured with him; but if called on to specify he will hesitate, and then invent his grounds as he goes along—if able—and forget them on the next occasion. Another paretic dement will allege that he is five thousand three hundred and seventy-two feet high, his actual height being rather under five feet (actual observation). If placed side by side with a taller man, and asked to estimate his size, he correctly assigns, six feet. If asked whether he has to look up or down to measure his neighbor he unhesitatingly admits that he has to look up. But on being confronted with his inconsistency, how it is possible that a man over five thousand feet higher than the tallest giant can possibly look up to an ordinary-sized mortal, he simply reiterates that he is so many thousand feet higher than the interlocutor or any other man. A third person claims that he is General Grant, the week before he claimed to be Rothschild the banker, but abandoned that idea when told that the latter was dead. He is unable to say when the war began, what his business was before he became a general, what battles he fought in, and finally what land he is president of. A systematic delusional Ulysses Grant would, in marked contrast with the paretic, be a walking history of the war and of Ulysses Grant; he would very probably content himself too, with the more plausible delusion that he was a brother or intimate friend, or a subaltern of the general, to whom the latter owed his "inspiration" and success.

While, on the whole, the organization of a delusion reflects the essential type of the insanity of which it is a manifestation, it must be admitted that there are special delusions or groups of delusions, in which the formal contents and the superficial guise of these conceptions have a diagnostic value which is lacking with others. The delusions of marital infidelity and sexual mutilation, when combined, suggest the existence of alcoholic insanity. The combination of unsystematized sexual with religious delusions of a hallucinatory tinge is characteristic of certain forms of epileptic insanity.

The current notion, expressed in several treatises, that delusions are independent of sensorial impressions, is erroneous. Aside from the fact that systematized as well as unsystematized delusions are sometimes accompanied, and are often determined and modified, by hallucinations,* it is a common observation that visceral illusions, if they do not cause, modify the character of delusions. Thus, a patient, in the writer's observation, who believed that he had a doctor in his belly, was found to have extensive peritoneal adhesions; another, whose case has been cited from Esquirol in most elementary treatises, had the notion that he had a

---

* A case was recently discussed and recognized to have been correctly observed in the Medico-Psychological Society of Berlin, where a systematized delusion had developed in consequence of a dream, no pre-existing mental taint being discoverable.

council of bishops in the same cavity, which after death revealed the same pathological condition. The common delusion that some one or other organ or member has been destroyed or petrified or turned to wood, cork, or glass, or is infested by noisome parasites or reptiles and maggots, is probably based on partial anæsthesias and paræsthesias. Here the delusion is as much in disproportion to the exciting cause as the idea of a dreaming person that he is falling down a great height is to the dropping of his arm over the side of the bed; or the dream of Captain Parry, that he was standing upright with his feet frozen into the ice of Baffin's Bay, was to the fact that his feet were uncovered at the time he had this dream.

Here, however, as elsewhere, the essential element is the cerebral fault (the logical failure to correct the delusion, with the weakening of the *ego*) that permits the development of this insane symptom. The visceral disease or the sensorial disturbance is an accidental factor; the patient would be insane with or without it, in the vast majority of instances; and in its absence his defective mind would fall a prey to some other delusion, under the assaults of some other incidental occurrence, which, equally with the visceral disorders alluded to, would fail to provoke delusions in a healthy brain.

The superficial character of the delusions of the chronic insane, interesting as it is to the laity who visit the " showwards" of an asylum,* is ordinarily of lesser value to the alienist, for reasons similar to the ones just assigned. When the fiend Thomas, exploded a dynamite battery, killing over a hundred persons at Bremerhaven, a lunatic in a German institution developed the delusion that the clockwork of just such a battery occupied the place of his heart, and at a certain moment would explode the charge and destroy the asylum. When Nobiling shot at the German emperor over a dozen lunatics accused themselves of complicity, or offered to give information of alleged conspiracies which they had unearthed. When the Pope referred in his allocution to the Star which was to relieve the Church of its persecutors, more than one insane devotee thought he was the one referred to. Whenever a political movement, like

---

* It was sufficiently evident to a lunatic described by Kiernan: that the grosser the delusion the greater the interest of the visitors, and the more likely their donations of tobacco, etc.; that he shammed having the delusion of "tar and grease filling his head" to accommodate them.

the war of the Commune, that of the Rebellion, an election, or an assassination, occurs, it is reflected, caricatured, and travestied in the delusions of asylum inmates. The enumeration of features due to such transient influences, and revealing so little of the true foundation of a symptom, may advantageously be waived at this point.

CHAPTER III.

IMPERATIVE CONCEPTIONS AND MORBID PROPENSITIES.

There are certain mental phenomena, reflections and suspicions arising in lunatics, which, although they differ from the delusion, in that the patient is able to reason himself out of them and to recognize their absurdity at times, yet tyrannize the patient's thoughts, and sometimes his acts, as markedly as the most firmly-rooted organized insane idea. These phenomena arise suddenly, without any obvious connection with previous thoughts; they appear like spontaneous explosions of some uncontrolled segment of the nervous system, and are aptly called IMPERATIVE CONCEPTIONS.

On analyzing these remarkable symptoms more closely we find that they sometimes arise by suggestions, but suggestions quite inadequate to produce such impressions in a healthy state. Just as sane persons in looking over a precipice feel tempted to precipitate themselves down its dizzy height, just as an eminent *savan* in crossing a bridge was so strongly tempted to push a boy sitting on the parapet into the water below that he felt constrained to turn and walk away; so the sight of a conflagration or an execution or the recollection of a suicide excites in the minds of the insane, imperative conceptions dictated by the impressive spectacle or event. A morbid impressibility of the nervous system * may be reasonably assumed to exist in these cases, and this assumption is borne out by the fact that imperative conceptions are more common in females than in males, in youthful and imbecile than in aged and

---

* The writer may make the reservation here that he is not disposed to admit that the imitation of great crimes or of bloody deeds, in itself, and unaccompanied by other evidences of insanity, is a proof of its existence.

strong-minded persons, and under such conditions as pregnancy, menstruation, and the convalescence from fevers.

Many of the imperative conceptions remain *in statu quo* for years, as, for example, the morbid fear of places (*Agoraphobia* of Westphal), that of narrow quarters (*Claustrophobia*), the "*Grübelsucht*" of Griesinger, and the interesting *Mysophobia*, or fear of defilement, described by Hammond. One of the patients on whom the latter author based his description has since been under the writer's care; she was in much the same condition as when the author referred to saw her years before; she did not venture to handle many articles at all for fear of poisoning first herself and then her children by contact; nor could she look at the wall without speculating as to the dangerous influences which might emanate from it. The sight of crockery, of food, of the floor, of water, in short, of anything connected with her household affairs gave rise to painful conceptions, to rid herself of which she struggled in vain. In this case the theory that the imperative conception is a *rudimentary delusion* seemed to be borne out; for the patient was speculating at the time on the possibility of the existence of an inimical influence against her. In more than one instance the writer has observed that the announcement of delusions of persecution at the hands of Jesuits, of Free-masons, and of other secret societies was supported by similar imperative conceptions, against which the patient had ceased to struggle, and which became in consequence fixed ingredients of the delusive belief.

Just as the insane delusion determines the development of those morbid projects, or insane enterprises, so characteristic of many chronic delusional lunatics; so the imperative conception often leads to the IMPERATIVE ACT, or, as it is more commonly called by English writers, the MORBID IMPULSE. But the imperative act is directly determined by the overpowering conception suggesting it, and is not the result of a linked reasoning, as the morbid project is.

In some instances the same morbid impulse recurs again and again, or continues throughout the life of the patient. There are subjects actuated by a homicidal impulse, slaying when and where they can, while under the immediate influence of the imperative conception. Such persons have been styled homicidal monomaniacs, unfortunately, for there is no such form of insanity (see Monomania), and patients with this symptom should be designated: lunatics with *in-*

sane homicidal impulses. The frequency with which subjects at or about the time of puberty, epileptics and imbeciles, develop the impulse to incendiarism has led to the erection of the group *pyromania;* and the thieving impulse, when it exists independently of any of the commonly recognized divisions of insanity, is similarly known as *kleptomania*. These and other terms of the kind may continue in our nomenclature as symptom designations only, with the restriction that they are to apply to lunatics in whom the morbid impulse is the most prominent and continues to remain the characteristic feature of their disorder. These terms should not be applied to sufferers from well-marked forms of insanity, such as, for example, terminal and paretic dementia, which may manifest themselves in episodial thefts and acts of incendiarism or destructiveness, and which latter are then not necessarily impulsive in character.

PYROMANIA, or the morbid impulse to commit incendiarism, is a symptom which may occur in epileptics and menstruating girls, in consequence of fiery visions or fluxionary congestion; it may develop in any lunatic with destructive impulses, and is sometimes found in pregnant and hysterical females. It is usually exhibited at or shortly after the period of puberty, and as a rule with weak-minded or imbecile persons affected with an hereditary taint or a neuropathic constitution. The commission of a large number of incendiary acts in a given neighborhood, when not due to an organized movement, to the desire of malicious persons to revenge themselves, or to some other criminal motive, may be generally traced to a patient presenting this dangerous symptom.

Pyromania, like kleptomania, may be a leading manifestation of periodical insanity, and indeed one morbid impulse, dipsomania, has given its name to a variety of that disorder. (See Periodical Insanity.)

Needham reports a case where an educated lady, while in the climacteric, developed an impulse to murder on seeing some knives. Recognizing her condition she voluntarily went to an asylum, and stayed there five years.

The suicidal and homicidal impulses, based on the imperative conceptions of melancholia, are among the most dangerous manifestations of that disorder. Whole families have been immolated in obedience to these terrible suggestions. The sight of a revolver, a knife, a rope, or the reading of accounts of death by any strange or unusual method, is followed by the adoption of similar means calculated to accomplish self-destruction or the death of others.

Pregnant women develop the strangest conceptions and impulses, often blended with the morbid appetites of their condition. It seems that here the murderous and cannibalistic impulses are usually directed against those nearest and dearest to them, on that same basis of contrariness which appears to govern the imperative conceptions of the insane generally, and the analogous impulses of the sane, illustrated in the irresistible tendency to laugh while attending funerals, and to use profane expressions in the midst of solemn surroundings, which tendencies are far more frequent than is ordinarily believed.

A sharp distinction should be made between the homicidal and other destructive impulses, and the *morbid propensities* which result in similar acts. The former are isolated phenomena, incongruous with the rest of the mental sphere, and the criminality of the acts, morally and legally considered, is recognized after their perpetration. The latter are exaggerations of the propensities which are either less markedly present, or lie dormant in the sane mind, or represent psychical atavisms. The lunatic affected with the *quasi*-instinctive propensity to murder, to eat human flesh, or to indulge in abnormal sexual acts, if confronted with the enormity of his inclinations, may recognize their technical illegality, but in his own mind he feels no contrition for the act in itself, and considers himself justified in following his inclinations whenever he can safely do so, and finds a favorable opportunity.

Morbid propensities and impulses are sometimes combined, as in the cases of pregnancy alluded to.

The morbid propensities are frequently confounded with the imperative conceptions, because they lead to similar results, and because, as in the cases referred to, they occasionally coexist. They probably depend on analogous fundamental states, but differ in being more firmly rooted in the patient's organization; they are perversions of those instinctive tendencies which are common to mankind, and they control the individual in an analogous manner to, but usually more intensely than the physiological desires.

The morbid propensities are perversions of the two main instinctive tendencies of the human race: the desire for food and the sexual appetite. In maniacal and hysterical conditions the sexual appetite may be exaggerated to a remarkable degree, constituting *satyriasis* in the male, and *nymphomania* in the female. The patient suffering from the former condition may commit rape, resort to indecent exposure of his person, and make insulting proposals to females; while the nymphomaniac, aside from her solicitation of male persons and masturbation, may reveal her state of mind by extravagances in dress, courtesan-like behavior, and a tendency to make obscene accusations against other females, less for the sake of injuring these than for the purpose of suggesting to the interlocutor what her own desires are.

It should be recollected, however, in estimating the motives of a female patient manifesting exaggerated sexual ideas, that these are not necessarily of a character looking to animal gratification. In insanity we find every analogy of mental health; and just as a large number of perfectly healthy women are devoid of a sexual appetite, or manifest but feeble indications of it, so we have occasion to observe female patients whose conduct, while corresponding in the main features to that of the nymphomaniac, does not culminate in attempts at sexual acts. Such patients are very likely to develop platonic admiration for male and even female* persons, or to become religious devotees or enthusiasts. All through the history of insanity the student has occasion to observe this close alliance of sexual and religious ideas; an alliance which may be partly accounted for because of the prominence which sexual themes have in most creeds, as illustrated even in ancient times by the Phallus-worship of the Egyptians, the ceremonies of the Friga cultus of the Saxons, the frequent and detailed reference to sexual topics in the Koran and several other books of the kind; and which is further illustrated in the performances which, to come down to a modern period, characterize the religious revival and "camp-meeting," as they tinctured their mediæval model, the Münster anabaptist movement.

The most important morbid propensities, from a medico-legal point of view, are the *sexual perversions* and *anthropophagy*. Instances of these remarkable conditions are collected in Caspar's treatise on forensic medicine; but our

---

*The case of an acutely maniacal patient who fell in love with her nurse, related by Krafft-Ebing in the *Archiv f. Psychiatrie* seems to the writer to belong to this category, and not to be strictly one of sexual perversion.

more perfect knowledge of their pathological nature we owe to Krafft-Ebing.

While sodomy, pederasty, and other disgusting abuses of the sexual apparatus were common, and even cultivated by more than one nation of antiquity, and are such everyday occurrences in Armenia, Syria, the home of the ancient corybanthism, and in other Oriental lands to-day as to be there considered legitimized by custom—not to mention the fact that they are too common in our large cities to be considered anything beyond the outcome of salaciousness, idleness, and opportunities created by that over-fed luxury which tires of natural gratification—there are individual cases where these acts must be regarded as the imperative results of a faulty organization.

In these instances there are sometimes physical signs such as asymmetry or other malformation of the skull, deformities of the ears, and notably of the sexual organs, which indicate the organic nature of the perversion; and, it may be added, which exemplify the close relation existing between the development of the nerve-centres and the sexual apparatus. The existence of these signs in some cases points to a deep error in development, dating from an early embryonic period. Such a developmental fault may be assumed, by analogy, to exist in a lesser degree in those subjects which do not exhibit external signs. Searching inquiry generally reveals other anomalies in the nervous functions, and although these sexual perversions have occasionally been noted in persons of fair and even good intellectual powers, they are generally associated with more or less mental weakness.

A curious contribution to the discussion of this question was made by an intelligent judicial officer uamed Ullrich, who himself afflicted with that form of this perverse tendency, in which sexual love is displayed toward the same sex, wrote a book in its defence, claiming that it is justified by natural laws, inasmuch as it occurs as a zoological phenomenon in certain insects;* and he fortified his position in a most elaborate manner by the precedents furnished by the classical nations of antiquity. Such cases as his strongly sustain Westphal's claim that sexual perversion is not nec-

---

* Among Coleoptera. It may be not unrelated to this question that the phenomenon of a union of both sexes, in such a way that one lateral half of the animal is male and the other female, or where alternate quarters are of opposite sexes, is most frequent in this order of insects.

essarily and by itself a proof of insanity, as some have believed.

The most common form of pathological sexual perversion is the love of the same sex. The male subjects are as a rule peculiar from boyhood up, manifest feminine tendencies, have a mincing gait, prefer dolls and girlish toys to boyish sports, delight in assuming the female costume, and develop platonic and sexual love for persons of the same sex. Seminal emissions and even the full orgasm are pro-

FIG. 1.

voked by a grasp of the hand or an embrace of the adored subject; and so far as the recorded cases go, the subjects repudiate the idea that they are guilty of sodomy with disgust and indignation, their *modus operandi* consisting in mutual titillation. Cryptorchidism, hypospadias, defective development of hair on the pubes and face, club-foot, and other signs of imperfect development, are often found here, and a feminine expression is very common.

An historical instance of this form of perversion is that of Lord Cornbury, a cousin of Queen Anne, and, as the son of

Lord Clarendon, afterwards a member of the House of Lords, also at one time governor of the colony of New York. This person, whose picture, preserved in the library of the N. Y. Historical Society, illustrates the asymmetry and feminine appearance natural to and the costume adopted by such subjects (Fig. 1), was, according to his historian, a degraded and hypocritical being, utterly devoid of a moral sense, and so thoroughly mean and contemptible that in a short time all classes of the population were arrayed against him, compelling his removal. His greatest pleasure was to dress himself as a woman, and the good citizens of New York frequently saw their governor, the commander of the colonial troops and a scion of the royal stock, promenading the walls of the little fort at the Bowling Green, with all the coquetry of a woman and the gestures of a courtesan.

We have less knowledge of sexual perversion of this variety in the female sex; though the accounts of "Lesbian love" would seem to indicate its existence in antiquity. The writer has now under observation a lady suffering from periodical insanity, who has an intense but pure affection for a lady companion whom it is necessary to allow her to have to manage her outside of an asylum.

The second variety comprises cases where sexual gratification, while resorted to with persons of the opposite sex, is accompanied by cannibalistic or analogous perverted desires. Several of the Cæsars, a family which presented numerous examples of transmitted mental disorder, delighted in seeing maidens slaughtered from sexual motives. More recent instances have been adduced by Lombroso, and at the present moment the province of Westphalia is excited over the commission of more than a score of murders, performed in the most revolting manner, on young girls who had been previously violated, every indication pointing to the same person as the perpetrator.*

---

* One of this class of instinctive butchers made the following statement regarding the murder of one of his long series of victims, who was first violated and then treated in the manner detailed by himself: "I first opened her chest, and divided the fleshy parts of her body with a knife. Then I dressed this person, as a butcher dresses cattle, and chopped her body into pieces, so as to get them into a hole which I had dug on the mountain. I can say that while opening the body I felt so ravenous that I trembled, and cut out and ate a piece." Others have torn out the heart, and drunk the blood of their victims. and Tirsh (reported by Maschka) cut off the breasts and genitals of an old woman whom he had killed and violated, and cooked and devoured these parts,

It is to be insisted here that even these terrible sexual aberrations may exist as combined results of a vicious inclination and cynical brutality in persons not insane. The term *Anthropophagy*, as indicating a morbid perversion of the sexual appetite, calling for the satisfaction of murderous and cannibalistic desires, should be limited to those cases where there are signs of heredity, somatic evidences of degeneration, and other manifestations of a faulty nervous system. In one such case, the executed monster Leger, Esquirol found gross brain disease of the kind sometimes discovered in the insane.

Necrophilism is a name given to the propensity to violate dead bodies, which in very rare instances is found as a manifestation of periodical insanity. The French Sergeant Bertrand, who broke into churchyards, under great risks, and dug up female bodies to violate and mutilate them, is a case in point. In this instance the hereditary history, the periodical recurrence, and the association of this frightful propensity with signs of maniacal excitement demonstrated the insanity of the individual. But there could be no better exemplification of the doctrine that few acts, however extravagant, are in themselves signs of insanity, than that this, the most incomprehensible of all conceivable crimes, like Anthropophagy, has been committed by persons whose sanity could not be disputed.

CHAPTER IV.

HALLUCINATIONS AND ILLUSIONS.

Hallucinations and illusions are erroneous perceptions sometimes occurring within the limits of health, but in their more intense development characteristic of certain forms of mental disturbance. Here, as elsewhere, it is the manner in which the symptom originates, the basis on which it develops, and the relation it assumes toward the *ego* that determines its interpretation by the alienist.

An hallucination is the perception of an object as a real presence, without a real presence to justify the perception. The perception of a heavenly vision by the religious lunatic, the odor of putrefying and other disgusting substances complained of by hypochondriacal and demented paretic patients, the voices commanding the sacrifice of relatives or self to insane fanatics, and those driving the melancholiac to the commission of suicide, are among the more characteristic examples of hallucinations in the insane. In all these cases there are no objective grounds for the perception.

An illusion is the perception of an object actually present, but in characters which that object does not really possess. The delusion of a hypochondriacal patient that he is decaying, because he detects a decaying smell, while there is *in reality* a chronic catarrh of his naso-pharyngeal passages, resulting in a bad though not a putrid odor, is founded on an illusion; the patient who, hearing the wagons rolling on a street imagines that his name is being repeated in regular rhythm, has illusions; and the melancholiac who, seeing the apprehensive glances of his relatives, imagines that they sneer at and threaten him, is suffering from the same class of symptoms. In short, while the hallucination, as far as the outer world is concerned, is a creation out of nothing and entirely fictitious, the illusion is but a misinterpretation of a physiological impression and has a partial basis in fact.

In the strictest psycho-physiological sense an hallucination is a morbidly intensified memory. Gray nervous matter, wherever situated, if connected with a sensory periphery, is capable of receiving and retaining impressions. The length of time through which the impression may be retained differs in different segments of the central nervous apparatus. When, for example, a person looks at a window with a black figure standing in front of it distinctly outlined against a very bright sky, and, after regarding it for a certain length of time fixedly, shuts his eyes and turns away, he will find the image still continuing. It gradually becomes fainter, but may last for several minutes. The fact that the retinal "after-image" follows the movements of the eyeball shows that it has its seat in the retina, and proves that the ganglionic cell layer of that ocular tunic is capable not only of receiving impressions but also of retaining them for a brief period of time. Now let a person be immured in absolute darkness for a week; let him, without exposing his retina to any other impression, repeat the experiment, and it will be found that the retinal after-image will last very much longer! In short, the less the nerve-cells of the retina are subjected to the rapidly-succeeding, quickly-interrupted, and ever-changing impressions of the outer world, the better becomes its registration power. It is because of the multitude of impressions crowding and jostling each other and involving the same retinal elements over and over again, that it becomes unfitted to retain impressions longer, and the incapacity is in this case a beneficial one.

But there is an area in the brain—the cortex of the calcarine fissure, its neighborhood, and a considerable extent of the convexity of the occipital and possibly of the parietal lobes—which is known to be the central abutment of fibres connected with the optic tracts. A host of anatomical and pathological observations support this assertion. Here we find that the cortex has a structure differing materially from that of the rest of the cerebrum, and indeed, as Meynert pointed out, one not without some analogies to that of the retina itself.

Here the impressions occurring in the retina and transmitted through the optic nerves, tracts, and radiations, are registered as memories. It is because the cortical nervous elements are much more numerous than those of the retina * and that the same cells and groups of cells are not called upon to serve as recipients so often, that *visual memories* are more permanent than the after-images.

Memories depend for their fixation on three factors: the intensity of the impression, the frequency with which it recurs, and the functional disposition of the central nervous elements at the time of the impression. The same laws that govern the after-image also govern its more stable analogue, the memory. Just as a brilliant illumination is more certainly followed by an after-image than a feeble one; so a prominent fact, overwhelming or crowding out less important collateral circumstances, is more apt to become fixed in the memory than the latter. Just as the retina, after a long period of inactivity, is more sensitive in relation to after-images; so the cortex of the youth, of him who awakes from a refreshing slumber, or who has been idle, is better fitted to register new impressions than the cortex of the aged, of him who is fatigued, or has been straining his memory for some time.

If a person presses a pin's head against any part of the closed eye he will see a corresponding spectre, apparently

---

* And that the numerous associating tracts present a basis for a greater number of association possibilities.

on the other side of the field of vision. If he uses a triangular, a circular, or a perforated object, the image will have the corresponding shape; and in the same way the apparent situation and number of the impressions may be varied. This experiment, like the fiery visions noted by patients who are suffering from pathological changes in the retina, shows that impressions may be produced simulating those ordinarily resulting from the impressions of the outer world, by *non-physiological irritation* of gray matter.

Just as the spectre produced by pressure on the eye is an artificial imitation of the retinal after-image, so the hallucination is a morbid intensification of a memory, produced by the irritation of the registration field of that memory.

Just as the irritation whose result imitates the true retinal light impression is applied at the normal site of such light impressions, so the irritation resulting in an hallucination must be directed to the normal centre of the memory registrations—in other words, to the brain cortex or its subsidiary depots.

Just as the physiological irritation is competent to reproduce old recollections, through a healthy use of the association tracts, and the consequent increase of the nutritive and bio-chemical changes in a given cortical or sub-cortical nervous area; so a meningitis, a toxic agent, a relative hyperæmia, or some other profound nutritive disturbance of the same area, will produce the same result *in kind*, but so much more intense *in degree*, that the reawakened impression, instead of being of the ordinary intensity of the memory, becomes lifelike and simulates a real impression.*

Within the limits of health hallucinations are usually limited to those occurring during sleep, in the course of dreams,

---

* According to Meynert a reversal of the normal course of impressions takes place in addition; the pathological irritation extends centrifugally toward the periphery, and, transmitted to the peripheral sense-organs, determines the lifelike reality of the hallucination. There are grounds for this belief; but the secondary irritation of the sensory periphery is not essential: some of the most elaborate hallucinations occur in persons who have had their eyeballs extirpated, and the writer has made the *post-mortem* examination in a case where the most marked and complicated hallucinations of vision, the Deity, angelic processions, worms in the food, etc., occurred *sub finem*, the optic nerves being found cut in two by the pressure of a tumor and the thalamus softened, leaving the cortex alone to serve as the basis of the symptoms. Hallucinations may originate in thalamic disease or disturbance anywhere along the course of the optic fibres, but their entry into the consciousness can only take place in the cortex.

or in the intermediate state of sleeping and waking. If the observation of Treviranus, made on a patient whose brain surface was exposed, that the brain surface is paler in sleep than in the waking condition, but richer in blood during sleep interrupted by dreams than in dreamless sleep, is confirmed, we shall have to look to irregularities in the blood-supply as the source of the hallucination in the insane. The fact that musicians are more apt to have hallucinations of hearing, painters and sculptors those of sight, microscopists those of the objects which they study, would seem to indicate the potency of excessive physiological irritation to predispose to the production of the same phenomena as those resulting from pathological irritation. Here again the analogy between the extra-physiological phenomena occurring within the limits of sanity and those characteristic of insanity is manifested. The hallucinations of the insane, generally relate to familiar objects, to those which the patient has been in daily contact with, or which are connected in some way with his dominant conceptions. The hallucinatory religious monomaniac sees the gates of heaven ajar, St. Peter beckoning, processions of angels and saints, the crucifixion; or, on the other hand, the devil with myriads of imps dancing round and mocking him, the visions of St. Anthony repeated, or the strange beasts described with so little reference to zoological harmony in the "Revelations." The insane claimant of a throne hears encouraging or threatening comments on his claim, sees detectives dogging his footsteps, or has visions of great state receptions. The lunatic with sexual ideas sees images of the most voluptuous kind. The hypochondriacal lunatic sees his skull open, fungous growths on his brain, his viscera exposed, the liver riddled with abscesses, and other morbid conditions, varying with the extent of his reading and experience of diseases.

It is in consonance with the same fact that insane hallucinations vary in different races and communities, and with different periods of history, like the dreams and delusions of the day and the people. The visionary lunatic of ancient Greece saw the gods and goddesses, or satyrs, dryads, and nymphs peopling the forests; he of the middle ages communed with saints, or saw the devil, in the fashion of that day, with horns, a goat's beard, a barbed, arrow-headed tail and a pitchfork; to-day the visions of heaven and hell are more after Bunyan's style or Miltonic in character; and with the practical tendencies of our growing civilization they are undoubtedly becoming rarer than they once were.

The discrimination between the insane illusion and the insane hallucination cannot always be sharply made. Indeed, within the realm of sanity the two are noted to run into each other. What is generally considered the best example of hallucinations in the sane, the vision of a dream, is probably in part illusional. Scherner has shown that with the majority of dreams involving visions and conceptions of water the urinary bladder is full; where there is a feeling of looking or falling down a dizzy height the stomach is disturbed, and here a centripetal influence analogous to that transmitted by the vagus and determining a subjective dizziness in gastric disorder, is undoubtedly produced. The dream-visions of luxurious banquets are generally referable to a sense of hunger; and, as showing that the mild irritations which determine the illusional hallucinations of sleep when exaggerated may provoke them in the waking state, we need but refer to the numerous instances where shipwrecked or other travellers, after a long period of starvation, fell into a delirium tinged by similar visions.* Among the insane there are similar transitions between the pure hallucination, the illusional hallucination, and the pure illusion.

While disturbances of the visceral or general sensations may determine hallucinations having a certain semblance of the real about them, like those just cited, there is another class which presents less of the systematic and plausible character. These are visions of innumerable objects of the same kind : insects, fish, worms, maggots, snowflakes, or sounds of a hundred bells, a thousand whispers. In the dreamy state such hallucinations are probably referable to "luminous dust" impressions flitting before the closed' eye and tinnitus aurium as their starting point; in a diseased state they imply a deep nutritive or toxic disturbance in the

---

* A few years ago several lumbermen lost their way on the ice of Lake Ontario, and were found on a little island, from which it was difficult to remove them owing to the attractive vision of a splendid feast and a warm fire which occupied the minds of these persons, who had almost reached the point of death from cold and starvation combined. This winter two teamsters lost their way in the snows of Montana. Both were saved by the pluck and wood-craft of one of them, but he experienced great difficulty in preventing his comrade from darting away into the woods, where he claimed stood a man with a basket of provisions and a house with lights from which proceeded the noise of a carousal.

brain itself. Hallucinations of this kind are common features of opium and alcoholic delirium, of the delirium of meningitis, of the visions of the paretic dement, and are in these two latter disorders of grave import.

It is difficult to determine to what extent hallucinations in the insane are determined in their origin and recurrence by the faulty conceptions and reasoning of the patient. There are undoubtedly patients suffering from delusional insanity whose hallucinations are a sequel of their delusions. Just as some artists possess the power of almost hallucinating a face once seen, and faithfully transferring its lineaments to canvas, or as Goethe, absorbed in selfish reveries, suddenly saw himself before him on the road; so the delusional lunatic, intensely concentrating his morbid ideation in a religious, an erotic, an ambitious, a hypochondriacal, or a persecutory notion, develops corresponding hallucinations. Inasmuch as the hallucination strengthens and confirms the delusion opinion out of which it grows, the grave signification of this complicating symptom in the chronic insane becomes evident.

Again, there are cases where the hallucination determines the character of a subsequent delusion, just as there are a few—a very few—authentic instances where a dream has produced an insane delusion. Insanity of this character, as a rule, offers a better prospect than that where the delusion is a primary factor, and determines the development of, and becomes in turn confirmed and more firmly organized by the hallucinations it provokes.

Hallucinations are found with most forms of insanity, but, being often of a more evanescent character than delusions, they are not as frequently noted; they are also less readily announced, either because the patient imagines that the visions and voices are of a sacred character, and that it would be therefore a sin to communicate them to the "uninitiated," or because some unwise person, having previously ridiculed him, the patient, insane as he is, apprehends that his tale will be treated as a fiction again. There is, however, one form of insanity in which hallucinations are so prominent and numerous that it has been termed "hallucinatory confusion." Hallucinations are also characteristic of insanity developing in prisoners, and are here doubtless due to the favoring influence of darkness, solitude, and silence in producing hallucinatory phenomena. The same remark applies to the insanity of blind persons.

The true signification of insane hallucinations and illusions, like that of the insane delusion, lies in the fact that the logical apparatus, instead of correcting the error of the perceptions by the inhibitory influence of collateral perceptions and reasoning, incorporates them in an insane belief, or bases insane actions on them.

In briefly enumerating the varieties of insane perceptions, it is best to consider hallucinations and illusions together.

Visual hallucinations, or insane *visions*, may vary in intensity and distinctness from mere blurs, clouds, or haloes, to flashes of light, bright color perceptions and *fac-similes* of the reality. The perception of single faces and figures often engaged in some occupation, the details of dress and of the features being distinctly reproduced; or of noisome reptiles or dancing devils in larger numbers and lesser distinctness serve to characterize different varieties of insanity. Sometimes the same hallucination continues for years. A patient may see the same person or animal following him wherever he goes. At others a fixed hallucination may become the basis for transitory ones. A patient sees himself in heaven or in hell continuously, but the other inhabitants of the hallucinated locality are continually changing. In their complexity and expansiveness such hallucinations sometimes approach the abstract visions of the poet. This variety is found in chronic insanity, and is less favorable in its prognostic indications than the more confused and evanescent hallucinations of some acute derangements.

Illusions of sight generally relate to persons, and are like the corresponding hallucinations determined by pre-existing delusions. Either the patient refuses to recognize his relatives (generally one), and asserts that they are strangers, or —and this is a not uncommon occurrence in the asylum ward —greets and embraces others as his or her children, sisters, brothers, or parents. These combined illusions and delusions of identity are so powerful, that in one instance a patient who asserted that the writer was her son, though properly recognizing all her other surroundings, and very unmanageable generally, followed his directions and voluntarily went to the asylum, because she could not believe her own son would advise her against her and his interests.

Hallucinations of hearing are more frequent in chronic delusional insanity and in melancholia, than those of sight, and are particularly common in persecutory delirium. Not infrequently, hallucinations of vision and those of hearing coexist. A melancholiac sees and hears those who are mocking and pursuing him; the chronic delusional lunatic sees and hears the God, or his messenger, who commands him to slay himself or others. Often the patient has a sort of recognition of the fact that the voices and other sounds he hears are not in the outer world, and assigns their seat to his own head. One of the writer's patients heard the same melody indefinitely repeated at a spot under the left parietal boss. Others hear the voices as if these originated in other parts of the body ; if rhythmical, in the chest; if less constant and regular, in the abdomen. In the former case an illusional transformation of cardiac sounds, and in the latter of intestinal rumblings may be sometimes demonstrated to be the exciting cause.

There may be the greatest variety in aural hallucinations; they may be limited for years to the repetition of the patient's name. A lunatic with this symptom, whom the writer exhibited before the New York Neurological Society, and who heard his name called by every drayman in the street, after assaulting persons in consequence of the imaginary insults, travelled from New York to Philadelphia, and thence to St. Louis, with the object of getting rid of the " persecution." Finding that the voices continued, he went to Canada, thinking that in another country, at least, he would be let alone. Of course the symptom continued and it continues to this day. The otherwise intelligent patient had read treatises on insanity, admitted that the writer was correct in classing hallucinations as signs of insanity, admitted also that the voices on the street were not real, but persisted in declaring that the voices he heard at his place of business were genuine. As the writer walked with him to the place where the meeting of the medical society was held, he turned around and said, "That driver called out my name now; you must have been abstracted not to notice it." A half a year later he sent a telegram requesting the writer to come to his place of business and to convince himself, as well as to remedy the matter by presenting to his persecutors the inhumanity of calling out the name of a person who was as insane as he admitted himself to be. *

---

* This case is cited because it is the purest example of a single hallucination and consequent delusion continuing for a long period (over ten years) in the writer's experience. It aptly illustrates, too, that the doc-

Hallucinations of hearing are more frequently secondary to delusions than hallucinations of sight. This is in harmony with their relatively greater frequency in chronic delusional and melancholiac derangement, of which persistent delusions or delusional states are characteristic. Sufferers from hallucinations of hearing under these circumstances are particularly dangerous. They hear the command to kill some one, to throw their children into the fire, they hear promises of eternal bliss to follow as a reward of some other terrible deed, or they hear themselves accused of foul crimes, or the pursuing footsteps of officers of the law, and in subjective self-defence will kill or maim others; or, to escape their persecutors, may resort to suicide.

In other cases the hallucinations are more agreeable. Such are generally more unfavorable as to prognosis than the last-mentioned cases, although their episodial results may be less serious from a medico-legal point of view. The patient hears the "music of the spheres," celestial harmonies, angelic anthems, the tributes of his grateful subjects, or, like a paretic demented merchant from Chicago, he listens at the keyhole to the telegraphic messages of John or Jack, and replies, directing what disposal is to be made of his extensive domains, trunks of gold, and stock investments in that city. The conversations which patients hold with imaginary persons are, of course, entirely the creation of their own minds, and the replies attributed to others are nothing but the thoughts of the patient apperceived as spoken speech by him.* It has even been supposed that the mental processes of one hemisphere may reach the degree of hallucinations, while the other functionates in a less exaggerated manner, and registers the "hallucinated thoughts" as the spoken speech of an imaginary friend or enemy. Support has been supposed to be derived in favor of this theory from the fact that hallucinations of hearing are sometimes unilateral.

It is an interesting observation, the analogue of which was referred to in speaking of visual hallucinations, that

---

trine of a "partial punishability" of the so-called "partial lunatics" is an anachronistic absurdity. This patient committed a series of insane acts, more or less directly based on his hallucinatory delusion, several of them conflicting with the law, for all of which he was, in equity, as irresponsible as the paretic who commits a theft or an incendiarism.

* The paretic dement in question enunciated both the queries and replies himself.

while in earlier times the aural hallucinations of the chronic insane more frequently related to celestial voices, the introduction of the speaking-tube was followed by a transferral of the voices to imaginary systems of tubing in the walls of the patient's apartment. Later, the introduction of the telegraph was followed by the idea (heraldic of the telephone) that the voices were carried to the patients by systems of wires; this delusion latterly has been modified by the adoption of the name of the instrument.

Auditory illusions are generally based on entotic sounds; the tinnitus of cerebral anæmia * is frequently interpreted falsely by the patient. The distinguished comparative anatomist, Leuret, having been bled on one occasion, complained of the carelessness of his attendants in spilling acid on a costly marble table; he hearing. as he claimed, the effervescing sound thus produced. Moos, the Heidelberg otologist, reports a case where the *bruit*† produced in the *bulbus venae jugularis* was interpreted by the patient as the noise of railroad trains and factories, to escape from which the patient committed suicide.

Hallucinations of smell and taste usually coexist, a fact attributable to the close association existing between our conceptions of taste and smell, and an additional proof, if it were wanting, that hallucinations essentially originate in the brain.

Patients having hallucinations of sight or hearing, sometimes have the corresponding hallucination of smell or taste. Thus, patients who in the middle ages saw the devil, smelled his brimstone; those who hear conspirators suggesting poisoning, thereupon smell or taste poison in their food. False taste and smell perceptions are common in hypochondriacal insanity and in paretic dementia. In the latter disease they have in the writer's experience a

---

* That extreme variation in any direction of the bio-chemical state may produce disordered sense perception is illustrated by this case where anæmia is the fundamental cause, and by the occasional observation, that hallucinations of hearing appear on assuming the horizontal position or even on lying on one side, that is, with a relative hyperæmia.

† The importance of this *bruit* in producing hallucinations and insanity has certainly been overestimated by Moos. The ear probably accommodates itself to all vascular sounds, in a state of health; as it must do with regard to the pulsations of the intrinsic arteries of the labyrinth; just as a miller after a period ceases to hear the noise of his mill, and distinguishes the lesser sounds through it, which are inaudible to the newcomer.

bad import, and indicate rapid deterioration. This applies also to masturbatory insanity, of which disorder such hallucinations are almost characteristic. In these three forms of mental disorder the patients smell dead bodies, putrefying and filthy substances, noisome gases, seminal discharges, etc.

It is not uncommon to find that alleged hallucinations of taste disappear with the relief of a dyspepsia, showing that in reality the false sense perception is illusional.

False perceptions of general sensory impressions are probably always illusional in character. Of this kind was the illusion of a paretic dement in the writer's observation that he was infested with vermin, of a patient reported by Mickle whose skin was "taken off and hung up to dry," and of others who imagine their body infested by maggots, or sprinkled over with gold coins, or turned to some other foreign—generally valuable—substance. The degenerative changes in the spinal cord noted in such cases furnish the basis for the false perception. Hypochondriacal and hysterical lunatics find a rich field for false beliefs in illusions of touch and of the visceral sensations. The demonomaniac interprets borborygmi, pulsations in the head, intermissions of the heart beat, and heaviness at the pit of the stomach, as the assaults of the demons within. The delusional lunatic with a hypochondriacal tendency believes that his limbs are made of glass, stone, wood, or cork. The hysterical delusionist imagines that she feels the movements of a child or an animal, on the strength of her tympanitic distension and intestinal movements.

Even here, the influence of the age is felt, and those disordered perceptions which were interpreted as the result of supernatural influences in the middle ages are to-day described as electrical sensations, ascribed to magnetic and planetary influences, or attributed to the intrigues of secret societies.

CHAPTER V.

THE EMOTIONAL DISTURBANCES.

Disturbances of the emotional state are almost universal in insanity. With the exception of the congenital forms of derangement and those dating from childhood, all insanities exhibit as an initial feature a change in the emotions. Either these are more readily aroused than in health, as in the pathological fury * which is manifested by patients suffering from head injuries, epilepsy, and neuralgia, and particularly in the prodromal period of paretic dementia; or they are less easily aroused: the emotions are blunted, as in stupor, in the emotional indifference to friends and relatives of the melancholiac, in the last stages of paretic dementia, in the terminal deteriorations, and, above all, in alcoholic and epileptic insanity.

There are two forms of emotional disturbance which are pathognomic of the antithetical groups: mania and melancholia; and which, constituting as they do their intrinsic features, will be discussed together with these.

Great confusion has arisen from the use of the term EMOTIONAL INSANITY. If this term were to be limited to those forms of derangement in which the perversion of the emotions, in the strictest sense, is the essential and unvarying feature, it would be a useful designation for that general class of simple insanities comprising mania and melancholia. But, unfortunately, it has been applied in the very widest sense, and stretched to cover the conception of moral disturbance as well.

There are several psychological as well as psychiatrical reasons for separating the *sentiments* from the *emotions*. It may suffice to point to the fact that there are many patients in whom there is a painful emotional state who do not become less benevolent or less kindly disposed, though perhaps more indifferent to the world at large or their relatives, just as there are patients developing an expansive emotional state who do not entertain more elevated sentiments than they did before, and often, indeed, become more profane, vulgar, and cruel, presenting a great moral contrast with their emotional exaltation. The moral sen-

---

* "*Affect*" of the Germans; this condition has been frequently confounded with "*transitory mania s. furor*" from which it differs in the fact that while the memory for the acme of the former state may be abolished, it is not destroyed for the whole period of the disturbance as in transitory furor. In our forensic annals many cases of pathological (and non-pathological) fury may be found, where the plea of emotional insanity, transitory mania, moral insanity, homicidal mania, or epilepsy, has been made, with the result of confusing the criminal jurisconsulist and throwing discredit on forensic psychiatry.

timents may be unaffected while the emotional ones are disordered, and *vice versa*. While there is a bond of connection between them, it is in the alienist's experience not an absolute one.

Disorder of the moral sentiments may be congenital, and equivalent to a partial imbecility, as the father of American mental science, Rush, first pointed out. The memory and the reasoning powers may be so slightly affected in this condition that their deficiency is practically unnoticeable; or the reasoning processes—and this is more frequent—may be as perverse as the moral state. Moral perversion may be also acquired. It is a common accompaniment of advanced epileptic, and it is constant in masturbatory insanity. In paretic dementia and other terminal states it is also observed, but its mechanism is here somewhat peculiar, and the moral perversions of the deteriorating forms require a separate consideration. (See Part II.) In traumatic and alcoholic insanity, moral perversion, or a blunting of the moral sentiments, seems to grow out of the corresponding emotional obtuseness.

It may be advanced as a cardinal canon of psychiatry that in insanity the moral feelings are usually more or less dulled or perverted.* The deficiency of the moral feelings may, however, be of a different kind; in certain cases it may be a necessary result of intellectual enfeeblement, it may be due to an obtuse emotional condition, or it may be an original deficiency analogous to that lack of musical sense, or color-blindness, which may coexist with a fair faculty of language and good contour perception, just as that moral imbecility which authors call moral insanity may be found associated with fairly good logical powers in the abstract. An intense egotism is sometimes found to lie at the root of a constitutional inability of the individual to recognize any moral obligation to others. In such cases, abstract moral conceptions may be inculcated by education. But the subject of this condition, while he may be able to guide his conduct by these, regards them as purely utilitarian con-

---

* It would be unnecessary to emphasize that moral perversion even to the extent of criminality is consistent with insanity, had the contrary not been urged by medical witnesses on more than one occasion. It has never been asserted, I believe, that insanity improves the moral sentiments; yet the conviction of insane criminals has been effected by parading their previous records of irregularities and crimes as destroying the claim of insanity.

ceptions. They never become as organically fixed as where they have been developed from that inborn tendency which is the common heritage of normal mankind.

CHAPTER VI.

THE MEMORY AND CONSCIOUSNESS IN INSANITY.

Disturbances of the memory are common among the insane, either as temporary symptoms of the acute and transitory forms of insanity and the exacerbations of the chronic forms, or as progressive and permanent accompaniments of terminal deterioration.

AMNESIA, or loss of memory, is characteristic of the epileptic mental states, where, while the memory for all impressions prior to the explosion, as well as subsequent to it, may be unimpaired, the recollection of the events transpiring during the attack is either obliterated or clouded. It is found in degrees varying with the intensity of the intoxication in alcoholic and other narcotic insanities. It is more or less marked in the delirious phases of mania, melancholia, and the hysterical, parturient, and febrile states. Finally, it is a most important feature of paretic dementia and other forms of insanity with organic disease of the brain.

While the condition of the memory in the various forms of insanity can be more appropriately discussed in the second part of this volume, it is desirable to lay stress on two prominent facts at the outset.

In the first place, the current idea that all forms of insanity are accompanied by loss of memory is a false one; so false that the pretence of having lost the memory is under certain circumstances justly considered the strongest evidence of simulation. Quite aside from those sub-acute maniacal cases in which the memory appears to be morbidly sharpened for the time being, there are a large number of mild melancholiacs in whom the memory does not appreciably suffer. But, above all, there is a larger class of the chronic insane in whom for many years, and often through a long life, the memory remains as good as that of the average human being, and may even be noted for its faithfulness in regard to details. (See Monomania.)

Secondly, it is erroneous to infer that because a series of acts is carried out by a subject with an appearance of method and purpose, that these acts, their object and results, must necessarily remain registered in the memory. There are many automatic and even higher acts performed by epileptics in the post-epileptic state; but the entire period during which those acts are performed, may become an absolute blank in his mind when it returns to lucidity. In the case of a policeman, whose interesting history is detailed in the chapter on Epileptic Insanity, the patient went out imperfectly dressed, but donned his badge, seized his club, and patrolled a "beat"—the wrong one, it is true, but still he patrolled it in such a way as to attract no special notice. However, he knew nothing of it afterward.

Proof that still higher mental combinations and projects may be formed and executed, even in sanity, which afterward escape the memory altogether, is furnished by some of the rarer phenomena of somnambulism. Condorcet, the mathematician, solved a mathematical problem, which had worried him the day previous, during a somnambulistic state, and had no recollection of the fact the following morning. Similar observations are made on the insane.

Sometimes an entire period of life during which a patient has been insane appears blotted out of his memory, and in convalescing he may consequently consider himself to be still of the same age as when the disease began. But in most cases individuals recovering from insanity recollect what had occurred during their illness; and our juries have not unfrequently been moved to pronounce an unquestionable lunatic, claiming his discharge from an asylum on the ground of self-alleged sanity, a sane man, because of the detailed, circumstantial and connected account which the patient was able to give of the events leading to his commitment, and of those transpiring during an asylum sojourn extending over many months or years. This faithfulness of the memory is found in the milder cases of acute insanity and in monomania.* In more severe grades of acute

---

* Griesinger, confounding chronic mania with confusion of ideas or chronic secondary insanity with paranoia, claims that the memory is confused in the latter.

insanity, and even in some of the epileptic and particularly the toxic psychoses, the patient retains what the Germans term "*Summarische Erinnerung*," that is, a faint, sketch-like recollection of the main events occurring during the attack. Often the memory for past events is unimpaired, but the patient finds it difficult to reproduce his recollections, owing to a temporary depression of the associating apparatus. This is particularly common in melancholia. With such a condition the reception of new impressions is impaired, but it must be always borne in mind that the common statements of patients convalescing from an acute insanity, that they have no recollection of their insane period, must, with certain exceptions, be discredited, as these patients feel humiliated by their recollections, and waive every reminder by the easily-made claim of total amnesia.

A very curious disturbance of the memory is manifested in the condition known as *double* or *alternating consciousness*. It is exceedingly rare, and appears to be limited to the mental disturbances of menstruation and to periodical insanity.

This condition is characterized by the alternation of periods, in which the subject enjoys his memory and retains his sense of identity, with others, in which he fails to recollect the impressions of his healthy period, but possesses the faculty of learning new ones. In the next healthy period he recollects what occurred before the abnormal period, but does not reproduce every fact acquired in the latter. In short, the mental life of two distinct individuals seems to alternate in one person.*

It is the multitude of impressions and experiences registered in the organ of the memory that constitute the conscious *ego*. *Healthy consciousness is that condition in which the individual, while registering the impressions of the outer world to which his attention is directed at the time, correlates these*

---

*Some remarkable instances may be found related in Forbes Winslow's treatise on "Obscure Diseases of the Brain and Mind." It is a noteworthy circumstance that few of these are verified by more than one observer, and that few or none have been reported within more recent times. In one case, a girl who was outraged in the morbid period revealed this fact in the next morbid period, but had no recollection of it in the intervals. The occasional observation that epileptic patients whose attacks, manifesting themselves in automatic acts or in rhyming deliria, are repeated in the same way, as in a case reported by the writer in the *Medical Record* for 1881, presents some analogies to double consciousness.

with the summarized observations of the past. The sum of the observations constituting the *ego* is continually increasing; the scope of the *ego* becomes wider with every added impression, every correction of a perceptional or conceptional error, and with every new experience. The *ego* of the child is a different one from that of the adult; but it is the nucleus around which the *ego* of the adult is to gather. In other words, the *ego* of an individual is throughout the same complex unit, composed of the correlated and associated notional items collected up to a given time; but it is variable to this extent, that it is continually forming new bonds and adding new notional items to its mechanism in a healthy state. Consciousness is merely a single designation for that state in which this addition is being made. It involves, in the first place, the healthy memory of the past, that is, not a detailed memory of the entire past at the given moment, but that summarized sense of identity—the *ego*-consciousness which is as it were an abstract of the chief memories of the entire past; and secondly, the functional disposition to add to these memories and to incorporate them with the continuous *ego*.

So complex a mechanism as the *ego*, is, corresponding to its complexity, readily deranged, and it is its disturbance in insanity that constitutes the most essential feature of that disorder, the one whose analysis from a medical point of view is most important, and the one without which no medico-legal study of insanity can be reasonably attempted.

Self-consciousness may be disturbed in any one of its factors. It is always materially influenced by disturbances of the memory; and no grave disorder of the memory is supposable without a corresponding disturbance of the *ego*, for it deprives the latter of important component elements. But the intrinsically important disturbances of the ego-consciousness are of a different nature: they relate to the failure or the functional indisposition of the *ego* to incorporate the changing outer impressions, and to accommodate itself to them, or its inability to properly correlate these impressions with those already accumulated.

CHANGE OF THE SENSE OF IDENTITY is a result of the latter disability. The individual registering the impressions and conceptions of a given period of life, and combining them in a distinct union, without uniting them to those of the past, there result two *egos*, a present and a past one. Lunatics exhibiting this phenomenon may recognize in a shadowy way the conflict between the old ego and the spurious one; they then complain that they have been changed, and have become other individuals, in which case they will often speak of their former selves in the third person. More frequently the patient's spurious *ego* overwhelms the *ego* of the past, a complete change in sentiments and conceptions occurs, and a person previously a prosperous mechanic may so identify himself with the false character construed on a delusional basis, that it remains his supposed nature for life; he continues to his death to believe himself a king, a religious reformer, or a cheated inventor.

It has been occasionally observed where the patient has a divided identity that the conceptions of either the normal ego or the spurious one rise to the surface in the shape of hallucinations, and under these circumstances asylum inmates can occasionally be found sustaining dialogues between their older and their newer and morbid selves.

Sometimes the insane lose all sense of time, place, property, and propriety for a short period, to regain it again. This is particularly noticeable in senile and paretic dementia, in which disorders the patients may, even before serious illness is suspected, lose their way in the streets, go out without being properly dressed, leave restaurants without paying for their meals, post letters without any directions, sign documents or draw checks without method, and do a hundred other things which are like to those done by healthy persons in fits of abstraction; but they differ from them in this, that there is a real, substantial loss of consciousness, not a temporary blurring by an overwhelming but intense healthy mental process.

One of the most important states of consciousness in the insane is that which many melancholiacs and paretics manifest early in their disorder, namely: the consciousness of an impending loss or disturbance of reason. It is probably less the result of a true appreciation of the subjective mental phenomena than the direct result of strange intracranial or perhaps of visceral sensations, due to the nutritive or destructive lesion of the brain. In other forms of insanity the appreciation by the patient of his morbid state as a morbid one is the best augury of a happy termination of his illness.*

---

*The various forms of disturbance of consciousness, such as stupor,

It is a noteworthy fact that patients who have entirely lost their normal self-consciousness, and whose mental mechanism has been used to construct a false and absurd identity, are yet able to appreciate their objective surroundings in their real characters, while utterly unable to regulate their subjective beliefs by them. Thus a paretic dement with the most extravagant delusions looked down with contempt on another patient suffering from a mental obtuseness following cerebral hemorrhage as a "poor, paralyzed fool," and the writer has not unfrequently witnessed the badinage which patients indulged in against their fellows on a basis of mental infirmities actually observable in the latter. The case related by Zacchias, observed in a Roman asylum, can probably be duplicated in every larger institution for the insane. A visitor entering the one in question was shown around by a very intelligent cicerone, who explained the nature of each patient's insanity, drew out delusions, and otherwise made the visit of the stranger so interesting and profitable, that the latter thought him to be one of the asylum authorities, if not the superintendent himself. Finally they came to a little chamber, where the "cicerone" said (pointing out a patient), "Here is a sad case—the worst one in the asylum. This poor fellow thinks he is the Redeemer of mankind—the fool, when it is I who am the true Redeemer!"

The insane are not blind to the merits and demerits of the sane either; a fact important to bear in mind in their treatment. In more than one instance has the writer known the patients of an asylum to expose the ignorance of persons in authority with a sarcasm and wit which was far from unjustified by the facts of the case. A paretic dement, whose disorder was a complication of a pre-existing monomania, and who had earlier in his career been a proficient surgeon and dentist, successfully defied a colleague to enumerate the cranial nerves or their foramina of exit; burst out in a torrent of invective against such ignorance, and, without a penny in his pocket, announced the project to build a large school adjoining the asylum in which to instruct physicians in the rudiments of medicine

---

ecstasy, and those associated with delirium and epileptic states, are discussed in the second part of the book. The same disposition has been made of certain special disturbances of the will, and other signs which would otherwise require repeated discussion.

and the anatomy and physiology of the brain, "according to Bell and Magendie." A female monomaniac, who had escaped or been "liberated" from several asylums, ridiculed the faulty logic and ignorance of an asylum officer who was an advocate of phrenology, and who actually claimed in one of his annual reports that his great success in the treatment of the insane was due to his phrenological attainments. Shortly after the testimony of the medical witnesses for the prosecution in the Guiteau trial was given—an important feature of which was the unanimous statement by these gentlemen that insanity was never hereditary—the male patients in one ward of a large State asylum, whose superintendent had testified in addition, " disease is never transmitted," held a meeting and passed certain resolutions relating to the superintendent—grounding them on the fact, that almost the first questions asked of a patient's relatives on his arrival at that same asylum, related to the hereditary history. When the writer adds that the member of this "convention" who reported the incident to him was a patient who had claimed and obtained his liberty by legal methods, and who was undoubtedly insane still, also admitted that his fellow patients who had been "liberated" in the course of a liberation epidemic were still insane—which was true, though sane juries and judges saw it not—and that he himself suffered from an organic affection of the brain which had affected his mind—which was also true—the reader may form some idea of the erroneous nature of the views on the strength of which physicians unfamiliar with the insane pass on the question of mental soundness, believing that a lunatic must be a lunatic to his fingers' ends, and must be at all points and at all times unreasonable in his thoughts and absurd or extravagant in his acts.

The mind of an insane person may be, and in some instances is, a more elaborate mechanism, one with more individual components, and of consequent wider scope as to its combinations, than the mind of many a sane person. But the mental factors of a healthy, inferior mind, if fewer in number and simpler in their associations, are well arranged and united into a consistent and continuous *ego*. It is the failure to effect this union that constitutes the chief element of what is so aptly called *derangement* and *alienation*.

CHAPTER VII.

THE WILL IN INSANITY.

There is probably not a single case of insanity in which the will in the widest sense is not at some time or other in the course of the illness more or less disturbed. The free determination of the will in certain directions is perverted by delusions; it is hampered by delusive interpretations and by depressive emotional states; finally, it is disturbed by delirium, confused by incoherence, and abolished by dementia. The patient who, apparently of sound mind to the laity, in other directions, believes that if he looks out of a window he will poison the world with a glance, and consequently persists in staring at a wall or in keeping his eyes closed, has his will perverted by the delusion. But in all these and similar cases the will is not intrinsically deranged, but misapplied. As far as its strength is concerned, and irrespective of its delusive basis, the will of an inspired or calculating lunatic is analogous to that of a fanatic or a determined adventurer: it may indeed be remarkable for its firmness, but for firmness in a direction which is determined by disease.

It is thus often seen that, as far as the deliberate planning and execution of an insane act is concerned, it differs in no wise from the acts requiring planning and deliberation in the sane. Yet the original motive may be the most extravagant delusion or project, the most disgusting morbid propensity, or the most horrible imperative impulse.

In those forms of insanity in which the *ego* is exalted there is a corresponding exaltation of purpose. The maniac, thinking himself to be physically and mentally powerful, becomes as extravagant in his undertakings as he is in his self-esteem and his word-delirium. In other words, he becomes *hyperbulic*. But, owing to the multitude of conceptions crowding on him, and the corresponding multiplicity of purposes, some of which may even conflict with each other, the projects and their execution in the maniac and maniacal paretic appear to be unsystematized and boyish.

In insanity with depression, the opposite condition of affairs, or *abulia*, is found. Here the patient is unable to exert his will, either on account of the arrested and impeded ideation, or because everything appears blank and hopeless to the patient. In melancholia, however, of that form in which physical depression is least marked and in which delusions are absent, the most appalling forms of combined suicide and homicide, revealing a most determined if a morbidly based will, are common occurrences.

The inability of certain patients to rid themselves of absurd thoughts, of aberrant trains of ideas—*Grübelsucht*—and of morbid fears—*Mysophobia, Agoraphobia*—is the expression of a disturbance of that higher phase of the will by which the healthy human being is able to limit his attention to the real, or to let it travel off in the realms of fancy as his choice may direct. In the insane with imperative conceptions this choice is rendered less free or impossible by the dominant idea.

The French have given the name *manie* or *folie raissonante* to a form of insanity in which the reasoning powers in the abstract suffer little or not at all, while the will is greatly disordered. Under this unfortunate term a chronic disorder properly classed under monomania, as well as the lighter form of acute mania, have been confounded. It is observed that such patients can reason plausibly, speak intelligently, but commit absurd acts, whose absurdity they may recognize and apologize for afterwards. These subjects may even be noticeable for their keen wit, apparent shrewdness, and the skill with which they seek to conceal or to palliate their acts. But, in studying the career of such individuals as a whole, it is discovered that their whole life history consists of a series of such absurd and insane acts; and a closer analysis shows that where these are not the outcome of imperative conceptions and morbid propensities, they are based on a reasoning which, however superficially bright and to that extent misleading to the laity, is as faulty logically as the acts which it provokes.

CHAPTER VIII.

THE PHYSICAL INDICATIONS OF ACQUIRED INSANITY.

The great central organ whose disordered states are sometimes manifested in mental alienation is not only the mirror of the outer world, as Wundt aptly expresses it, but in its ganglionic masses and expansions it receives fibres connected with every visceral and other somatic periphery, and consequently reflects and influences the activities of the body itself. The impressions which are utilized in the building up of conceptions, the motor innervations which in their intricate association mediate the expression of the will and of symbolized thought, and the visceral and vasomotor innervations which stand in such intimate relations to the emotional states, are all projected in the organ whose diffuse disease processes may determine mental alienation. It is not surprising, then, that functions mediated through the same organ should in its diseases be disturbed together; in other words, that physical indications should mark the development of mental disorder.

The special interest which attaches to the physical indications of insanity is twofold. In the first place, as they are more tangible to our instruments of precision than the mental signs, they are of great value in indicating the nature of the presumable physical disorder underlying insanity, in the frequent absence of positive *post-mortem* evidence. In the second place, the physical signs constitute, when well marked—as they are in certain forms of insanity —such unfailing criteria of those forms, that they are not only of great aid in the diagnosis of these affections, but also of efficient service in the detection of simulation.

The physical signs of insanity naturally fall into two groups: first, those which are initial or accompanying phenomena of an insanity affecting previously healthy individuals; second, those which are indications of a transmitted or acquired constitutional taint; in other words, signs of a predisposition to insanity. These will be considered separately.

THE CONDITION OF THE RETINA IN INSANITY has been extensively studied, and the introduction of the ophthalmoscope was followed by the most extravagant claims with regard to its importance in the diagnosis of insanity. It is generally recognized that this instrument is of great value in diagnosticating other cerebro-spinal affections, such as posterior spinal and diffuse sclerosis, tuberculosis of the pia, and tumors of the brain. It is an invaluable adjunct in determining the existence of those phases of kidney disease which are most likely to be associated with mental troubles, and it should always be employed in suspected paretic dementia, syphilitic and other forms of insanity with organic disease, because these may be accompanied by changes in the ocular background hinting at if not demonstrating the nature of the disorder. (See Paretic Dementia.)

Aside from insanity with organic diseases, the psychoses do not manifest such regularity in the retinal condition as to justify the loudly-trumpeted claim of Bouchut, that the ophthalmoscope is also a "cerebroscope," nor the fanciful and bold assertions of those who have actually attempted to exalt this instrument to the position of a medico-legal test of insanity! In maniacal and delirious conditions generally, there may be a hyperæmic flush of the optic disk, and in stupor and passive melancholia this region may be very anæmic. But these deviations from the ideal healthy state are not more considerable than those encountered in any large eye-clinic whose material is derived from a sane population. The writer has observed an extreme hyperæmia of the disk in three cases: one that of an acutely delirious maniac, who died of maniacal exhaustion, and whose brain exhibited commencing organic changes; the second, that of a paretic dement, the inception of whose disorder was marked by overwhelming hallucinations of fiery visions, and whose illness ran a furibund and rapid course; the deep flush found in the latter case was exceeded in a sufferer from a saturnine encephalopathy with mental derangement which terminated in recovery. In all such cases other and more conclusive physical and mental signs are abundant; and while the ophthalmoscope will continue to remain a valuable instrument of research, it is in no sense an absolute test, and only in insanity with organic disease a collateral sign of value.*

The observation of THE ELECTRO-MUSCULAR REACTIONS has led to many interesting and to some contradictory results. In the present state of science it has, like the retinal examinations, a positive value only in insanity with organic disease. It is found, aside from this, that all forms of insanity in which irritation and excitement predominate, are marked by exaggerated irritability of the muscles, which, however, may pass to the other extreme in maniacal exhaustion. Exaggerated reaction is found in mania and

---

* In pubescent insanity and other forms of alienation, with which masturbation is a marked factor, the writer has found extreme pallor of the optic disk. But this pallor he has found in a much more intense development in five sane masturbators who had such a degree of retinal anæmia that they had gross limitation of the field of vision, without structural change.

agitated melancholia, diminished reaction in atonic melancholia, dementia, stuporous insanity, and katatonia. But these reactions may vary in different parts of the body, and it has even been noticed that the reaction differs on opposite sides of the body in melancholiacs. The finer questions of electro-diagnosis are covered by the question of the existence or non-existence of central organic disease, and of certain metallic intoxications which are sometimes complicated by insanity (saturnine insanity) and, as thus far determined, they have no unquestioned bearing on the problems of insanity proper.

DISTURBANCES OF SENSIBILITY are frequent in insanity. Anæsthesia and analgesia may be so pronounced in demented patients, that severe burns and traumatic injuries are not appreciated by them. Hyperæsthesia and paræsthesia are often found in mania; hyperalgesia harmonizing with the painful emotional state is sometimes a feature of melancholia.

Jolly, Benedict, and others have vainly endeavored to associate the existence of hallucinations with disturbed reaction of the auditory nerve; but nothing could be very much more fallacious than the theory of the former, that the hallucination of whole sentences heard in the voice of some person known to the subject during an electrical examination, indicated a disturbance of the auditory nerve. Such disturbances essentially involve the organ where complex hallucinations can be alone evolved—the cortex of the cerebral hemispheres.

In paretic dementia of the "ascending type," the sensory disturbances of locomotor ataxia are appreciable, even before the mental symptoms are well developed. The well-known laryngologist, Elsberg, passing through the writer's office, recognized a patient whom some years before he had treated for a throat catarrh, and remarked that he had rarely seen an instance of such pronounced laryngeal anæsthesia as this patient then presented. The patient later showed signs of the incipient stage of a second exacerbation of paretic dementia, and there was no other anæsthesia than this one discoverable then; subsequently general anæsthesia and ataxia became marked.

In many instances the determination of sensibility is rendered impossible by the dementia, the atony, the restlessness, and sometimes the delusional mutism of patients.

Howard, of Montreal, claims that anæsthesia is pronounced in epileptic and other forms of dementia.

A great deal has been written about the COMPOSITION OF THE BLOOD AND THE URINE IN INSANITY, but little of this literature is of interest to the practitioner. While there are those whose enthusiasm for new therapeutic measures, and whose ignorance of the true nature of insanity, led them to believe that the vitiated blood composition, which they supposed the foundation of all forms of mental disorder, could be removed together with the insanity by transfusion; scarcely less absurd propositions have been made, based on the alleged fact that the phosphates are continuously being drained from the brain in insanity, and pass out in the urine.*

It is true that changes in the blood composition occur in insanity. Without committing ourselves to the view of Sutherland, that the non-agglutination of the blood-corpuscles in rows is an indication of certain forms of alienation, and, remarkably enough too—as that author claims—more constantly an evidence of insanity in the male than in the female patient, it may be conceded that anæmia of an extreme kind, sometimes reaching the degree of a pernicious anæmia, is a feature of insanity with depression and stupor, as well as of apathetic dementia. But even in melancholia, particularly in that variety which is termed *melancholia sine delirio* (see Melancholia), the writer has failed to find any anomaly of the general nutrition, the blood composition or the secretions in several—exceptional—instances. Excluding the fatal forms of insanity and those above mentioned, the anæmia of asylum inmates is not so much more marked than in the corresponding poor-house or tenement-house population, as to constitute a valuable criterion of insanity. It is when present a noteworthy collateral phenomenon, and rarely anything else.

---

* The suggestion of a writer in the *American Journal of Insanity*, that phosphorus is to the brain in insanity what iron is to the blood in anæmia, is an outgrowth of the obsolete theory that a diminution of phosphorus in the brain is characteristic of insanity. Brain-starvation is not regarded as the ever-present danger, it was once considered; indeed Siemens (*Arch. of Psychiatrie*, XV.) showed that forced feeding was carried too far under the old view. To regard insanity as an exclusively nutritive disorder, was consistent with views formerly held at Utica, that insanity can never be hereditary, and that mania and melancholia are identical psychologically and pathologically,

The solid constituents of the *urine* are often increased in insanity, and it is particularly in the exacerbations of paretic dementia that this increase becomes so marked as to suggest the possible explanation of the loss in weight observed at these times, on the ground of renal elimination. The old view that it is particularly the phosphates which are increased in the urine of patients suffering from excitement, and that this is the expression of a chemical exhaustion of the brain, is not borne out by more recent and careful examinations. In the writer's experience it is unusual to find the increase of phosphates in the urine of the insane to exceed the limits found in patients suffering from spinal irritation and the masturbatory neurosis.

In those forms of insanity and their episodial outbreaks in which there are marked changes in the blood-pressure, albuminuria is frequently and hyaline casts are occasionally observed. This is the case with the maniacal and apoplectiform phases of paretic dementia, the post-convulsive epileptic states, and acute alcoholic insanity.

While the examinations of the various organs and fluids above alluded to have rather a collateral than a fundamental value, those of the circulatory apparatus, though they have not yielded any indications of insanity destined to occupy a place side by side with what are after all the important and essential signs of insanity: the mental disturbances; they have at least pointed out paths of research which if conscientiously followed will enlighten us as to the real nature of the maniacal and the melancholic attack, and of the exacerbations of paretic dementia. It is well known that the pulse-trace is modified in health by various emotional states, and it is in harmony with this that the SPHYGMOGRAPHIC EXAMINATION in insanity connected with disturbed vaso-motor states reveals such suggestive changes from the normal pulse character.

In the secondary and terminal deteriorations the pulse-trace is that designated by Wolff as the *pulsus tardus*: the ascent is not marked by a sharp apex as in health. This expression of the normal vascular tone is lost—in other words, there is a degree of vascular paresis. In mania there is as a rule normal and variable tension, while in acute delirium there is a rapid deterioration of the pulse-trace, which ultimately becomes monocrotic, with the apex of the ascent or percussion stroke greatly rounded, thus resembling the pulse of the terminal states with cardiac anenergy or enfeeblement, and with vaso-motor paresis. A similar condition is found in the later stages of paretic dementia, in the earlier stages of which illness the pulse-trace has a different character, to be hereinafter referred to.

The TEMPERATURE OF THE INSANE does not present any constantly marked variations except in melancholia, where it is commonly what is termed sub-normal, that is, between 96·5° and 98° Fahrenheit (36·2°—37° C.), and may in stuporous insanity and the cataleptic phases of katatonia sink to 95° and less, gradually diminishing with a fatal termination. In paretic dementia it varies within narrower limits and in different directions in the various types and phases of that disease. Howard, of Montreal, claims that the lowering of the temperature in insanity generally, is more common than the writer is able to admit; but it seems from his figures that prolonged insomnia may lead to a lowering even in maniacal disorders. As a rule the temperature is altogether normal in simple mania, and even in the most violent frenzy of that psychosis when the forehead to the touch appears to indicate a fever-heat, the thermometer may not show a higher temperature than that found in the sane. The not inconsiderable variation of the temperature in the sane,* and its quite constant decrease in old age, should be taken into account in attributing a value to the temperature variations in insanity. Marked variations in this disorder are probably due to involvement of the cortical thermic centres.

THE FUNCTIONS OF THE INTESTINAL TRACT show great disturbances in some forms of insanity. Indeed, the ancients and those recent alienists of the somatic school of Jacobi, according to whom the os sacrum covered more of the seat of insanity than the os frontis, were inclined to believe that because insanity was sometimes averted or otherwise favorably influenced by a brisk cathartic, that visceral disorders might be the primary factor in the production of insanity. Others, like the renowned Van der Kolk, were induced to divide all forms of insanity into two groups, idiopathic and sympathetic, according as their primary origin was supposed to be in the brain, or in some other organs, notably the viscera. It is now generally admitted that in the vast majority of cases the essential foundation of insanity is in the brain, and that only exceptionally do peripheral disturbances cause or materially modify the disorder. When the

---

*Determined by Landon Carter Gray, of Brooklyn.

same set of hypochondriacal delusions is found in two brothers, both of whom are found in the autopsy to present the same constriction of the intestinal canal, the hasty guess might be made that a common cause —a visceral disorder—had produced the same mental disturbance in both, whereas in reality insanity was hereditary in the cases alluded to, and the visceral disturbance was as much a modifying factor as any other accidental circumstance, affecting the brothers equally, might have been.

In atonic mental states the atony of the general muscular periphery is shared by the intestinal muscular tube as well as by the œsophagus. Under these circumstances difficulty or inability to swallow, tympanitic distension and constipation are present. These conditions may alternate with an adynamic form of diarrhœa, a most undesirable complication of the stuporous and melancholiac states: while in mania a slight diarrhœa, due to a different condition; overactivity of the normal peristaltic movements; is neither an uncommon nor an unfavorable symptom. In that variety of melancholia in which there is atonic contraction of all the muscles, manifesting itself in a "frozen attitude" and tetanic contraction of the facial muscles, there exists a similar tonic spasm of the intestinal tube and even of the urinary bladder; the former leading to constipation, the latter to spastic dysuria.

Anomalies of the appetite, such as anorexia, which is characteristic of the atonic and depressive forms of insanity, and the *bulimia*, or ravenous appetite of mania and the maniacal phases of paretic dementia, are partly of direct psychical origin, or the indirect results of the mental state as manifested in secondary atony or anæsthesia of the stomach.

THE CONDITION OF THE SKIN AND ITS SECRETIONS is sometimes greatly altered in insanity. The skin may be dry and parchment-like, particularly in melancholia; in all atonic states it has a clammy feel, and in dementia may be exceedingly oily. In mania the skin is more apt to be moist. The so-called *effluvium* of the excited insane is due to their ex- increased cutaneous secretions and the decomposition of the sudor and macerating epithelium. It has not—in the writer's experience—anything characteristic; and probably those observers who consider themselves able to diagnosticate mania with their regio olfactoria have done so on the strength of an unanalyzed experience in the maniacal wards. Here there is indeed an odor rarely encountered elsewhere, but it is due to a combination of the odors of other excretions with those of the skin, which is rarely encountered in ordinary hospitals. Keep a maniacal patient clean, give him a frequent change of linen, and he will not present anything specially indicating his disorder to the sense of smell! With certain idiots a musky odor has been observed by an Italian alienist, and it is more reasonable to suspect a specific change here than in the acute psychoses.

Some believe that the hands of patients addicted to masturbation are remarkably clammy and blue in contrast with the rest of the body. This fact, which is borne out by the writer's experience, is probably due to the enfeeblement of the distant capillary circulation in these patients, and is not peculiar to this condition, but is found in stupor and melancholia as well.

THE MOTOR DISTURBANCES in insanity, intimately allied as these are with the mental symptoms, and indeed often rather of a psychical than of a strictly somatic nature, are among its most prominent indications. In the case of at least three forms of insanity the absolute diagnosis cannot be made when certain motor symptoms are absent, *i.e.*, in paretic and syphilitic dementia and in the alcoholic forms. And in many other instances, as in the motor agitation of mania, the atony and catalepsy of other forms and the automatic movements of the states of mental enfeeblement, they constitute characteristic features of the psychoses.

Paresis of a peculiar type is found in paretic and syphilitic dementia, and also after acute delirium, as the result of destructive lesion of the nerve-centres. Contractures may similarly result, while convulsions may be among the first indications, as they are almost constantly found to be among the last accompaniments, of paretic dementia. The convulsions found with epileptic, the peculiar movements occurring with choreic and hysterical psychoses, as well as the motor and co-ordination disturbances of insanity resulting from sclerosis, tumors, and meningitis, are to be excluded as not being genuine evidences of the insanity itself, for they are all found in epilepsy, chorea, hysteria, sclerosis, tumors and meningitis without insanity.

Ataxia is a frequent phenomenon of paretic dementia, and characterizes one of its types; aphasia and anæsthesia are likewise common in this disorder, which is a perfect repertory of all possible motor disturbances.

The muscular over-activity of mania and the muscular passiveness of melancholia are psychical expressions of these mental states, and will be considered with them.

A tetanic condition of the muscles is found in a number of atonic melancholiacs, and its disappearance without co-incident amelioration of the mental symptoms is an indication that the patient will not recover, but is passing into pa sive dementia.

The cataleptic state of the muscles is one of the characteristic features of katatonic insanity.

A special interest attaches to the innervation of the facial muscles in all forms of insanity; to the experienced alienist there are no more suggestive signs of mental disorder than the *insane expression, manner and attitude*. This is not to be wondered at. 'What is a more faithful indicator of a condition—which, however morbid, is at bottom a mental one—than the thoughts and their expression in the patient? Just as the play of the facial muscles, the expression of the eye, the movements and attitude of the body, are indications of the emotions, and to experienced mind-readers of the very thoughts of the sane; so the animated or angry expression of the maniac, the sad and thoughtful gaze of the melancholiac, the blank face of the stuporous and the silly countenance of the pubescent lunatic are indications to the practical alienist of the form of insanity with which he has to deal. Indeed, in the absence of written or spoken mental symbols in certain cases, the expression and attitude of the patient are the leading indications of his insanity. They acquire a special value from the fact that the most expert actor finds no sign of insanity so difficult to feign as the insane expression. Insane thought documents itself not in words and writings alone. The monomaniac who believes himself a king betrays it in his vain attitude and supercilious scowl; the paretic dement who thinks himself a millionaire assumes a lofty step; the erotomaniac exhibits an affectionate look, the religious lunatic that of a fanatic, a zealot or a mystic, while the nymphomaniac betrays her condition, if not in every gesture, in every glance.

In those forms of insanity which will be considered under the head of terminal deteriorations, in some imbeciles, and more rarely in monomaniacs, peculiar movements are observed, which are without purpose, but which it seems the patient has to go through with, as if driven by some unseen power. Samuel Johnson, who was a case of monomania with imperative conceptions, was thus found one day in Twickenham meadow, surrounded by a laughing mob, amused at the great lexicographer's sitting astride a book and imitating the motions of a jockey riding at full speed. On other occasions it was noted that this man could not go through a certain street without touching every post on the way, and if he forgot one, he turned back to make good the omission. On crossing a certain threshold he regularly turned a pirouette, etc. Asylum inmates exhibit similar odd movements; some will go through movements very like the genuflections and crossings or self-castigations of a religious devotee, without being able to assign any reasons for doing so; others will walk in a certain circle, triangle, or square, and bring down their feet with special emphasis at given intervals; still others appear to be continually brushing away something. As all these movements are disconnected in their motive from the thoughts of the patient, and are thus analogous to the imperative conception (Chapter III.), they are termed *imperative movements*. Some believe that they have their origin in delusions, which are obliterated by dementia, only the movements remaining and becoming a habit. But this explanation certainly does not cover many cases, particularly such as Johnson's.

Another class of movements is found in idiotic, imbecile and extremely demented patients: these are the automatic *rhythmical movements*. Patients of the kind mentioned may for months exhibit the same movement when awake, either in oscillation of the entire body from side to side, or from before backward, or of the head, the eyes, the mouth or the hands alone.

The *pupil* exhibits marked deviations from the normal standard in insanity. In passive melancholia, katatonia, pubescent and stuporous insanity, it is equally dilated, sometimes to a maximum degree. A dilated and mobile pupil is also very generally met with among epileptics. In aged patients and with certain anomalies of refraction this may be greatly modified, and alcoholic epileptics and alcoholic melancholiacs may have medium-sized or even narrow pupils. In mania the pupils are either normal or contracted, and only in mania from starvation are they dilated. Kirn has observed an initial pupillary spasm in periodical mania, while in katatonia, the writer has observed a disappearance of the dilatation as the patient passed out of the cataleptic state, spontaneously, as well as when this occurred under the influence of nitrite of amyl.

While in all the conditions thus far named the pupillary changes are not much greater than those occurring in the non-insane population, in paretic dementia they have a high diagnostic value. Extreme initial myosis and irregular contraction, followed by paresis of the sphincter, are very frequent accompaniments of this disorder, and sometimes indications of the precise phase of its progress.

Speech disturbances in insanity, aside from those due to organic disease, involve changes in the rhythm and manner of speaking, and, like the facial expression of the patient, indicate the nature of his disorder to some extent, aside from the actual contents of his sayings. Thus a rapid flight of words is a marked feature of mania and the maniacal phases of paretic dementia, while a slow, impeded speech is observed in melancholia and dementia, and this sinks to absolute mutism in the stuporous, atonic, and cataleptic states, as it also does under the influence of overpowering delusions and in hysterical insanity.

The "verbigeration" of katatonia and the "echolalia" of imbecility are rather psychical than motor manifestations in their essence and will be considered with the special forms.

Just as the contents of insane documents are of great value in determining the nature of a suspected mental disorder, so the *handwriting* of the insane in itself exhibits many suggestive peculiarities. In those forms of insanity which are marked by motor disturbance, a peculiar tremulous character of the handwriting is manifested; in monomania the paper is covered with underlining marks, queries, exclamation points, dashes, and strange symbols. Some patients write almost microscopically fine, others very coarsely, still others combine these extremes in one and the same document; while one patient in the writer's observation used the same page six times, writing once transversely, then longitudinally from top to bottom, then longitudinally from bottom to top, then upside down, and twice diagonally. In another case, that of an insane author who printed his own book, every form and variety of type was employed by him; and a good example of the manner in which the symbols and italics are abused by the insane is furnished by a poem of which the following is the conclusion:

PHYSICAL INDICATIONS OF ACQUIRED INSANITY. 77

> So take this advice before I go;
> (¡ Ah every MISER here below!)
> And save yourself from the horrid fate—
> When you think to enter Heaven's gate;
> To be told to stand aside and wait—
> And *let the poor little child pass by*,
> ¡ That he left *to freeze, and starve, and die !*
> While he had hugged the dross of EARTH;—
> And *without pity saw it* die!
> And heedless saw its mother's eye,
> Sunken in sorrow and dimmed by grief,
> When he would offer no relief;—
> *Hoarding life's blessings like a thief!*
> ¡ Ah! C-H-I-N-G-O C-H-I-N-G-O L-I-N-G O L-I-N-K—
> ¡ *Now pitches him down to* HELL, I think!—
> Where let the miserable MISER go!
> Only fit for horrid HELL below,
> 'Tis the doom he sought—so let him have it !
> The FIRE OF HELL—so let him brave it !
> ¡¡¡¡ C-H-I-N-G-O! ¡C-H-I-N-G-O! ¡ L-I-N-G-O! ¡ L-I-N-K !!!!!—
>   [—¿¿¿ 13034 ???—]

The writer has also observed analogous oddities in the compositions of an insane musician, who, writing a "money-polka," had a peculiar symbol inserted among the notes, indicating that at such and such a point bags of money must be jingled to the tune. But these formal aberrations in writing are motor expressions of morbid projects, and rather belong to the strictly psychical indications of insanity. They are here referred to only because it is customary to regard the handwriting and its mannerisms in the insane together.

THE TROPHIC DISTURBANCES OF INSANITY in the widest sense comprise the congenital anomalies to be considered in the next chapter, as well as that general nutritive disturbance of insanity, which manifests itself as a rapid loss of weight in melancholia and stupor, a less considerable loss in mania, as an alternating loss and gain in circular insanity, and in rare instances as a progressive pernicious anæmia. The congenital anomalies are indications of the evil influence of arrested or perverted development of the brain in peripheral development, and the acquired trophic disturbances are the outcome of the profound influence which the matured brain exerts on the bodily states. Recently the term trophic disturbance has come to be used in a more limited sense, excluding the above, and in this sense as applying only to positive and local manifestations of deranged nutrition and growth the term will be here considered.

Like other gross signs of nervous disease the trophic disturbances are most varied in character and constant in occurrence with insanity associated with organic disease. A peculiar change in the bones, by which these either become unduly brittle, or so soft that a knife suffices to divide them (in one instance in the writer's experience, to cut the calvarium in places and to open the thorax) is found in about a quarter of the patients suffering from advanced paretic dementia. Not unfrequently charges of violence and abuse are based on these spontaneous conditions, which latter are entirely dependent, like the changes which occur in the articular extremities of the long bones in locomotor ataxia, on irritative and destructive lesions of the brain and spinal cord. Almost the entire train of trophic disturbance found in disease of the spinal cord—muscular lipomatosis and deuteropathic muscular atrophy, phlegmonous ulcers, bullæ, atonic decubitus, and that malignant decubitus which spreads with such remarkable rapidity as to defy all efforts of treatment—are all found in paretic dementia.

Herpetic and other eruptions, peculiar forms of pigmentation, premature grayness, and particularly the blanching of the hair in unilateral or symmetrical patches,* have been observed in acute insanity, and in one instance in the writer's experience, that of a negro paretic, a blanching of the skin in irregular anastomozing patches, was observed just prior to an accelerated progression of the disease.

A noteworthy trophic disturbance is that known as OTHÆMATOMA. This consists in the formation of a sanguineous tumor of the external ear. Anatomically it consists of a hæmorrhage into the substance of the aural cartilage proper, and is probably due to a diseased state of this cartilage or its vessels; though the possibility of its production by violence alone cannot be excluded, for this event has been recorded by good observers as occurring in sane subjects. The writer has been struck by the fact that in large pauper asylums with few attendants, and those of an undesirable class, and with large numbers of ill-fed and irritated patients crowded in the wards, the physicians were able to show numerous cases of othæmatoma; while in private institutions and continental asylums there was sometimes not a single recent case to be found, and the

---

* Also observed in spinal irritation with predominant vaso-motor symptoms.

scars of those who had had othæmatomata frequently dated from a sojourn in some infirmary or other less well-managed institution. Whatever the fact may be, it is generally admitted that if violence is a factor in producing othæmatoma, it is a subsidiary one to the structural predisposition of the patient's tissues. Injuries which would fail to produce any noticeable reaction in the sane, produce othæmatoma in certain lunatics.

This symptom, also termed the "insane ear," was found in 48 out of 2,297 patients by Kiernan. He found it with every form of insanity except with monomania, which was not at that time differentiated from secondary confusional insanity by this observer. It seems to be least common in melancholia and most common in epileptic, paretic and terminal dementia, and has been occasionally observed to herald the approach of a delirious exacerbation. Othæmatoma, from its common occurrence in incurable forms of insanity, has been considered of grave import, but it is not in itself of prognostic value, and a few cases of acute insanity have been reported to end in recovery, although this symptom was present.* It is now generally considered to be a collateral trophic disturbance of insanity which may require special treatment, but which has no intrinsic signification. If uninterfered with, the tumor undergoes shrinkage, the blood is absorbed, and great deformity results. This sequel may be to a great extent avoided by making numerous small punctures over the most prominent portions of the tumor, and cleaning out the half-clotted, half-fluid dark blood and tissue *débris* constituting its contents. If it should be shown that othæmatomata, when found in terminal or paretic dements, imbeciles, and epileptics, are as constantly associated with blood cysts of the arachnoid, as the author has found to be the case in a limited number of examinations, this sign may attain a relative value of a kind which cannot now be assigned to it.

With regard to the sudden visceral complications of insanity occurring without any external influence or local

---

* It is significant that regarding the three recoveries which Kiernan observed among forty-eight patients with othæmatomata, that author, in a copy of his paper, sent the writer two years subsequent to its publication, inserted marginal manuscript notes to the effect that all three patients subsequently returned to the asylum; one as a paretic dement, a second as a melancholiac dying in an attack of *raptus melancholicus*, the third as a (probably) periodical maniac.

cause, and frequently without a local predisposition—such as pulmonary gangrene, infarctions, pleural ecchymoses, hæmatoma of the lower bowel, necrotic changes in other organs distant from the brain—these are almost without exception the result of gross vaso-motor and trophic influences provoked by exacerbations of the central nervous affection. They hence follow the apoplectiform and epileptiform phases of paretic dementia, the seizures of epilepsy, and the hæmorrhages and other acute cerebral complications of terminal dementia. In some instances, as in the pulmonary gangrene of acute insanity, other factors, such as general disturbances of nutrition and pulmonary atelectasis from respiratory stagnation, may participate in the production of the lesion.

It is of considerable medico-legal importance that in the insane, black and blue marks of a kind resembling bruises may appear spontaneously. In some cases these marks approximate a marbled character; they are not as linear in direction as marks made by blows with a stick or strap, and hence cannot be readily confounded with the marks produced by such implements. But trifling bruises and mere pressure may produce marks resembling the latter in demented patients.

In one form of insanity—*delirium grave*—the trophic disturbances are so marked in all patients who live any considerable length of time, that they may be considered among the characteristic signs of that disease. They resemble the changes noted by Mitchell after peripheral nerve-injuries, and by others after cerebral hemiplegia. (See Delirium Grave).

In some phases and forms of insanity, notably in the earlier stages of paretic dementia, it has been found that wounds and ulcers heal with remarkable rapidity, more quickly by far than in health. With this the general nutrition of the patient appears to be in an excellent, if not an extravagant, condition. Later in the same disease the reverse occurs: slight injuries are followed by rapid and destructive sloughing, and with this there is a general deterioration of the bodily state.

In maniacal conditions a favorable influence of the insanity on existing bodily diseases has been frequently observed; a convincing proof, if such proof were needed, that the cant that all lunatics are "sick men," in the sense that their bodily mal-nutrition, irrespective of the condi-

SIGNS OF THE INSANE CONSTITUTION. 81

tion of the cerebral organ, is the foundation of their insanity, is not justified by conscientious observations.

While—as repeatedly insisted on in this chapter—no even approximately scientific study of insanity can be made without a consideration of the bodily states often accompanying it; yet, for the purposes of diagnosis and psychological, as well as medico-legal analysis, the mental symptoms and those somatic signs which are directly indicative of or parallel to the mental symptoms will forever occupy the position of the essential signs of alienation. A delusion, an imperative conception, a faulty logic, a disordered emotional state, will always prove a more profitable discovery to the diagnostician than an anæmia, a constipation, a bed-sore, or an excess of chlorides in the urine. And among the physical signs those, which, like the movements, the handwriting, and the facial expression, are the direct outcome of the mental state, have a more convincing diagnostic and prognostic value than the disturbed phenomena of vegetative life.

CHAPTER IX.

SOMATIC SIGNS OF INSANITY INDICATING THE EXISTENCE OF A CONSTITUTIONAL TAINT OF, OR A PREDISPOSITION TO, INSANITY.

Individuals are sometimes born with so defective a nervous organization that as soon as the first manifestations of higher mental life appear, or should appear, the defective or perverted anatomical foundation exhibits its nefarious influence in the defective or perverted state of the mind. Such persons may properly be said to be born insane, and though the quibble may be made that a being cannot be said to be of non-sane mind before its mind is developed, it may be safely assumed, that where the essential foundation of insanity is present the elements of insanity positively exist; just as club-foot, although perchance the child may never have attempted to nor ever will walk and exhibit the functional manifestation of its organic difficulty, is properly considered a disorder of the organs of locomotion.

But even the quibble alluded to can be met by facts

Just as children may be born affected with chorea, eclampsia, hydrocephalus, and a large number of other diseases; so cases have been recorded, however rare they may be, where mental derangement was observed from birth. As might be anticipated such derangement is manifested in the instinctive and reflex spheres. . Idiots are observed to have a more stupid look than other children, and sometimes to be unable to suck; others are found to exhibit odd motions or fits of furious violence, and Greding records the instance of the child of an insane woman which had fits of destructive mania at its ninth month, and died with them before the end of its first year of life. As a rule, the functional expression of the insane brain in childhood is in convulsions; a fact in parallelism with the daily observation that the febrile deliria of the adult are in infants often replaced by convulsive spells.

Organic diseases of the brain, such as meningitis, hydrocephalus, and cerebral syphilis, developing either before or after birth, lead to irregularities in development which functionally may manifest themselves in insanity, and organically in anomalies of the skull shape, of dentition, and of peripheral development. The earlier the time of the injury the more serious its results, and this applies above all to those errors or flaws of development which occur in members of a degenerating family line; that is, one in which insanity of a certain type or any grave neurosis has already been developed.

Both paternal and maternal influences are active in determining the conformation of the offspring, but the relative preponderance of maternal influence in hereditary transmission is almost a dogma of Natural History. Maternal influences being concerned in the maturation of the ovum, a process lasting many years, and the development of the impregnated ovum to maturity, it is easily understood why a healthy male influence, limited to a briefer period of the germ-history, should be less potent to neutralize the evil influence of a vitiated female parentage, than a healthy maternal influence is to neutralize the influence of a morbid or aberrant male ancestry. This finds a confirmation in the conclusions of Richarz.

1. The chances of the transmission of insanity are greater if the mother is mentally affected, than if the father is insane.

2. The chances of transmission are greater for that child

which is of the same sex as, and which resembles the insane person.

The same author, after a careful study of numerous cases, established the following order of liability to insane inheritance:

Mother insane.   1. The daughter resembling the mother(?)
"   2. The son resembling the mother.
"   3. The daughter resembling the father.
"   4. The son resembling the father.
Father insane.   5. The son resembling the father.
"   6. The daughter resembling the father.
"   7. The daughter resembling the mother(?)
"   8. The son resembling the mother(?)

It is scarcely necessary to add that the liability to insanity is still greater when both parents are insane or have a nervous disorder; and that conception during an exacerbation of their illness * or in the event of consanguinity almost renders its transmission a certainty.

That the transmission of many of the cerebral defects, with which the alienist is concerned, occurs in some way at the moment of conception, although the exact manner of the transmission will never become known, is the general surmise of those who have studied the subject. Far more certain is our knowledge concerning the embryonic mechanism of these defects, and of the influence which fœtal and maternal impressions and injuries exert on the development of the nerve-centres. The latter knowledge furnishes valuable arguments by analogy in support of the accepted conclusions regarding the hereditary group. Modern embryologists have gone so far as to imitate the known natural teratological malformation of the nerve-centres through artificial methods. By wounding the embryonic and vascular areas of the chick's germ with a cataract needle malformations are induced, varying in intensity and character with the earliness of the injury and its precise extent. More delicate injuries produce less monstrous development, and it is particularly the partial varnishing or irregular heating of the eggshell that results in the production of anomalies comparable to microcephaly and cerebral asymmetry. It is this latter fact, showing the constancy of the injurious

---

* Just as children conceived by a drunken father have been repeatedly found to be epileptic, imbecile, deaf mute, or insane.

effect of so apparently slight an impression as the partial varnishing of a structure not directly connected with the embryo, that may suggest the line of research, or rather of reasoning, to be followed in seeking for a plausible explanation of the maternal and other impressions acting on the human germ. What delicate problems are to be solved in this connection may be inferred from the observation of Dareste, that eggs transported in railroad cars and subjected to the vibration and repeated shocks of a railroad journey are checked in development for several days. It requires no great stretch of the fancy to imagine a less coarse, molecular transmission to take place during the maturation of the ovum, or its fertilization, or, finally, the embryonic phases of the more complex and, therefore, more readily disturbed and distorted human germ, and thus to account for the disastrous effects of insanity, emotional explosions, and mental or physical shocks of either parent, on the offspring.

All doubts as to the potency of maternal impressions to affect the shape of the fœtal body and its organs must be dispelled by such positive evidence as is furnished by the two following cases: one selected from the domain of zoology (because of the absence here of many of the objections which might be made to cases observed in the human species), the other from that of an alienist's experience. At the meeting of the Zoological Society of London, held February 24, 1863, Dr. C. E. Gray, the curator of the British Museum, presented the body of a chicken whose beak and feet closely resembled those of a parrot. The sender of the specimen reported that several such instances had occurred in his poultry yard, and he attributed the monstrosities to the fact that one of the hens had been frightened by a parrot which was kept in a cage in the same yard, and which had the habit, when the hens approached its cage for food, of screeching violently at them. More remarkable in many respects is the case described by Wille. A healthy woman, while pregnant by a healthy husband, experienced a sudden fright at seeing a man without a nose. When her child was born its nose was found flattened, there was harelip, and, beside other evidences of an early defect in the germ axis epiblast, the cerebral hemispheres were found confluent in the middle line. The writer had under treatment a child which, when first seen at its nineteenth month, seemed to be a forcible example of the influence of maternal visual impressions on the conformation of the child. Its head was noticed from birth to have the following peculiarity: the frontal region being excessively narrow, the greater part of the skull cavity seemed to be crowded behind the ears. The narrowing of the frontal region was unsymmetrical, the left frontal bone appearing as if not half the area of its fellow, and exhibiting a depression extending into the left temporal region, while the ocular aperture on that side was less wide than its fellow. Before the writer's assuming charge of the case it had had for three months from ten to thirty epileptiform attacks daily, which ceased after five days, under the use of bromides and proper dietetic treatment. The child was then absolutely idiotic; it was at last accounts weak-minded, and altogether an "enfant arriere," had not learned to walk at its twenty-second month, and exhibited violent outbursts of temper at times. Owing to strabismus (deficiency of inner and inferior *recti*) it did not then appear to use its eyes at all. At last reports it managed to render its ocular axes parallel by carrying its head in an oblique position toward the object it desired to fixate. The only cause the healthy mother was able to assign was a sudden fright experienced by her during a panic on board a steamer a few weeks before the infant was born. But as the deformity in question must have originated at a much earlier period of gestation, and as the whole physiognomy of the child was an almost caricature-like reproduction of one of its father's features, it seems more reasonable to attribute the deformity to a visual impression. The father sustained an injury of the left eye years ago, and the palpebral aperture being closed, the whole side appeared contracted, though in reality there was no cranial asymmetry in his case. In all these and similar cases the embryonic disturbance extends far deeper than the deformity of the subject seen by the mother—an evidence that the injurious influence has partially revolutionized the germ-layer arrangements and proportions. A striking example of the influence of maternal conditions of another kind is furnished by Herbert Sankey, where, in a somewhat neurotic family, all the children were normal except two, with whom the mother had had severe epileptiform convulsions while *enceinte*. (Appendix I.)

One of the strongest arguments in support of the view which refers the hereditary and congenital constitutional insanities to anatomical defects is furnished by the great analogy existing between the transmission of these insanities and the known phenomena of inheritance studied by zoologists. Some of the resemblances have been already referred to; a most important one is the frequent intensification of the malady as well as of the anatomical defects. For example, in an observed case, a father was merely morbidly eccentric, and presented no cranial anomaly indicating an anatomical mal-development. His child, however, became an original monomaniac, and had asymmetry of the cranium and defects of innervation, indicating such maldevelopment. In another instance a monomaniac, of very good education and marked ability in many directions, who is now out of the asylum and engaged in chemical experimentation and the occasional issue of insane pamphlets, had an idiotic daughter, with a deformed cranium and strabismus, who died early with convulsions. It must be admitted that such cases strongly support the suspicion that the anomaly visible in the descendant pre-exists in the parent in so slight a degree that it remains impalpable to our imperfect methods of examination.

While the misshapen bodies, the deformed ears, the anomalies of the sexual apparatus, and of dentition, sometimes found in lunatics of the " degenerative series," have a bearing on the question of the inheritance of insanity and properly rank as among the somatic *stigmata* and diagnostic marks of that disorder; special interest attaches to the peculiarities of the insane skull, as an index of the disturbed development of the mental organ contained within.

The attention of observant visitors passing through the wards of an asylum for the insane is often directed to the large proportion of the inmates exhibiting anomalies of the cranial configuration. Aside from a number of cases where the skull is either perceptibly above or below the average in all its dimensions, and another series in which the cranial asymmetry is grossly exaggerated, there are found others where, either separately from these aberrations or coexistent, there are discovered extreme brachycephaly or dolichocephaly, flattening of one or the other end of the cranial ovoid, disproportions between the dome and the base, promontory-like extensions of the vault, in short, every conceivable deviation from that standard which the experience of anatomists with sane subjects justifies us in considering as the normal one.

The question naturally arises, whether a relation can be

SIGNS OF THE INSANE CONSTITUTION. 87

established between these anomalies of the cranium and the presence of their bearers in an asylum. Most of the classical writers assume the existence of such a relation, and content themselves with noting the coincidence of mental alienation in general, with cranial malformation. Foville the elder, after excluding cases of idiocy and cretinism from the list, found that one sixth of his insane patients presented marked cranial malformation. Morel was the first to refer to an intrinsic relation between the character of the insanity and the associated cranial anomalies. He established the special frequency of the latter with the hereditary and degenerative forms. The same anomaly, a peculiar flattening of the occiput, was found in thirteen subjects presenting the symptoms of the *manie raisonnante* of the French. A different malformation, characterized by flattening and depression of the forehead, with an increase of the transverse diameter of the skull, has been noted as being of special frequency with insanity manifesting itself in homicidal, suicidal, and other dangerous impulses. The stenotic skull has been as frequently observed with congenital weak-mindedness, associated with mental and motor agitation. Among special symptoms none appear to be so commonly associated with a particular cranial deformity as pyromania. The Germans have found osteophytes on the *sella turcica*, and abnormal prominency of the clivus projection in the dead, and distortions of the cranium in the living subject, which pointed to a basilar deformity in several of those exhibiting this impulse. The *crania progenia* have been found in subjects representing connecting links, as it were, between idiocy and monomania. But pronounced asymmetry of the skull is found to be less constantly associated with special forms of derangement; very frequent with epilepsy and epileptic insanity, it is a not uncommon accompaniment of monomania and imbecility.

Lasègue and Garel established the existence of a close relation between idiopathic epilepsy and faulty ossification of the cranial sutures with resulting asymmetry. Morel found the latter as frequently in the degenerative forms of insanity. It is quite common in chronic insane epileptics.

The combination of cranial deformities with such different forms of insanity seems little calculated to explain, *per se*, the character of the connection between the visible anomaly and the mental aberration. That the cranial deformity is merely the expression of an error in development, simultaneously affecting the organ of the mind, and that herein its signification really consists, is almost self-evident. But before this line of inquiry can be profitably followed certain clinical relations must be clearly grasped whose accentuation at this point hence becomes necessary.

The defective states of the mind, which are the frequent manifestations of an hereditarily transmitted taint, may be ranged in a serial chain, whose links are constituted by different forms of alienation, and merge gradually into one another. One end of this chain is constituted by idiocy, the other by that systematized perversion of the intellect, which, classed by the great Esquirol under the too often misinterpreted designation of monomania. is now known variously as megalomania, chronic delusional insanity, primary intellectual vesania, and by a host of other terms, more or less correctly expressing some prominent feature of this form of mental alienation. On first sight these two conditions would appear to be separated by an impassable chasm, and this from a psychological, as well as from a strictly somatic point of view. No greater contrast could be exhibited within the walls of an asylum than by placing side by side an idiot and a lunatic with systematized projects and delusions: on the one hand there is a state characterized by an utter absence of every higher mental co-ordination; on the other, one which manifests intricate and varied combinations of the mental mechanism, which are analogous to those of the normal mind. In the former case, the gross anatomical defect, which is in such perfect parallelism with the mental hiatus evident, is usually perceivable on the first glance; in the other, no deviation from the normal physique may be discoverable with our available methods of examination.

On studying, however, other cases of insanity arising in the course of hereditary transmission, there are found certain groups, for which it is justifiable to claim a place, intermediate to that occupied by the two extreme forms mentioned. They are composed of cases of imbecility, with or without morbid impulsiveness, and others with primary delusional insanity united with weak-mindedness as a prominent factor. On inquiring yet more profoundly into the mutual relation of these different forms, and including in the analysis those transmitted insanities associated with epileptic manifestations and with moral imbecility, observers have perceived that not only is there an uninterrupted single line of gradation running from idiocy to primary insanity with systematized delusions, but that, in addition, there are numerous collateral branches springing from this parent stem at certain altitudes, which crop out into more or less independent developments. It is, hence, not possible to draw an absolute line of demarkation between the different forms comprising this series. The more extensive observations become, the clearer is it recognized that there exist transition forms uniting into one great class all the hereditary insanities. Startling as these propositions seem when advanced thus categorically, they are not in their essence new ones. Cloaked in less decided language, their expression may be traced in the writings of the earlier masters of mental pathology.

As far as positive data have been obtained in the pathology of hereditary and congenital insanity, many of the cases coming under this head constitute a united group, characterized and held together as an anatomico-pathological unity by the presence of a teratological anomaly. Morel recognized the affiliation existing between most of these groups, and identified their somatic resemblance in the common presence of the hereditary *stigmata*. His researches did not extend to a minute study of the brain. Le Grand du Saulle, in a masterly series of papers, affirmed the propositions of his great predecessor, and fortified them chiefly by clinical observation. Krafft-Ebing, in his earlier monograph on the criminal jurisprudence of the insane, laid great stress on the importance of the stigmata and especially of cranial deformity in the recognition of certain hereditary and doubtful cases of insanity. More recently he has drawn the line very sharply and happily between the "psycho-neuroses," as he terms the ordinary forms of insanity developed in individuals in whom no degenerative taint exists, and the "psychical degenerative states" which comprise the greater part of the cases with which the stigmata are found.

The conventional notion of idiocy and imbecility—the forms which stand at the lower end of the degenerative series—associating these conditions with a simply quantitative deficiency of the fore-brain, is a very imperfect one. The researches of numerous observers have shown that qualitative defects, using the term "qualitative" in its wider sense to cover both morphological and histological aberrations, are at least as common and perhaps more characteristic features of the idiotic and imbecile brain, and are for the most part as closely related to the cranial malformation. These defects may be enumerated under the following heads:

1. Atypical asymmetry of the cerebral hemispheres, as regards bulk.
2. Atypical asymmetry in the gyral development.
3. Persistence of embryonic features in the gyral arrangement.
4. Defective development of the great inter-hemispheric commissure.
5. Irregular and defective development of the great ganglia and of the conducting tracts.
6. Anomalies in the development of the minute elements of the brain.
7. Abnormal arrangement of the cerebral vascular channels.

All of these conditions are separately, or in the combination of several of the features above mentioned, occasionally found in the brains of lunatics, appertaining to the group of monomania, or so-called "partial insanity." They are also and more constantly found in that mixed and unclassified group of cases clinically occupying an intermediate position between the higher systematized perversions of the mind falling under the designation of monomania, and the lower group of congenital imbecility. Inasmuch as the latter cases almost universally exhibit a pronounced malformation of the cranium, and in all the instances which have been thoroughly examined—with the result of revealing cerebral defects—such malformation was a prominent feature, it may not be unwarrantable to anticipate the existence of cerebral defects, where similar external malformations are discovered during life.

There are undoubtedly cases of insanity of inherited origin, in which cerebral defects are not discoverable. Instead of conflicting with the views to be expressed in the present chapter, this fact furnishes on the contrary a strong support of them. What is more natural than that in a series of cases, ranging between idiocy with its gross and palpable physiological defects, and monomania with its relatively close analogies to physiological cerebration, a corresponding structural gradation, insensibly approaching the normal limits, should exist? The general conclusion that the explanation of the existence of a certain class of the hereditary insanities and the basis of the manifested symptoms is to be sought for in morphological, that is, in *quasi*-teratological faults of structure and not in post-natal changes, happens to harmonize with every clinical experience.

Authors have given much prominence to the alleged fact that a *transformation* in the type of a given hereditary neurosis is the rule. It is true that such a transformation frequently occurs; that an epileptic parent may have an imbecile child, that a monomaniacal ancestor may have an idiotic or a deaf-mute descendant, while a father suffering from alcoholism may have an epileptic son. It is also true that the superficial features of neural disorder in the child may be the very opposite of those found in the parent. But a closer study shows that the occasional observation of the latter fact has been made the basis of too wide a generalization. It is particularly in the case of monomania that a great similarity in the manifestations is found in an entire insane family branch. Suicidal morbid impulses have been observed to be transmitted with such constancy, that although the fact that the suicide of his ancestors at a certain age had been faithfully kept from a patient inheriting this tendency, and the possibility of imitation was thus excluded, he committed suicide at about the same period of life which had been the fatal one with his ancestry.

The readiness with which any one of the various mental and nervous disorders may undergo a metamorphosis into another form, in the course of hereditary transmission, seems interpretable in only one way. It is to be assumed that a change in the nature of the transmitted structural defects or their intensification or mitigation may account for such a transformation. Thus asymmetry of the brain and skull are found in hereditary epilepsy and in hereditary monomania, but it is not yet known in what particular direction the asymmetry must be manifested to determine one or the other of these conditions.

The frequent association in a given family line of the periodical psychoses in some, and epilepsy in other members, is a suggestive one, as both disorders are manifestations of vaso-motor states.

The erroneous idea has been inculcated in several treatises, that as soon as the degenerative psychoses have once manifested themselves in a given family line, the course of that line is infallibly in the direction of further degeneration and extinction. This, while true in a large number of cases, is not so with regard to all. For example, the Austrian philosopher, Schopenhauer, had imbecile relatives in his ancestry. This instance, like the sometimes noted appearance of a one-sided talent or genius in a mentally unsound family, illustrates the occasional triumph, however partial that triumph may be, of the conservative and progressive over the degenerative tendencies.

CHAPTER X.

The Morbid Anatomy of Insanity.

The interesting question whether there are pathognomonic signs of insanity in the dead as well as in the living body, is a natural sequel to the two foregoing chapters. The examination of the living patient reveals the *manifestations* of a morbid state; the existence of somatic signs and stigmata *suggests* its nature; but it is only the anatomical and physiological methods which are calculated to *prove* the *intimate relation* between this state, if demonstrable, and the symptoms depending on it.

Let the reader reflect for one moment on the presumable degree of intensity which a lesion must reach in order to affect the mental faculties! Let him bear in mind that the majority of the symptoms of insanity are merely excesses or deficiencies or perversions of functions—excesses, deficiencies, and perversions whose very existence implies that their organic basis must be *anatomically* intact! He will then recognize the futility of searching for coarse formative changes in such subtle affections as most of the simple primary pyschoses are. It is well established that fatal brain intoxication can be produced by conium, opium, and chloroform, without producing any changes in the nerve structures proper which the microscopist is able to detect. And we need not wonder that with the presumably grosser dissociation of the nervous molecules which takes place at the moment of death, some of those finer changes which are compatible with life, though sufficient to produce mental disturbance, should be obliterated, just as similarly delicate changes are obliterated in death from acute hydrophobia, tetanus, and strychnia poisoning.

In a large number of cases where the functional disturbance reaches a high degree, or continues for a long period, tissue changes occur, either in the way of a primary lesion of the nerve elements, or of vascular degeneration with its ensuing result: tissue mal-nutrition. These changes when found often furnish valuable indications as to the original nature of the insanity with which they exist. It is to be regretted, however, that the study of these lesions has been hampered by premature theorization and erroneous observation to an extent which is unparalleled in medicine.

On account of the dearth of constant pathological findings in cases of insanity, the earlier, and some contemporary pathologists have been reduced to remarkable extremities in endeavoring to assign insanity to material causes, or to explain certain of the symptoms of insanity on material grounds. Some of the resulting errors are no doubt due to the fact that those who have committed them have not had as large a material of normal subjects at their disposal as would have enabled them to determine the extent of normal variation. One alienist, for example, claims as a pathological fact, that the sciatic nerve is flattened in general paralysis of the insane; yet this nerve is always a flat cord, and variably so, in normal subjects. Two other writers find certain cells of the parietal region and its neighborhood to be "pathologically enlarged" in special forms of insanity; yet over eight years ago Betz demonstrated that just such cells constitute the nests of gigantic pyramids in that very region of the normal cortex. Their absence, not their presence, would be abnormal!

Much as is to be eliminated from our pathological records, as accidental or artificial, those records become still further reduced, when we attribute a proper value to the registration by alienists of certain vascular and other histological appearances as lesions of insanity which are found in every healthy adult. The more delicate changes in the adventitia of the cortical blood-vessels, deposits of hæmatozin and granular matter in their neighborhood, and pigmentation of certain nerve-cells, it can be positively affirmed, are constant occurrences in subjects dying sane, and who have never been anything but sane. Only when these changes pass a certain degree can they be attributed to insanity. It is also to be borne in mind that many actually morbid appearances in the insane are not necessarily related to the mental disorder as such. These are prominently the atheromatous change of the blood-vessels and the so-called miliary aneurism. Both occur at least as frequently and in the same distribution in the sane population as in the insane, and the clinical signification of both changes is so thoroughly well known and appreciated that their potency to produce those symptoms of mental alienation, which occur equally well in insane subjects with healthy blood-vessels, may be questioned.

In many cases of advanced insanity, particularly with the so-called secondary states and terminal dementias, numerous morbid appearances are found, which, while they bear an unquestionable relation to the mental state of the patient as it was preceding his demise, have no real etiological signification; established years after the inception of the insanity, they must be considered as secondary, and frequently as passive results. The so-called arachnoid cysts, and many instances of a moderate degree of cerebral atrophy, are of this character.

The value of our psycho-pathological literature is also impaired by the fact that many of the "lesions" recorded by observers have been, not the result of disease, but of imperfect manipulation. A recent German observer, for example, describes gaps in the white substance of the cerebral hemispheres of paretics, which the writer, with Mendel, can only consider to be artefacta.* Another source of error is the action of frost on the partially hardened brain. When a high degree of cold is reached, the entire organ becomes riddled with cracks and falls apart; with lesser degrees, the natural spaces, such as the pericellular and perivascular gaps, become much exaggerated by the expansion of the fluid within them in the freezing process, thus simulating the results of long-continued obstruction to the lymph outflow.

But the most remarkable error that has been committed in this direction is one which, from the fact that it has been made the basis of a very sweeping and misleading generalization, deserves special mention at this point. Several of the earlier pathologists described bodies known as "colloid

---

* Both Plaxton and Savage have confirmed the author's views on this subject. They have been verified by a number of continental critics, and appear to be generally accepted.

spheres" and "miliary sclerotic patches" as found in the insane brain. These bodies, as the writer has shown, can be produced by the action of alcohol.

When the brain or cord of any animal is submitted to the action of this fluid a sufficient length of time a peculiar series of changes occurs, simulating lesions, but which can be produced at will in the healthiest human and animal brains. In the first place, alcohol has been long known to extract *leucin* from nervous tissues, and the writer has found crystalloid or sub-crystalline spheres of *leucin* scattered more or less regularly through the tissues or accumulated in the perivascular areas of healthy brains submitted to the action of alcohol. By varying the length of exposure, or by immersing the brain, or different portions of the same brain, at different stages of *post-mortem* change, in alcohol, these artificial "lesions" may be pleasingly varied. That these bodies are *leucin*, or at least chiefly consist of *leucin*, is demonstrated by their chemical reaction. Among the tests which can be conveniently applied are their insolubility in ether and chloroform, and difficult solubility in alcohol (requiring 1,040 parts of cold and 800 parts of hot alcohol); the bodies do not stain in carmine and are refractive,[*] besides they are either hyaline or show a radiatory striation.[†] While the observers in question committed an error in manipulation merely, it remained for two later writers to build up theories on these artificial precipitates, which deserve notice, if for no other reason than because they have been considered the semi-official utterances of a body of alienists, and have misled the profession not a little. The fact that these bodies were found or rather manufactured equally in "general paresis, subacute mania, and chronic melancholia," is responsible for the theory, that in insanity the alienist has to contend with but "one diathesis;" and the same fact doubtless also furnished the grounds for the dogma emanating from the same source which regards "mania and melancholia as essentially the same psychologically and pathologically." The "lesions" in question, varyingly designated as "granular bodies," "grumous granular matter," "encysted morbid

---

[*] This feature varies, probably with certain admixtures.
[†] Overstaining sections containing these bodies in carmine, and particularly in hæmatoxylin, does stain the latter: ordinary staining fails to do so.

products," "colloid bodies," "globules of a fatty nature," "nitrogenous plasma, proteinaceous in its nature," "miliary sclerosis," and "spots," have been utilized to sustain the claim, that the microscope "unerringly points out the seat of disease" in insanity.* It may be safely predicted that these and similar views will retain a certain degree of value, of at least a historical character.

The views of sound authorities are unanimously to the effect that the connection between the symptoms and lesions of insanity is obscure, and that the cerebral disease underlying insanity is in many cases undiscoverable. "Thirty years ago I would willingly have written on the pathological cause of insanity; to-day I should not attempt as difficult a task, so much of uncertainty and contradiction is there in the results of *post mortems* made on the insane to this very day." Thus wrote the great Esquirol half a century ago. One† of the contemporary French alienists, after citing this passage, adds: "What Esquirol said at the time he wrote, that might he have adhered to to this day, notwithstanding the incontestable progress which has been made more recently in the pathology of mental diseases. Although, during the last few years, experimental physiology, pathological anatomy, and histological researches have yielded valuable acquisitions to the pathology of the nervous system, this cannot be equally claimed in relation to insanity properly speaking; the clinical study, the direct observation of patients, have contributed much more to advance this branch of science. It is indeed to-day possible, with a large number of diseases of the nervous system, to establish the relation between lesion and symptom; this approximation is, however, even to this day almost impossible with the greater number of mental diseases." While the two extracts here quoted fairly represent the views of the classical and modern French authorities, the following almost identical citation indicates the state of opinion in Germany, that land which divides with France the honor of standing in the front rank of mental medicine. "In turning to pathological anatomy for enlightenment, it cannot be suppressed that in a certain number of *post mortems* of the insane, palpable morbid appearances of the brain are ab-

---

* Report of the New York State Asylum at Utica.
† Dagonet: "Nouveau Traité Elementaire et Pratique des Maladies Mentales."

sent. . . . Experience teaches that it is almost exclusively the primary forms, the first stages of insanity, in which we find nothing palpable in the autopsy, and must content ourselves with assuming anomalies of innervation, blood-distribution, and chemical composition."* Quotations of a like character might be multiplied.

In recent mania, properly so called, the acute mania of English writers, a form of insanity comprising from ten to fifteen per cent of the admissions to a large number of asylums whose statistics the writer has collected, no deviation from the normal structural conditions has been found. The nerve-cells and their fibres show absolutely nothing that can distinguish them from the nerve-cells and nerve-fibres of healthy persons. The only appearance revealed with any degree of constancy in uncomplicated cases is a fulness of the cerebral blood-vessels,† which in extreme instances may be accompanied by hæmorrhages or hæmorrhaginous transudations. In these and other cases metamorphosed elements of the blood are found scattered through the brain cortex and sometimes in the white substance. In those cases which the writer has himself examined *post mortem* he has found nothing ‡ beyond a very slight degree of cortical hyperæmia, and one not differing from the hyperæmia found in sane persons dying with certain lung diseases, and not equalling that hyperæmia sometimes found in the brains of those dying of fevers. In not a single accurately studied case, reported by trustworthy observers, has a characteristic lesion of the essential mental apparatus been found, nor has any doubtful appearance of any kind been discovered that has not been found, in lesser degree, in sane persons also. The latest and most thorough writer § on mania concludes his chapter on pathological anatomy, naturally the most meagre in his work, with these words: "From all this I draw the

---

* Krafft-Ebing, Lehrbuch, I. p. 11. This quotation is well supplemented by Schüle (*Seq.*).

† Calmeil, Article Manie, Dict. de Méd., etc.; Meynert, Psych. Centralblatt, 1871; Westphal, Charité-Annalen, I., 1876; Ripping, Die Geistesstörungen der Schwangeren, Wöchnerinnen und Säugenden. Stuttgart, 1877; Mendel, Die Manie, 1881.

‡ The writer excepts from these remarks his findings in a single case of *delirium grave*, a condition which he does not agree with Mendel in classifying under the head of "Mania," properly so called.

§ Mendel : Die Manie 1881.

conclusion that mania is a disease, whose patho-anatomical basis we have thus far not been able to discover in the brain. In this light we must look upon it as a functional disorder, because those alterations (of the brain) which must be supposed to determine the disturbance of the cerebral functions are undemonstrable with our present means of investigation."

When we recollect that only from two to five per cent of maniacs die, and that, consequently, chiefly those subjects reach the *post-mortem* table in whom, either through complications or an intensification of the maniacal excitement to the degree of maniacal exhaustion, the brain disturbance must be assumed to have reached a far higher pitch than in the ninety-five or more per cent surviving; when we bear in mind that Esquirol, Griesinger, and others consider the possibility of and cite cases to prove the causation of maniacal symptoms by processes associated with anæmia instead of hyperæmia of the brain, it must be clear that the evidence concerning the tangible signs of a form of insanity comprising a large proportion of asylum admissions is fragmentary and inconclusive where it is not entirely negative.

Nearly the same remarks apply to simple melancholia; the most complete series of investigations—that of Meynert—yielded no other result than the conclusion that cortical and meningeal hyperæmia are far rarer and lymph-space dilatation more frequent here than in mania, and on this he was enabled to base some ingenious hypotheses. The writer's own investigations, including the examination of several melancholiacs dying in and out of asylums, have satisfied him that, with our present means of investigation, a change in the essential nervous structures is not demonstrable.* Even in melancholiacs whose symptoms have reached the degree of atonic stupor, the morbid appearances, such as œdema of the membranes and pallor of the cortex, are possibly referable to starvation from prolonged refusal of food, and to the lung diseases which are so frequently concomitant to this affection. These differ so little from the similar appearances found in ordinary hospital patients dying from inanition and phthisis, that we can only conclude, as in the case of maniacs who recover, that the larger proportion of melancholiacs exhibit nothing posi-

---

* Katatonia is not included here.

tively demonstrating their disease in the brain structure. How uncertain the anatomical basis is in those forms of insanity for which the writer defends the retaining of the term "monomania,"* forms comprising as much as twenty-five per cent of some asylum admissions, cannot be better illustrated than by a citation from the author of the article "Insanity," in Ziemssen's Cyclopædia. "The pathological anatomy of monomania is the most meagre chapter of our generally extremely moderate knowledge of psychological morbid anatomy. In the case of some 'original monomaniacs' (*Originär Verrückte*) Sander and Muhr have called attention to interesting anomalies of the skull and brain, in part congenital and in part acquired in early infancy. But, for the great majority of those later attacked by monomania, no gateway promising to lead with any certainty to the laboratory of these serious and deeply grasping changes, which attack the very essence of the *ego*, has been so far discovered."

In epileptic insanity, sometimes comprising eight per cent of asylum admissions, no morbid appearances can be found in a large number of recent cases, and those found, such as asymmetry of the brain and skull, rather indicate the frequently hereditary and congenital origin of the disorder than that they explain the exact basis of the insanity. Even in advanced cases of epileptic insanity it is not the exception to find but very slight and doubtful indications of that disorder *post mortem*. In some instances, atrophy of the brain, vascular kinking, cystic degeneration of the cortex, and extreme pigmentation of the nerve-cells, particularly in the medulla, are found. With chronic secondary insanity, dementia, and other terminal deteriorations, the brain is more constantly diseased; it is diminished in weight, the convolutions appear shrunken, the cells of the brain-rind wasted, the blood-vessels show the vicious influence of long-continued disturbances of the blood-supply; but even here there are cases where the *post-mortem* evidence is not conclusive. With imbecility and idiocy anatomical evidences of imperfect brain development are common. Finally, in progressive paresis, or paretic dementia, the *post-mortem* record is exceptionally satisfactory. It is safe to say that if, in a case where this disease is claimed to have lasted longer than a year, certain changes in the nerve-

---

* See Paranoia.

centres are not found, the justice of the claim is seriously impaired.

Summarizing the teachings of the master-minds in pathology, of reliable observers generally, and the writer's own experience, he considers himself justified in saying that positive and indisputable evidence of insanity cannot be found in more than thirty per cent of the insane; that in another thirty per cent slight changes are found, not differing in character, though perhaps in extent, from what is observed in some sane subjects; while, in the remainder, there is no visible deviation from the normal standard of any kind. To be more specific in regard to the various forms of insanity, the likelihood of finding characteristic structural changes in the insane brain may be represented as follows: In true and recent mania that likelihood is as 5 : 100; in acute melancholia it is almost zero; in epileptic insanity it is as 20 : 100; in monomania it is as 5 : 100; in the terminal states, it is as 60 : 100; in imbecility and idiocy, as 80 : 100; in paretic dementia it reaches the figure 99+ : 100, and here alone and in insanity with organic diseases does the autopsy approximate the dignity, from every point of view, of a scientifically positive test. The writer does not give these figures as exact, but only to indicate as nearly as figures derived from a comparison of authorities and experience can, the degree of certainty with which a *post-mortem* finding may be anticipated, and the extent to which it may serve as the test of an opinion.

The various genuine changes which are found more or less constantly associated with the ordinary forms of insanity may be considered together. There is one form of insanity, paretic dementia, which is a pathological entity; its morbid anatomy is as well known as that of other organic nervous diseases, and the relation between its morbid changes and the symptoms thereon depending is sufficiently constant to justify the establishment of a pathogenesis of this malady. For these reasons the morbid anatomy of this form of insanity will be separately considered; and the subjoined remarks are not intended to cover the pathology of that disorder, nor of those rarer forms of insanity whose anatomical features require to be specially referred to in the second part of this manual where these forms are separately considered.

The greatest interest attaches to the changes in the essential apparatus of thought and action—namely, the brain;

and in this organ it is particularly the cortex or brain-rind of the cerebral hemispheres which has been made the subject of the closest study. It is in the nerve-cells of this cortex that the individual elements of thought and action are supposed to be seated, being represented by force oscillations in their complex protoplasm; and it is in the nerve fibrillæ connecting these cells, and in the coarse fibre-bundles associating neighboring as well as distant cortical areas that the physiologist is compelled to locate the correlating and co-ordinating mental faculties, in short those without which no logical comparison nor conclusion can be made.* From all this it is evident that a rational psycho-pathology is really nothing more nor less than a branch of general cerebral pathology; or, as Meynert expressed it, that the pathology of insanity is merely a chapter in the morbid anatomy of the fore-brain.†

CHANGES IN THE CEREBRAL TISSUES.—In the majority of the recent cases of insanity no coarse change of the cortex can be detected which is not dependent on the degree of vascular injection. It is due to the arrangement of the cortical capillaries that an unusual degree of injection marking itself in the deep and superficial cortical plexuses obliterates the normal lamination of the cortex to the naked eye. On hardening specimens of such a cortex the lamination returns, and under the microscope the distinctness

---

* It has been not inaptly said that the mental co-ordinations acquired in the course of a higher civilization are the ones which are most often defective in the lunatic. It may be presumed that the sub-cortical nerve tracts which are the latest in attaining their full development, and which indeed have not attained that development at birth, are the essential mediators of those intricate co-ordinations. As we proceed from the lower mammalia to the higher, and as we follow up the successive stages of the human embryo and fœtus, we find that the isolation of conducting tracts, and the related overgrowth and emancipation of the associating systems, appears to be the ideal aim of the highest development. What more imperative conclusion is there in the entire field of mental science than that the source of the more complex disturbances of the mind must be sought for right here?

† This term has been erroneously used by some writers as a synonym for the frontal lobes. In reality it is an English rendering of the German term "Vorderhirn," or the Latin equivalent "Prosencephalon," and in its widest sense includes the cerebral hemispheres and the thalami. One of our best comparative anatomists, Wilder, thinks the vernacular term "fore-brain" unnecessary. It is, however, a convenient one for the alienist, as, in the sense here set forth, it is the nearest anatomical equivalent for the physiological conception of the "organ of the mind," which ought—provisionally at least—to include the thalami.

of the cells and the layers in which they are arranged is found to be as great as in the normal brain, thus showing that the obliteration of the coarse lamination is only a seeming one and has no histological significance. The essential cortical structure is similarly unaffected by the extreme anæmia found in melancholia and stupor. But in cases in which a transition to dementia is likely and which reach the autopsy table, through some inter-current disease, the perivascular and periganglionic spaces, which are supposed to appertain to the lymphatic system of the brain, are found to be unusually large. This phenomenon is probably the expression of a retarded lymph out flow. In some extreme cases an accumulation of lymph-bodies in and near these spaces occurs; this appearance was noted in a pronounced degree by the writer in the single case of katatonia, examined by him *post mortem*. It is susceptible of the same interpretation as the perivascular dilatation with which it is usually connected.

In chronic insanity a number of changes are noted which may be regarded as partly the secondary results of the vascular lesions, to be later considered, partly as the anatomical expression of a premature senescence of the nervous tissues. The closer the insanity approaches the degree of a dementia the more apt are these changes to be marked and characteristic. They are hence prominent and frequent in that secondary form of insanity which is commonly called chronic mania (which the writer proposes to term secondary or chronic confusional insanity), and most prominent and frequent in dementia; while they are far less frequent in monomania, though they are found even here as an expression of the partial deterioration observed in this affection. The periodical and epileptic forms of insanity, marked as they are by recurring vaso-motor crises, are often accompanied by changes showing the destructive influence of the latter.

The various vicissitudes to which the healthy brain is exposed, its physiological hyperæmias, and probably also the febrile disorders and emotional strain through which its bearer has to pass, register their influence in pigmentation of the nerve-cells, in accumulations of blood pigment and granular material in the adventitia of the blood-vessels, and these changes, gradually intensifying with age, lead to the vascular degeneration and nerve-cell atrophy sometimes found in advanced life. The influence of the more intense vaso-motor cataclasms of insanity, and the greater functional abuse of the organ of the mind in this disorder lead to a correspondingly more intense and more rapid tissue deterioration in the insane. It is for this reason that the lesions of the forms of chronic mental disorder here referred to do not generally differ in kind from certain appearances found in most healthy persons who have passed the twentieth year, though they may differ greatly in degree.

In extreme cases of chronic insanity, particularly where marked mental deterioration has permanently set in, atrophy of the essential nerve elements is found. To the naked eye this is made manifest in diminution in size of the entire brain, in enlargement of the ventricles, in shrinking of the convolutions, and in consequent gaping of the sulci.* By the balance it is made evident as a general decrease in weight, particularly of the cerebral hemispheres. With the scalpel it is recognizable by a greater firmness of the tissues, due to the relative preponderance of the connective over the strictly nervous substances. Finally, the microscope shows that the nerve-cells are diminished in number, and either are shrunken in all dimensions or present appearances of passing through the successive stages of destructive degeneration. More or less change is observable in the white substance: gaps form more readily in hardening than in the healthy brain, showing that there is a rarefication of tissues; the myelin is not as well marked, and various fine shades of color are noted which are not observed in the the healthy brain; strands of nerve fibres are seen which have a grayish or brownish tinge, and others which have a dead-white color, and these are usually found to be connected with those cortical fields where atrophy is most marked. These changes in color are not the expression of a genuine sclerosis, but of some finer disease process whose real nature has yet to be determined.

CHANGES IN THE NERVE-CELLS.—The more violent the manifestations of the psychoses were during life, the more likely are the nerve-cells to show the effect of rapidly produced nutritive disturbance. Their protoplasm in that case is cloudy, their processes stain poorly, and the nucleus is

---

*It is due to this fact that the atrophic brains of the chronic insane exhibit the typical sulci much more satisfactorily for the beginner than healthy brains.

ill marked. But it is unusual to find this change in other mental disorders than acute delirium, and in those following fever and accompanying meningitis. In acute mania and melancholia the nerve-cells are perfectly healthy to the eye, and typical specimens for anatomical demonstration have been obtained by the writer from the brains of those dying of either of these disorders.

In chronic insanity of long standing there are particularly two series of changes which engage the attention of the pathologist. The first seems to be a rather *passive atrophy;* the cells appear to be diminished only in size, the smaller nerve-cells are indistinct, and the nerve processes seem to be fewer. In extreme degrees of this change some of the cells scarcely take carmine staining at all, although side by side with them are found others showing the normal degree of imbibition of the staining fluid. It has even been claimed by Förster and others that calcification occurs under similar circumstances. All this suggests that a gradual deterioration from the highly protoplasmic character of the nerve-cells has taken place. This condition is found in that apathetic dementia which follows melancholia and stuporous insanity.

The second series of changes is of more common occurrence. It is an intensification of that normal involution of the nerve-cell which takes place with advancing years. While the pigmentation of the nerve-cells of the healthy cortex is in middle life limited to the larger pyramidal cells, it extends to the smaller ones in chronic insanity. While the pigment granules of the normal brain-cell are exceedingly fine, and do not appear to affect the fibrillary structures of the cell protoplasm; in the diseased condition the pigment is in clumps, consisting of coarse granules, obscuring or crowding away the nucleus, and obliterating the fibrillary structure referred to. This change seems to produce different results when it affects the smaller nerve-cells. These are often converted into an apparently homogeneous yellow substance, without any alteration in shape; the nucleus shares in this change in a lesser degree, but does not lose its outline or distinctness. Inasmuch as with this class of changes it is common to find morbid states of the blood-vessels, it is reasonable to regard the recorded pigmentary-granular change of the nerve-cells as the result of active nutritive disturbance. The final phase of this change, a destruction of the nerve-cell,* is noted in the larger pyramids in extreme dementia, just as it is a regular occurrence in advanced senility. It manifests itself by the obscuration or disappearance of the nucleus, and the separation and disappearance of the cell processes. Finally, nothing but an irregular mass of granules and pigment serves to mark the former site of the ganglionic element. Observers have spoken of this process as a fatty degeneration or a fatty pigmentary change. But micro-chemistry, while it has not yet revealed the exact nature of these bodies, has demonstrated that they are not of a fatty nature. And it may be a proper place to insist here that the current statement, repeated in more than one text-book on physiology, and copied into several treatises on insanity, that the active physiological ingredient in the brain is a " phosphorized oil or fat," is absolutely erroneous. The healthy brain contains no true fat, and the various reports of fatty degeneration of the diseased brain will have to be very carefully sifted in the light of modern chemistry before they can be accepted. The recorded proofs of fatty degeneration seem to rest on the reaction of the alleged fatty material to ether. But ether affects other substances than the fats in a similar way.

CHANGES IN THE NEUROGLIA.—The so-called nuclei of the neuroglia, small round bodies which are derived from the formed elements of the blood, are frequently increased in number in insanity of long standing, particularly when it is accompanied by excitement. Unfortunately not enough is known of the true nature of the neuroglia to enable pathologists to explain a familiar change, common in advanced deterioration. It is remarked that the longer a specimen of the brain cortex is hardened, especially if it be hardened in chromic acid or its salts, the more decided becomes a certain *fibrillous texture* of the neuroglia. It is not determined thus far whether this network of fibrils is of a nervous or a connective-tissue character ; it certainly is the expression of a normal structure. The longer the hardening processes continue the more does this network contract, its meshes then widen, and numerous small roundish

---

* It is customary to speak of nerve "cells," although it should be borne in mind that, in the strict histological sense, the ganglionic bodies of the cortex are not cells in the common acceptation of the term.

or oval cavities appear. Evidently the hardening process is accompanied by the solution and removal of some substance and the condensation of the fibrillar network. Now what hardening does after a long period in a healthy brain that will it do in a much shorter time in the brains of some of the chronic insane, particularly in senile dementia. Indeed, a similar condition may be here determined in the *fresh* tissues, and is, therefore, undoubtedly pathological.

CHANGES IN THE BLOOD-VESSELS.—The cerebral cortex is one of the most vascular tissues in the body. Among parenchymatous tissues it is exceeded in vascularity only by the blood glands. It is, consequently, but natural that some of the more important pathological changes of insanity should be found to commence in the intricate nutritive plexus of the cortical blood-vessels. The changes in blood-vessels may affect the structure of the blood-vessel itself, its contour, and its appendages.

Structural Changes of the Vascular Wall Proper.—Proliferation of the nuclei in the vascular tissues and a fine granular or a colloid change of the muscular coat are common appearances. The term "colloid" here used has reference only to the optical appearance of the morbid deposit; it is not intimated that the latter has the composition of what is ordinarily called colloid substance. It is also to be insisted on here, that the so-called colloid transformation of entire blood-vessels reported by some observers is due to the methods of hardening employed, and is therefore not a *bona-fide* lesion. A general sclerosis of all the cortical vessels is a common condition in advanced insanity, and may be regarded as a sequel of the nuclear proliferation alluded to. In extreme cases this sclerosis reaches the degree of a fibrous transformation, the nuclei, previously abundant, now disappear as such, and the whole vessel may degenerate into a fibrous filament, perhaps even devoid of a lumen.

Changes in the Adventitia.—Much controversy has grown out of the claim that there are two lymph spaces around the cortical blood-vessels, one which is sub-adventitial and one which is extra-adventitial. There seems to be no doubt possible that while the space of His—that is, the perivascular space—does not exist in health, it certainly exists in disease, probably as a dilatation around the adventitia, compensating for the impeded lymph outflow through the natural sub-adventitial channel. This dilatation is accompanied by a dilatation of the peri-ganglionic spaces, and this latter reaches so extreme a degree, not only in paretic dementia (where it is most common), but also in advanced epileptic and periodical insanity, that the spaces resulting sometimes become visible to the naked eye. (See Fig. 2). This is the origin of the so-called *état criblé* of the cortex, which will be described in detail in the chapter on Paretic Dementia.

The most constant morbid appearance in insanity is the deposit of granular and pigmentary material in the adventitia. This is to be regarded as the result of repeated fluxionary states, and is probably one of the earliest changes in progressive insanity, as indeed it is but an intensification of the same appearance, found on a greatly reduced scale, in every healthy adult. The granular pigment, usually of a yellowish, and rarely of a decidedly brownish tinge, is found accumulated in scattered foci or in larger patches along the course of the vessel, and particularly at the vascular bifurcations. While the arterioles which are visible to the naked eye show this change very markedly, the smaller vessels may be quite free from it.

In advanced insanity, particularly of the epileptic and periodical types, and in chronic confusional insanity with excitement, profound changes in the direction and shape of the blood-vessels are frequent. The blood-vessels, which in a healthy state are straight or slightly undulating, in these conditions become tortuous, twisted, and redoubled on themselves, so that in exceptional instances, instead of a straight vascular tube, as in health, we have a pseudo-glomerular coil. The calibre of the blood-vessel is also affected; it is not equal, but irregularly dilated and constricted. Sometimes the dilatation resembles a minute fusiform aneurism, and an almost varicose condition is occasionally observed to accompany this. But in not a single autopsy has the writer observed that true miliary aneurismal condition which has been claimed to be so common in insanity. He questions whether, in view of the fact that miliary aneurisms have been found by Virchow in healthy subjects, and unaccompanied by any structural change, those found in the insane should be regarded as more than accidental or collateral conditions.

It is a natural result of the weakening of the vascular walls, and the diminished resistance of the wasted surrounding tissues, as well as of the frequent fluxionary episodes of certain forms of insanity, that minute hæmorrhages and their traces are sometimes observed in the cortex of the insane. It is also not difficult to bring the changes in the neuroglia and nerve-cells previously noted, into relation with the impeded transudation of nutritive material, through degenerated vascular tubes, and the retarded removal of effete products through the overstrained and blocked-up lymph passages of the degenerating brain.

The observations made concerning anæmia and hyperæmia of the brain after death, if we exclude paretic dementia and acute delirium, have very little value. The real condition as it was during life is generally obliterated by intercurrent disease, or by the mortuary changes. In a body which has been placed on the table with the face down, the frontal lobes may be found to be more injected than the occipital; if it is placed on its back, the reverse will generally be found to be the case. All these considerations show how uncertain the interpretation of so-called congestive or anæmic conditions must be, and how little value can be attached to the observation of a number of authors, who claim that in maniacal conditions the brain is found hyperæmic, and in melancholiac ones anæmic, on the strength of the degree of injection found *post-mortem*. The condition of the brain as it was *intra vitam* can be better determined by an examination, during life, of the vasomotor system, with our instruments of precision; and after death, by the morphological characters of the blood-vessels and the lymph spaces observed under the microscope. Dilated lymph spaces signify obstruction to the lymphatic outflow; tortuous blood-vessels indicate a weakening of their coats through repeated overstrain. The former condition predominates with the atonic forms of insanity, the latter with those marked by excitement.

WEIGHT OF THE BRAIN.—It may be stated to be a general rule, that in chronic insanity generally the brain loses slightly in weight, and that the loss in weight increases with advancing deterioration, until in dementia it becomes considerable. In acute insanity and in early monomania there is often no perceptible difference from the healthy standard. Numerous brains may be found in the deadhouse of an asylum exceeding in weight those found in the hospital dead-house, though the *average* is lower in the former. In estimating the bearing of the loss in weight of the brain in insanity, the influence of intercurrent somatic affections must be also taken into account; for in ordinary chronic diseases without insanity a considerable loss in brain weight is often noted. Pfleger has found that the loss in weight in the sane dying of chronic diseases is about eight per cent, while in the insane the average weight is ten per cent below the normal average.

The same observer, in his very careful measurements, found that the male brain suffers a greater loss in weight in insanity than the female brain, while Meynert arrived at the opposite conclusion. It seems to the writer that Pfleger's statement may have been made without reference to the original weight of the organ in the insane of the two sexes. In many cases, particularly where there is a constitutional taint, there are evidences pointing to a primary deficiency in brain development. Inasmuch as the male brain is normally larger than the female brain, both absolutely and relatively, and would presumably suffer more in congenital deficiency, it is apparent that a serious source of error has been overlooked by the Austrian pathologist.

Neither the chemical analysis nor the measurement of the specific gravity of the insane brain has led to results of sufficient importance to detain us here.

CHANGES IN THE BRAIN MEMBRANES.—In chronic insanity the membranes frequently show a morbid condition. It must be borne in mind, however, that many of the membranous changes found in insanity are also found in the ordinary hospital population; while, on the other hand, brains may be found in persons who have been insane for a score of years, particularly among monomaniacs, whose membranes and tissues generally, may present a picture of ideal health!

Inflammation of the dura is rare in simple insanity, but is common in alcoholic, syphilitic, and paretic dementia. Its more characteristic features, as well as the morbid anatomy of the meningeal blood cysts, which are by some considered to be the result of pachymeningitis, need not be detailed here, as they will come up for consideration in the analysis of the morbid anatomy of the disorders mentioned.

A very familiar appearance to the alienist pathologist is a milky-white opacity of the arachnoid which may affect the entire expanse of that structure. It is found in a large number of the chronic insane, and is usually associated with epithelial granulations of the pia, and increase in the size and number of the Pacchionian bodies. All these conditions indicate the existence of long-continued vascular strain, with consequent exudation of formed elements of the blood, causing cellular infiltration and thickening of the arachnoid lamellæ on the one hand, and enlargement of the " safety diverticula " of the great cerebral veins, as which the Pacchionian bodies are to be regarded, on the other. All these conditions, however, occur in persons who never have been insane, and it cannot be maintained that they in any way demonstrate the existence of insanity, though where found it is justifiable to bring these morbid appearances into relation with the mental disorder, inasmuch as far more numerous and intense examples of these changes are found in the insane than in the sane population. Even in the latter group, those showing any considerable degree of these changes are generally persons who have had syphilis, rheumatism, gout, been addicted to alcoholic excesses, or have suffered from exhausting chronic diseases.

Although it is exceptional to find the pia adherent to the cortex of the insane, except in paretic dementia, in insanity with meningitis or syphilis, and in insanity following insolation, this condition may be found in the terminal period of any psychosis, and has been discovered even in monomania, particularly when the course of this affection has been marked by a number of *quasi*-maniacal exacerbations. There are sometimes observed in the insane, particularly in those whose disorder has been of a chronic character, a peculiar series of meningeal changes, which seem to the writer to be interpretable as the evidences of a chronic and subdued inflammatory process which is not the cause of the insanity, but the result of prolonged insane excitement combined with the alcoholic excesses to which this sometimes leads. In a case of this kind—an imbecile with moral perversion and feeble ideas of aggrandizement, and who died with symptoms strongly suggesting the existence of an idiopathic organic brain trouble,—the following conditions were found : On cutting into the dura of the right side a turbid fluid escaped. It was impossible to separate the dura from the other membranes over the frontal lobe, in an area corresponding to two thirds of its convexity. Here the leptomeninges and the dura were fused into a thick, dense common mass of membranous and pseudo-membranous layers, adherent to the cortex and inseparable from it. On vertical section, it was found that fibrino-purulent layers penetrated to the depth of the sulci filled the meshes of the pia and arachnoid, and merged with the thickened and infiltrated dura into a combined thickness of more than half an inch at the focus of the disease, which was over the posterior third of the middle and upper frontal gyri. The fibrino-purulent infiltration extended further backward than the adhesion of the meninges. The pia shows a distinct zone of a yellowish color, about half an inch in extent beyond the area of adhesion, and this zone detached peninsular processes along the chief vessels of the convexity. These, however, did not exhibit the creamy-yellow color of the chief disease area. On each side of the vessels a white or whitish-gray streak, varying from less than one to more than three millimetres in width, was noted. On cutting, this almost creaked under the knife; its consistency could be best compared to a well-sized parchment. In diminishing distinctness this peculiar condition could be traced as far back as the occipital lobe. It had every appearance in these districts of a lesion of ancient date, and the more active symptoms of focal cerebral disease immediately preceding death were evidently due to an intensification of the lesion in the way of an acute and local exacerbation; while the original mental deficiency was attributable to a congenital defect of the brain development, manifesting itself in gross asymmetry and atypy not only of the cerebral hemispheres but also of the peduncular tracts, and was altogether independent of the progressive morbid process discovered after death.

A number of morbid appearances are found in other parts of the body than the brain and its envelopes. But these are either concomitant lesions of other parts of the nervous system, such as the spinal cord, characterizing certain special forms of insanity, or trophic disturbances dependent on such lesions. In a comparatively small number of instances diseases of the viscera are found which have an intrinsic signification, and will hence be discussed with the forms of insanity dependent on such diseases. But it is to be insisted here that these cases are rather exceptional than the rule. The gross lesions frequently found in the chronic insane are usually the remote consequences of insanity or accidental accompaniments, and not its cause. For example, in hypochondriacal lunatics and dements, whose psychosis began as a melancholia, it is not rare to find a dislocation of the transverse colon, whose loop may hang very low in the abdominal cavity.

This condition is undoubtedly due to the intestinal inertia, the resulting coprostasis, and consequent dilatation and loss of elasticity of the gut. Other of the lesions found in the insane are rather signs than causes of their disease; that is, they are either evidences of the obscure influence exerted by the mind upon the body, or of the trophic *rôle* played by the brain in the nutrition of distant organs, or, finally, local expressions of a deeper general somatic state, on which the insanity and the somatic signs in question rest in common. For instance, amenorrhœa in women, and torpor of the enteric tract in all sexes are concomitant to, and not causative of, melancholia. It is hardly necessary to refer here to the grave visceral lesions found with the paralytic insanities, which are results and not causes of the cerebral disorder, and whose analogues have been produced through artificial cerebral lesions, in the lower animals;[*] nor to similar somatic changes, sometimes found in cases of advanced epileptic insanity.[†] Even palpable affections of the brain may be accidental accompaniments of a mental disorder, nay, even its indirect results! The cysticerci occasionally discovered in the brains of imbeciles, terminal dements, and paretics, developed from ova introduced with the materials devoured in obedience to the filthy propensities of such lunatics, are illustrations of the possibility of such a curious relation.[‡]

From the statements just made it follows, that before a preternatural appearance, found in the brain of an individual dying insane, can be adduced to explain the nature of the mental symptoms, or to illustrate the operation of

---

[*] By Brown-Séquard, Leudet, Ollivier, Dupuy, Eulenburg, and others.
[†] Dufour, "Annales Medico-Psychologiques," 1880.
[‡] Ullrich, "Allgemeine Zeitschrift für Psychiatrie," 1872, xxix. Wendt, "Fall von Cysticercus im Gehirn, als Folge, nicht als Ursache der Geistesstörung." Ibidem, 1874, xxxi. Meschede, Ibidem, 1873, xxv. The writer has frequently found cysticerci, single or in numbers not exceeding three, in monkeys of the genera macacus, cercopithecus, cynocephalus, and semnopithecus, occupying different gyri, including those most carefully studied by the localizationists. In none of these cases had any symptoms of cerebral disturbance been observed. It is, however, a beautiful illustration of the proposition that diffuse and multilocular lesions of the fore-brain always produce insanity, that in an earlier case of Wendt's ("Allgemeine Zeitschrift für Psychiatrie," xxv.), where beside numerous cysticerci found in the cerebellum and basilar parts, one hundred and thirty were found in the cerebral hemispheres, the insanity of the subject coincided with the probable period of their development, as it does in analogous cases found among domestic animals.

an originally remote cause, it should be submitted to certain tests. It must not only be shown to be independent of spontaneous changes occurring after death, and of artefacta produced by the investigator, but it is also essential that the lesion be not one of the same kind, extent, and location as one found as frequently, or more frequently, in subjects dying sane. It is needless to remark that the same requirements are to be met, when it is attempted to establish a relation between diseased conditions found in other organs than the brain and a co-existing insanity.

The connection between a given lesion and mental alienation having been rendered plausible after compliance with these preliminary demands, it remains to be ascertained whether the lesion as found is merely related to the particular phase of insanity as it existed before death, or whether it explains also the origin of the trouble. It is here to be insisted that, if the mental disorder is the terminal phase of a preceding series of mental symptoms, perhaps differing in character from their sequel, the discovered lesion can represent nothing more than the somatic basis of the terminal phase. It may enlighten us as to the basis of the preceding mental phases and the inception of the original evil in so far only as the principles of pathology permit us to infer from a *post-mortem* appearance what the initial steps of the morbid life-processes leading to it were.

CHAPTER XI.

THE CLASSIFICATION OF INSANITY.

While the proper classification of insanity which was initiated by the French and carried further by the German alienists may be said to be approaching a comparative state of perfection on the Continent generally, it is in America and England still in a very chaotic condition. Nearly every writer on insanity has offered a classification of his own. But while the student will find only slight differences between the classifications of a Marcé, Dagonet, and Esquirol, or a Krafft-Ebing, Schuele, and Meynert, and the general principle of their classifications is the same, he can only be confounded by a comparison of the systems of a Maudsley,

Hammond, Bucknill-Tuke, Skae, and Sankey. Suggestive hints and valuable demarkations may be found scattered here and there among the divisions of these authors; but, on the whole, the ambition to found new forms, often involving the doing away with the clinical principles which must underlie every practically available definition, and to establish formulæ in place of following observations as a basis of classification, impair their usefulness.

What could be more unfortunate than an attempt to classify insanity according to the faculties of the mind supposed to be affected in its various forms? Maudsley, one of the most progressive of English alienists, for example, divides insanity as follows:

I. AFFECTIVE OR PATHETIC INSANITY.
  1. Maniacal perversion of the affective life. Mania sine delirio.
  2. Melancholic depression without delusion. Simple melancholia.
  3. Moral alienation proper.

II. IDEATIONAL INSANITY.
  1. General.
     α. Mania.   } Acute and
     β. Melancholia. } chronic.
  2. Partial.
     α. Monomania.
     β. Melancholia.
  3. Dementia. { Primary.
               { Secondary.
  4. General paralysis.
  5. Idiocy and Imbecility.

Here we have mania and melancholia distributed under both heads; in one place as affective and in another as ideational insanity! An attentive observer would often find that the same patient presents mania or melancholia without delusion at one period of his illness, and mania or melancholia with delusion at another. In fact there are few cases of insanity which would not be found to occasionally occupy a place in either of Maudsley's great groups. Such classifications are faulty, because they lack clinical unity and consistency. The same applies to several other systems of classification which are constructed on similar principles.*

One of the most curious classifications is that which was recommended by a committee of the British Medico-Psychological Association in 1869, but not adopted by that body. The fundamental distinction is here made of "cur-

---

* Maudsley has been selected as the most modern representative of the school which, with Griesinger and Prichard, made the mental complexion of the disorder the main guide in classification. He does not, therefore, stand alone, but his error has been a very general one; and there are those who have not yet emancipated themselves from it.

able" and "incurable" forms. Let the reader imagine such a principle applied to any other class of diseases, say those of the intestinal tract ! The proponents of this classification seem also to have been aware of the existence of but one form of senile insanity, namely dementia.

Morel was the first to admit the etiological principle into the classification of mental disorders. He divided insanity into—1. Hereditary Insanity ; 2. Toxic Insanity ; 3. Insanity due to the transformation of other neuroses, such as epilepsy and hysteria ; 4. Idiopathic Insanity; 5. Sympathetic Insanity; 6. Dementia. While this classification is defective on account of the insufficient stress laid on the clinical features of insanity, it must be regarded as the first step toward that more perfect classification which has been recently proposed and defended by Krafft-Ebing and other Germans. But a long period passed before this principle was properly asserted, for it was carried to an extreme, and rendered the vulnerable object of much ridicule by the English followers of Morel. Skae's twenty and more etiological forms, containing as they do much that is useful, include also a "post-connubial insanity," which may be anything from Mania and Melancholia to Paretic Dementia; a Mania of Oxaluria and Phosphaturia, which no investigator has been so fortunate as to be able to confirm the existence of; and manias of "pregnancy" and "lactation," which may be as often true melancholias as manias. Unable to make even his accommodating system fit all cases, Skae was compelled to erect an "Idiopathic" group, which, in truth, contains over one half of all the known and accepted forms of insanity. There is naturally no correct balancing of the main forms here!

Dr. D. Hack Tuke improved greatly on the classifications of Skae and Morel, attempting to combine the metaphysical and etiological principles into one. But, by making the former the guiding and the latter the subsidiary element, he was compelled to place Mania under "Intellectual Insanity," and Melancholia side by side with "Exaltation," and "Moral Insanity" under the "Emotional Insanity." There is also no place in his groups for paretic dementia. On the other hand, "Moral Insanity," placed by Dr. Tuke under the second order of the second class, those occurring as an "invasion after development," may be also due to imperfect development, and therefore might fall under his first order as well. (Appendix II.)

The pathological principle has also been made the basis of classification. By none has it been carried further than by Voisin. He divides acquired insanity as follows:*
  I. Idiopathic insanity.
     a. Due to vascular spasm.
  II. Insanity dependent on appreciable brain lesions.
     a. Congestive insanity.
     b. Insanity from anæmia.
     c. Atheromatous insanity.
     d. Insanity consecutive to brain tumors.
  III. Insanity dependent on alterations of the blood.
     a. Diathetic insanities.
     b. Syphilitic insanities.

This classification requires no extended discussion. Voisin's views as to congestion and anæmia in insanity are utterly fanciful, and it suffices to characterize the unsystematic application made by this author of his adopted principles, that general paralysis is widely separated from the other "idiopathic insanities" with "demonstrable brain lesions." Besides, Voisin claims to have determined a pathological basis for nearly every form and variety of insanity, and has in this respect either anticipated his age or has fallen into such multitudinous errors that it may be safe to pass by a project which may or may not be realized in the future, but which at present is decidedly unfeasible.

One of the best classifications made within the last decade is that of Krafft-Ebing.† He divides insanity into two great

---

* Leçons Cliniques sur les Maladies Mentales, 1883. The classification offered by the same author in the edition of 1876 is more elaborate.

I. Acquired Insanity.
  a. Idiopathic.
    { Unaccompanied by demonstrable lesions.
      Accompanied by demonstrable lesions.
        1. Active congestion.
        2. Passive congestion.
        3. Simple anæmia.
        4. Secondary anæmia.
        5. Atheroma.
        6. Tumors. }
  b. Following the great neuroses.
  c. Sensorial insanity.
  d. Sympathetic insanity
II. Congenital insanity.
III. Toxic insanity.
IV. Cretinism, idiocy, and imbecility.
V. General paralysis.
VI. Senile dementia.

† "Lehrbuch der Psychiatrie," II., 1879.

THE CLASSIFICATION OF INSANITY. 117

groups, according as the disorder is the result of a disturbance of the developed brain or of an arrest of brain development. Under the first head he places insanity ordinarily so called, and subdivides further as follows:
A. Mental affections of the developed brain.
  I. Psychoneuroses.
    1. Primary curable conditions.
      a. Melancholia.
        α. Melancholia passiva.
        β.   "   attonita.
      b. Mania.
        α. Maniacal exaltation.
        β.   "   frenzy.
      c. stupor.
    2. Secondary incurable states.
      a. Secondary monomania (secundære verruecktheit).
      b. Terminal dementia.
        α. Dementia agitata.
        β. Dementia apathetica.
  II. Psychical degenerative states.
    a. Constitutional affective insanity (folie raisonante).
    b. Moral insanity.
    c. Primary monomania (primære Verruecktheit).
      α. With delusions.
        αα. Of a persecutory tinge.
        ββ. Of an ambitious tinge.
      β. With imperative conceptions.
    d. Insanities transformed from the constitutional neuroses.
      α. Epileptic.
      β. Hysterical.
      γ. Hypochondriacal.
    e. Periodical insanity.
  III. Brain diseases with predominating mental symptoms.
    a. Dementia paralytica.
    b. Lues cerebralis.
    c. Chronic alcoholism.
    d. Senile dementia.
    e. Acute delirium.
B. Mental results of arrested brain development : idiocy and cretinism.

The criticism to be made of this classification is threefold. In the first place, it does not accommodate itself to the fact that there is every connecting link between idiocy and imbecility on the one hand and monomania on the other. In the second place, the designation "Verruecktheit" is too generally used, being made to apply to two very distinct forms of insanity. In the third place it is unnecessary, as it is inaccurate to add the adjectives "curable" and "incurable" to the primary and secondary forms of the psychoneuroses, for in many cases the primary disorders are not cured by the very best treatment, and the stamp of incurability seems to be affixed to some cases from the start.

It may be readily surmised that where the best thinkers have failed to produce an unexceptionable classification, the failure must be due to some inherent difficulty of the subject. Few cases of insanity are exactly alike in all respects. Here we have a patient whose insanity is characterized by a deep emotional tinge, there one with moral perversion, another with morbid propensities, and still another with fixed ideas. Here is an entire group of asylum inmates without hallucinations, illusions, or delusions; there another without dementia. Here is a ward filled with patients whose mental symptoms are accompanied by somatic anomalies; there another, in which no patient may be found whose somatic state differs appreciably from that of ordinary hospital patients. The course of this psychosis is chronic, of that one acute; while in some it is even, and in others progressive. In certain cases we find characteristic evidences of insanity in the dead bodies, in others not. Sometimes the psychosis is primary in origin, at others secondary to another psychosis. Often the insanity exists by itself, and often it accompanies and is determined in its existence by some other complaint. Several forms are hereditary or congenital, and others are independent of congenital and hereditary influences. More than one form of insanity is intimately associated with developmental periods, while the majority may appear at any time of life. In short, there is every possible association of factors seeming to distinguish various groups of the insane, and none of these can be altogether ignored in classification. For this reason all attempts to classify the form of insanity according to any *one* given invariable principle are predestined to failure.

Let us walk through an asylum for the insane with an experienced alienist, and observe his method of diagnosticating and classifying the insane. He will not point out to the visitor any cases of "ideational," "atheromatous," or "post-connubial" insanity, but he will show him a monomaniac, a melancholiac, a paretic dement, or a stuporous lunatic; and he will be able to pick out a large number of patients, place them together, and demonstrate the general sameness in the symptoms of all monomaniacs, or of all paretic dements, or of all patients belonging to any one of the other clinical groups. If he make any finer distinction, he may remark that in such and such a patient the disorder is hereditary; that in this patient the delusions are depressive, in that one expansive, and occasionally he may allude to the fact that in certain patients the type of the disorder is determined by the etiology; that, for example, it may be alcoholic or epileptic in character. Let us now follow him to the "demented ward." Here he will exhibit a number of patients who seem to be all equally sunken into a condition of mental apathy and deterioration. But, on the alienist's directing attention to certain points in their past and present history, distinct types of "dementia" will become recognizable even to the novice. This patient who still exhibits a few faintly expressed delusions was originally a maniac, then his insanity became a chronic confusional delirium which has gradually passed into what is properly called a *terminal* dementia. A second patient has passed into a demented state by a gradual and progressive deterioration from a previous condition of mental health; his dementia is a *primary* * one, and not secondary to, nor the terminal epoch of some other form of insanity. A third has epileptic convulsions, and his dementia is the sequel of epilepsy and consequently an epileptic dementia. A fourth has had a hæmorrhage or other destructive lesion of the brain, and his feeble-mindedness is attributable to that; in other words, he suffers from *dementia with coarse organic disease* of the brain. A fifth presents a peculiar and characteristic grouping of motor and sensory disturbances, and intellectual and moral perversion combined with mental failure, which in their union constitute paretic dementia. In still another patient the dementia is *senile*, because

---

*This term has been used very confusedly; for explanation of the sense in which it is used here see " Primary Deterioration."

it is an expression of the involution of age. Finally, a patient will be shown who is apparently in the lowest depths of mental annihilation; but the alienist assures the visitor that the disorder is not a progressive but a temporary one; that it is an overwhelming of the mind by some emotional shock, or the result of an episodial brain-exhaustion, and that in nine cases out of ten the patient will emerge from his present state clothed in his right mind. Such a case is one of so-called *primary acute* dementia, better known as *stuporous insanity*. As the novice is made better acquainted with the psychical features of these groups, he will find that the distinctions on which they are based do not exist alone in the antecedent history of the patients; but that as a general thing the character of the symptoms of each of them has something specific: that the senile dement is miserly and suspicious, the paretic dement boastful and extravagant, while the terminal dement is either apathetic or agitated according to the nature of his primary mental disorder, and that he often exhibits the residual delusions developed with the latter.

If the origin and prospects of various cases of so apparently simple a disorder as dementia are so widely different as is here adverted to, it may be readily conceived that, with regard to the more positive manifestations of insanity, the distinctions must be still greater, and even more important as diagnostic and prognostic criteria. It would be manifestly improper to place all patients manifesting maniacal excitement in one common group. Maniacal excitement may be an indication of a disorder consisting of this excitement as the sole prominent symptom, namely, of *simple mania;* it may occur as an episode of *paretic dementia* and in *epileptic insanity;* finally, it may be the recurrent manifestation of a *periodical insanity*, or characterize the explosion of a *toxic* affection. As insanity is after all but the symptomatic manifestation of a brain disorder, and the pathological states underlying insanity are not well known, obviously the simplest and most profitable plan of classification will be the adoption of the clinical method as our main guide; then where etiology, pathology, and speculative psychology furnish valuable distinctions we may incorporate them as collateral aids in such classification.

The first distinction to be made is between those cases in which the insanity is the directly produced and most important disorder manifested by the patient, and those in which the insanity is an accidental and inconstant accompaniment of other diseases, and has its nature modified by these. The first group may be designated as that of the PURE INSANITIES, and the second as that of the COMPLICATING INSANITIES. Paretic dementia, simple mania, imbecility, monomania, and alcoholic insanity are the direct expression of disorders *primarily attacking the brain;* are *not necessarily* dependent on any other disordered bodily condition, and if attributable to such a condition are *not* thereby modified to any important extent. Rheumatic insanity, pellagrous insanity, and the post-febrile psychoses are *essentially* dependent on disorders which are *not primarily* cerebral in their location, and their symptoms are specifically modified by an *originally non-cerebral* cause.

In the class of the "pure insanities" two great divisions must be made. There is one group comprising mental disorders which affect persons previously of sound mind, somewhat after the manner in which a fever or a diarrhœa attacks a person of previously sound bodily health. The insanity here is not the explosion of a continuous morbid condition, but stands by itself *an isolated occurrence** in the midst of a relatively healthy career which it may check and end. The other group comprises disorders which are the explosions of a *continuous neurotic condition,* which may be inherited from a vitiated ancestry, acquired through intra-uterine or infantile brain disease, or developed under the influence of injuries to the skull and brain, or of excesses in the use of certain narcotics. In a rude way the first group corresponds to Krafft-Ebing's "Psychoneuroses" and "Brain Diseases with Predominating Mental Symptoms" united into one class; the second group is nearly equivalent to the combined "Psychical Degenerative States" and "Mental Results of Arrested Brain Development" of the same author.

PURE INSANITY NOT INTRINSICALLY DEPENDENT ON A CONTINUOUS NEUROTIC VICE is in turn divisible into sub-groups. The first is not associated with demonstrable active organic changes of the brain, while the second is so associated. Simple mania is a type of the first sub-group, paretic dementia of the second.

PURE INSANITY NOT INTRINSICALLY DEPENDENT ON A

---

* Isolated in the sense in which the term may be applied to fevers or other affections of the kind. A recurrence is not excluded, but it is never typical.

CONTINUOUS NEUROTIC VICE, NOR ASSOCIATED WITH DEMONSTRABLE ACTIVE ORGANIC CHANGES OF THE BRAIN is again divisible into sub-groups. In a first subdivision we find insanities which attack individuals irrespective of the physiological periods of development and involution, while in a second subdivision they are intimately connected with such periods. Simple mania and melancholia are instances of the first kind, while insanity of pubescence is an instance of the second kind.

PURE INSANITIES NOT INTRINSICALLY DEPENDENT ON A CONTINUOUS NEUROTIC VICE, NOR ASSOCIATED WITH DEMONSTRABLE ACTIVE BRAIN CHANGES, NOR RELATED TO THE PERIODS OF DEVELOPMENT AND INVOLUTION include most of the curable cases of mental disorder. They present themselves under two distinct forms, according as the disorder is primary or secondary to one of the other forms of the same series. Simple melancholia and acute confusional insanity are representative of the *Primary Forms*, and terminal dementia and chronic confusional insanity are examples of the *Secondary Forms* coming under this head.

The PRIMARY INSANITIES *not intrinsically dependent on a continuous neurotic vice, nor associated with demonstrable active brain changes, nor related to the periods of development and involution,* consistently with the dichotomous division which happens to mark this branch of psychiatrical classification, naturally fall into two categories. In one the insanity is always characterized by a fundamental emotional disturbance. To this category belong *mania*, marked by an exalted emotional state; *melancholia*, marked by a painful emotional state; *katatonia*, marked by a pathetic emotional state; and *transitory frenzy*. In the other category there is an absence of any profound emotional disturbance. To this category belong *primary dementia, stuporous,* and *acute confusional insanity*. The SECONDARY INSANITIES belonging to the same sub-group include *terminal dementia* and *chronic confusional deterioration*.

Pure insanity not intrinsically dependent on a continuous neurotic vice, nor associated with demonstrable active organic changes of the brain, BUT RELATED TO THE PERIODS OF DEVELOPMENT AND INVOLUTION, comprises only two forms: *insanity of pubescence* (the *hebephrenia* of Hecker-Kahlbaum), and *senile dementia*. It is a matter of doubt whether a number of cases of insanity occurring at the time of the second climacteric justify the erection of a special genus for their accommodation under this head.

The pure insanities which are not the outcome of a continuous neurotic vice, and ARE ASSOCIATED WITH ACTIVE ORGANIC BRAIN CHANGES, include *paretic dementia, delirium grave* (the *manie grave* of the French), *syphilitic dementia*, and *organic dementia*. Under the latter term the mental defects accompanying gross disease of the brain, such as cysticerci, tumors, hypertrophy, atrophy, sclerosis, and the ordinary cerebral vascular lesions are included.*

THE PURE INSANITIES WHICH ARE THE OUTCOME OF A CONTINUOUS NEUROTIC VICE OR TAINT are those to whose pathogenesis the conclusions of the ninth chapter apply in great part. The neurotic vice may manifest itself in a gross and general defect in brain development as in idiocy; in lesser defects associated with anomalies in the cranial shape and peripheral growth and innervation, as in cretinism, imbecility, and original monomania; in convulsions during childhood; or in a generally neurotic constitution and mentally abnormal character. The mental disorder may date from birth, from the period of puberty, or from the second climacteric, on the one hand, or it may be developed by exciting causes at any time of life, on the other. The neurotic vice is not necessarily transmitted; it may be acquired through traumatism, and the formation of the alcoholic or other narcotic habit; and it may also gradually develop on the basis of any of the constitutional neuroses, such as epilepsy and hysteria.

It is impracticable to separate the forms of these properly so-called *constitutional* insanities into subdivisions based on the intensity or character of the transmitted vice. Gross anatomical defects or lesser somatic indications of defective brain development are found indifferently in several of the forms of insanity belonging to this series; but they may be undemonstrable in the very same forms, and only the neurotic character may serve to characterize the predisposed person as a defectively organized individual. Attention has been already (Chapter IX.) directed to the fact that any form of insanity in this series may be transformed into another, and that the anatomical defects sometimes noted may be intensified in the course of hereditary transmission, or become potent in a descendant when they were absent in the ancestor. It was also observed at the same place that

---

* This form will not be considered in this treatise, except in its differential diagnostic relations.

science is not yet able to determine what kind of anomaly determines the existence of a given form of hereditary or constitutional insanity. The same kind of asymmetry has been found in congenital paranoia as in imbecility.

A very natural distinction can be made between those forms in which the insanity is associated with the great neuroses, and those in which there is no such association, and if there be, that association is accidental. An idiot may have epileptic attacks, but his idiocy is not dependent on these, but on the brain-defect with which he was born. A monomaniac may have hysterical or epileptiform symptoms, though these are accidental to, and do not essentially modify the monomania. But there are epileptic and hysterical patients who develop an insanity intimately dependent on the neurosis, and whose symptoms have a specifically epileptic or hysterical character. Similarly any lunatic belonging to these groups may become an inebriate, but it is either a result or an accident in monomania, imbecility, and periodical insanity, while in the alcoholic maniac the insanity is the direct outgrowth of and modified in its symptoms by the acquired alcoholic neurosis.

THE PURE INSANITIES WHICH ARE THE EXPRESSION OF A CONTINUOUS NEUROTIC VICE, BUT NOT DEPENDENT ON THE GREAT NEUROSES, comprise *idiocy, imbecility, cretinic insanity, paranoia,* and *periodical insanity;* those which ARE DEPENDENT ON THE GREAT NEUROSES comprise *epileptic, hysterical,* and *alcoholic insanity.*

The COMPLICATING FORMS are as numerous as the somatic causes which may determine the existence and modify the character of insanity. It is customary to designate that insanity following injuries to the skull and which has certain specific clinical characters as *traumatic insanity;* that following rheumatism and gout as *rheumatic* and *gouty insanity;* that accompanying chorea, as *choreic insanity;* that developing in the course of phthisis as *phthisical insanity;* and that due to powerful reflex influences as *sympathetic insanity.* Many of these forms are rare, others of only exceptional occurrence, but they deserve a separate place because their symptoms do not correspond exactly to those of the "pure forms," and their treatment is directly to be based on the etiology. Indeed, they may be called the etiological forms. One form appertaining to this series, *pellagrous insanity*, will not be discussed in this volume, as it does not occur in America, and is limited to such countries as Italy, where maize forms a staple article of diet, and where the disease known as *pellagra*, which is attributed to the living on spoiled maize, occurs in an endemic form.*

A glance at the subjoined table will give a better idea of the proposed classification than any further description. It will be observed that on adding the designation characterizing the species to that of the genus and the class, a definition of many of the enumerated forms can be compounded. Thus: Melancholia is a simple insanity, not essentially the manifestation of a continuous neurotic condition, not associated with demonstrable active organic changes of the brain, attacking the individual irrespective of the developmental and involutional periods, of primary origin and characterized by a fundamental emotional disturbance of a painful character. Insanity of pubescence is a simple insanity, not essentially the manifestation of a continuous neurotic condition, not associated with demonstrable active organic brain changes, attacking the individual in connection with the period of puberty. Paretic dementia is a simple insanity, not essentially the manifestation of a continuous neurotic condition, associated with demonstrable organic changes of the brain, which are diffuse in distribution, primarily vaso-motor in origin, and destructive in their results. As a rule much briefer definitions will serve the purposes of the alienist, but the fact that the proposed classification carries with it the terms of these definitions will seem to many the strongest proof of its consistency.

It is claimed in behalf of this classification, that while it is far from being above criticism in many particulars, it is calculated to meet the requirements of the practical alienist, in those respects in which the other classifications referred to fail.

It may be objected to on the following grounds, and it is proper to take up those objections which can be anticipated seriatim and at this point.

It may be claimed that the maniacal symptom group being found in certain cases of periodical insanity as well as in simple mania, the two should not be widely separated. To this it may be answered that in simple mania, the emotional disturbance is the sole essential feature, while periodical recurrence and a neurotic constitu-

---

* The other complicating forms will be considered incidentally in the chapter on Etiology.

INSANITY.

GROUP FIRST. PURE INSANITIES.

SUB-GROUP A.

Simple Insanity, not essentially the manifestation of a constitutional neurotic condition.

FIRST CLASS.

Not associated with demonstrable active organic changes of the brain.

I. DIVISION. Attacking the individual irrespective of the physiological periods.

a Order: Of primary origin.
Sub-order A. Characterized by a fundamental emotional disturbance.
Genus 1: of a pleasurable and expansive character............**Simple Mania**
Genus 2: of a painful character.
Simple Melancholia
Genus 3: of a pathetic character.
Katatonia
Genus 4: of an explosive transitory kind.
Transitory Frenzy
Sub-order B. Not characterized by a fundamental emotional disturbance.

Genus 5: with simple impairment or abolition of mental energy.
Stuporous Insanity
Genus 6: with confusional delirium.
Primary Confusional Insanity
Genus 7: with uncomplicated progressive mental impairment.
Primary Deterioration
β *Order:* Of secondary origin.
Genus 8: **Secondary Confusional Insanity**
Genus 9: **Terminal Dementia**

II. DIVISION. Attacking the individual in essential connection with the developmental or involutional periods. (A single order.)
Genus 10: with senile involution............................**Senile Dementia**
Genus 11: with the period of puberty.....**Insanity of Pubescence** (Hebephrenia)

SECOND CLASS.

Associated with demonstrable active organic changes of the brain. (Orders coincide with genera.)
Genus 12: which are diffuse in distribution, primarily vaso-motor in origin, chronic in course, and destructive in their results......................**Paretic Dementia**
Genus 13: having the specific luetic character..............**Syphilitic Dementia**
Genus 14: of the kind ordinarily encountered by the neurologist, such as encephalomalacia, hæmorrhage, neoplasms, meningitis, parasites, etc.
Dementia from Coarse Brain Disease
Genus 15: which are primarily congestive in character and furibund in development.
Delirium Grave (Acute Delirium, *Manie grave*)

SUB-GROUP B.

Constitutional Insanity, essentially the expression of a continuous neurotic condition.

THIRD CLASS.

Dependent on the great neuroses (orders and genera coincide).
I. DIVISION. The toxic neuroses.
Genus 16: due to alcoholic abuse............................**Alcoholic Insanity**
(Analogous forms, such as those due to abuse of opium, the bromides, and chloral, need not be enumerated here, owing to their rarity.)
II. DIVISION. The natural neuroses.
Genus 17: the hysterical neurosis...........................**Hysterical Insanity**
Genus 18: the epileptic neurosis............................**Epileptic Insanity**

FOURTH CLASS.

Independent of the great neuroses (representing a single order).
Genus 19: In periodical exacerbations ....**Periodical Insanity**
Order: arrested development { Genus 20: **Idiocy and Imbecility**
{ Genus 21: **Cretinism**
Genus 22: manifesting itself in primary dissociation of the mental elements, or in a failure of the logical inhibitory power, or of both.......**Paranoia (Monomania)**

GROUP SECOND. COMPLICATING INSANITIES.

These may be divided into the following main orders, which, as a general thing, are at the same time genera: Traumatic, Choreic, Post-febrile, Rheumatic, Gouty, Phthisical, Sympathetic, Pellagrous.

tion are necessary additional elements in periodical insanity. Inasmuch as these latter features determine the grave prognosis of periodical insanity, they are of greater practical import, as they are certainly of higher significance, from an abstract pathological point of view, than the symptomatic direction in which the disorder manifests itself. Consequently the neurotic predisposition and periodical recurrence of the malady must as criteria determining classification rank higher than the symptoms *per se*.

It may also be urged that heredity and the neurotic constitution, while they do not play as important a part in the simple as in the constitutional forms of insanity, yet they occasionally and in some forms, as in melancholia, quite frequently accompany the simple forms. But it is exceedingly rare for the patients suffering from a simple insanity to exhibit a *continuous* neurotic condition, and where they do manifest it, then "simple" insanity attacks them, as an acute disease may attack a previously healthy individual and evenly with one suffering from chronic disease or a constitutional vice but without any tangible connection with such chronic disorder or constitutional vice. An individual may inherit syphilis and become attacked by an acute pneumonia, but we do not speak of such a pneumonia as a syphilitic pneumonia; or another may have the tuberculous predisposition and die with a surgical affection, and while —this is said merely for the sake of offering an analogy —the statistics might show that more tuberculous subjects die of surgical affections of a given kind than non-tuberculous subjects, yet until a more intimate relation could be shown between the vitiated constitution and the local disease, we would hesitate to speak of surgical affections in these cases as of the "tuberculous" variety. It is the same with insanity: certain varieties like monomania and imbecility are almost invariably associated with an acquired or transmitted neurotic vice, and on comparing a large number of cases exhibiting these forms of derangement it is found that there is on the whole a sameness in the origin and nature of the symptoms. On the other hand, simple mania, stuporous insanity, and other of the simple forms are not as a rule associated with such taint, may attack persons previously healthy and free from hereditary taint, and are noted for an absence of those characters found with the constitutional forms. Syphilis and the tuberculous diathesis are undoubtedly transmitted, and the clinician is justified in characterizing certain medical and surgical affections from which subjects of such transmission suffer as the outcome of an hereditary or constitutional vice. But he will not place incidental affections of exactly the same character as those affecting the sane population—for example, ordinary catarrhs, attacks of indigestion, of diarrhœa, or the exanthemata—in the same category with the results of the constitutional affection. Just as a syphilitic subject may become affected with smallpox, so an imbecile may become a sufferer from acute melancholia; and just as a child afflicted with any hereditary cachexia may be carried off by a scarlatina or diphtheria, so a paranoiac may end his days as a paretic dement. It is needless to add that the occasional development of one form of insanity in a subject already suffering from some other form is no more a ground for considering the two affections to be inseparable, than it would be just to classify peritonitis and impaction of biliary calculi as varities of one and the same disease because the former may complicate the latter.

One of the strongest objections to be advanced against the proffered classification is that alcoholic insanity and senile dementia are placed remotely from the forms which like paretic dementia and acute delirium are associated with demonstrable active organic changes. It is true that considerable organic disease may be found in senile dementia and in alcoholic insanity. But in the former disorder these changes are the passive ones of involution, and not fresh processes attacking the previously sound brain. As to insanity developing on an alcoholic basis, those cases of it in which gross changes are found—changes which in that group of cases seem to stand in a constant relation to the symptoms—do not belong to alcoholic insanity proper, but constitute a variety of paretic dementia or insanity from coarse organic disease. The symptoms are then of an entirely different character, the morbid changes are both *demonstrable* and of an *active* kind, and the consistency of the classification proposed is nowhere better shown than here, where insanity which has a similar etiology, but a different clinical, pathological, and prognostic character from the alcoholic forms properly so-called, is removed from them by the terms of the definition heading the group in question.

A further objection may be based on the fact that dementia from organic disease and dementia from cerebral syphilis are not ranked with rheumatic, pellagrous, and post-febrile insanity. It may be alleged that they should be so ranked because they are all equally among the unusual manifestations of other diseases than those which fall within the ordinary ken of the alienist, and are hence true *complicating* forms. To this weighty objection it can be replied, that insanity from the physiological psychologist's point of view is a manifestation of brain disorder ; that we are correct in assuming that the ordinary psychoses are true cerebral affections, *primarily* of cerebral origin, and that it would be unwise from a patho-anatomical point of view—little as we actually know of mental morbid anatomy —to separate the organic affections of the brain producing insanity, even though they produce it but occasionally, from the known and hypothetical diseases of the same organ producing those symptoms more regularly. Besides, by retaining the distinction between those forms which *never exist* without an *essential* extra-cerebral disorder from those in which a cerebral disorder is the *primary* determining factor, attention is prominently directed to certain useful therapeutical purposes.

About one fact there can be no dispute, that, excluding the "complicating forms," the majority of the distinctions made will be recognized as necessary by the practical alienist. In a properly drawn up table of any asylum of over five hundred beds the reader will find that mania, melancholia, stuporous insanity, primary, terminal (secondary), senile, and paretic dementia, dementia from organic disease, acute delirium, alcoholic, hysterical, epileptic, and periodical insanity, states of arrested development, and monomania—possibly under the more popular though less exact title of "chronic delusional insanity"—all have a place. It may be assuredly claimed that these distinctions having stood the test of time must possess a practical value. The day is past when the asylum physician can content himself with such a classification as this one.*

Mania: acute, sub-acute, chronic, recurrent.
Melancholia: acute, chronic.
Dementia: primary, secondary, senile.
Amentia (!): idiocy, imbecility.
General paresis.

---

* Taken from the annual report of the New York City Asylum for the Insane, dated January 1st, 1879.

The average asylum attendants—and, in more than one instance noted by the writer, the asylum inmates themselves —are capable of mastering and—as far as an application can be spoken of—of applying such a system. When every excited patient is considered maniacal, every depressed one melancholic, every apathetic one a dement, and every stammerer a paretic, there is simply an end of scientific psychiatry, and if sight can be lost for a moment of the pathological and clinical aspects of the subject, the reflection remains that such a classification is equally unfortunate from a practical point of view. It leads to that dangerous routine which gives chloral and conium to the excited, opium to the depressed, and nothing to the apathetic patient, merely because they are excited, depressed, or apathetic.

While the same strong grounds advanced for the consideration, as separate forms, of those varieties of insanity mentioned at the opening of the above paragraph do not hold good with the others; that is, while the latter are not universally recognized to be as distinct as the former by eminent authorities, it is believed by the writer that they merit such consideration. For whatever disposition the future will make of them, it may be confidently predicted that the symptom groups of transitory frenzy, primary confusional insanity, katatonia, and the etiological forms will continue to be subjects for study, and present important problems of differential diagnosis, prognosis, and therapeusis to those alienists who analyze the symptoms of their patients not according to preconceived schemata, but in the light of the bed-side revelations. The clinical label may be changed, the clinical classification shifted or replaced by a patho-anatomical one; but the clinical picture will remain forever.

PART II.

THE SPECIAL FORMS OF INSANITY.

SUB-ORDER A.

The Simple Forms not Essentially the Manifestations of a Constitutional Neurotic Condition.

FIRST CLASS: THOSE NOT ASSOCIATED WITH DEMONSTRABLE ACTIVE ORGANIC BRAIN CHANGES.

I. Division. Attacking the Individual Irrespective of the Physiological Periods. Order a: *Of Primary Origin ; Sub-order* b: *Characterized by a fundamental Emotional Disturbance.*

SIMPLE MANIA.
SIMPLE MELANCHOLIA.
KATATONIA.
TRANSITORY FRENZY.

CHAPTER I.

MANIA.

Mania *is a form of insanity characterized by an exalted emotional state which is associated with a corresponding exaltation of other mental and nervous functions.*

THE TYPICAL CONDITION OF THE MANIAC * may be summarized in one phrase: loosening of the inhibitions, or checks, both those of organic and those of mental life. The perceptions appear more acute, the associations are quick, so rapid indeed that the ease with which the patient forms new and extravagant mental combinations, and the readi-

---

* It is not necessary to refer here to the fact that this word has been used in every possible sense, even in one equivalent to insanity as a whole. In these pages it is used in the limited sense: that is, "mania" without any qualifying clause refers to the condition treated of in this chapter alone.

ness with which novel suggestions present themselves, impress the novice as manifestations of a naturally quick wit, or of a talented and original mind. It is particularly that faculty termed the fancy which is extraordinarily active, and the images crowd each other in such profusion that the patient in endeavoring to announce them later becomes unable to keep step with his words, and although his speech is much more rapid than in health he is compelled to break off in the middle of a sentence to begin the next, and thus gives the superficial impression that his ideas are confused, when in reality they are not, at least in the earlier periods. There is a discrepancy merely between the rapidity of the conceptional and associating transits and those of which the speech tracts are capable.

Corresponding to the activity of the patient's thoughts he becomes declamatory in style, and his exaggerated manual gestures and the rapid play of his facial muscles indicate the nature of his disorder as well as his spoken words. He is not able to remain long in one place, as it is impossible for him to remain long silent.

The appetite and digestion are excellent, the sexual desires increased, the patient generally feels in high spirits; everything presented to his mind is *couleur de rose;* in short his whole condition resembles an intensified sanguine temperament. He forgets the cares and vexations which may have led to his illness; happy and contented, it is his desire to make others so. He scatters his worldly possessions among his friends and even among strangers; invites them to festivals or to banquets; and indulgences in drink so frequently resorted to in this condition, like the venereal excesses which are often its earlier manifestations, sometimes intensify the worst developments of the disorder.

If the maniac forgets the cares and troubles of this world, he becomes equally oblivious of the restrictions of its conventional and civil laws. All clogs and impediments are swept away by the rapid torrent of ideas and impulses overcrowding and jostling each other. Reflection has no time to exert its checking influence, and the beast in man comes to the surface. Men ordinarily reserved and women previously chaste display an animation in their looks, an obscenity in language, a lasciviousness in gestures and acts, and an obliviousness of propriety, shown in the publicity of the latter, which are among the most striking features of mania.

With all this the patient is quick at repartee, defends his acts with sarcastic retorts, or explains them in a *quasi*-plausible manner. The removal of the inhibitory faculty responsible for his violation of the laws of decency and sometimes of property, also enables him to defend those acts, and unfortunately for him, too, suffices to convince the average juryman and sometimes the wearer of the ermine, that it is not an insane man but a clever and amusing rogue they have to deal with; one who deserves not only punishment for his violation of, but additional penalties for his contempt for and trifling with the majesty of the law.

On the whole the patient in this condition greatly resembles a person slightly intoxicated. But just as the person who while slightly intoxicated is good-natured, generous, careless, mischievous, and perhaps lewd, when more deeply intoxicated becomes irritable, combative, and incoherent; so the maniac as his condition deepens exhibits a tendency to angry rather than to pleasurable excitement. Even in the lighter phases it is noted that he does not bear contradiction well; that trifling causes produce undue emotional reaction; that just as he laughs without cause, so he may become angry at an imaginary affront, at a mere interruption, or cry without due reason. His reproduction of impressions and association of conceptions now attains the rapidity of a delirium. Ideas which were previously followed out but not enunciated in words are now not even followed out in thought. The patient is unable to fix his attention long enough on a question to answer it even in his mind; the judgment becomes obtuse, dulled by the myriad of conceptions it is called on to control. Just as the beast came to the surface owing to the removal of the conventional clogs, egotism now asserts itself through the obliteraeration or weakening of the judgment. The patient believes he is rich, occupies himself with rapidly changing projects, is going to get a high office, or to sue the asylum authorities for a large sum as an equivalent for his "wrongful incarceration." A case of mania *in puerpero* which the writer observed at Meynert's clinic in the Vienna asylum exemplified at once the relative acuteness of maniacs and the unsystematized character of their delusions. Isolated on account of her violence, the patient tore every shred of clothing from her body, and then in an incredibly short space of time she picked the matting of the isolation-room to pieces, and made from the strands a most complete and tasteful dress, including every article of wearing apparel from the undergarments to a *cul de Paris*, a bonnet, stockings, and a satchel. This she wore for a long period; her abandonment of it was one of the first indications of recovery. At the same time she loudly asserted that she was an Austrian princess; on being asked which one she was she gave a name not on the list, and repudiated the attentions of an old deteriorated delusional lunatic who saw in the arrival of "her daughter" at an asylum the act of those conspirators who had deprived her, "the legitimate queen of Austria, of her birthright." Later on she accepted the relation, but this was evidently entered into in the same spirit displayed by children when, in play they assume the character of parents. of shopkeepers, or of Santa Claus. Her manner was proof of her insincerity, and she clearly despised and pitied her fellow-patient.

In this condition illusions are frequent and hallucinations sometimes present. A trivial resemblance of a stranger induces the patient to greet him as an old friend, or as some important personage who has come to confer an honor or to pay his respects to him. Like the mental tone and the delusions of the patient these disturbances of the perceptions are of a gay or expansive character; it is rare for an inter-current frightful or unpleasant hallucination to be noted.

It is frequently observed—in the writer's experience, in about one third of the cases—that hallucinations are striking features of mania at its inception. This feature characterizes the *hallucinatory mania* of Mendel, and is the ordinary form in which mania from acute diseases and puerperal states manifests itself. In a case of this character under the writer's observation—a patient suffering from mania *in puerpero*, who finally made an excellent recovery in spite of the fact that she had no asylum treatment—the first symptom of derangement noted was her taking a child's drum and belaboring it violently; on being remonstrated with she tore a number of dresses from the hooks of the wardrobe, because for several days—as she afterwards said—a disgusting odor had proceeded from these, which she found also tainted the food given to her.

The hallucinations in this form are generally present in all the senses. The patients hear military music, dogs barking, machinery, or obscene calls; they see marching regiments, processions, and priapic or heavenly visions;

they smell noisome or choking gases, and taste poisonous or fœtid substances in their food. But in so far as these hallucinations are intrinsically unpleasant they present this contrast with the similar hallucinations found in depressive forms of insanity, that the patient does not build up depressive but expansive delusions on them. Hallucinatory mania does not differ from ordinary mania except in the presence of the initial hallucinatory period, its further progress is exactly the same.

With the excitement in the sensorial, intellectual, and vegetative spheres, there develops a corresponding condition of the motor apparatus. Reference has already been made to the restlessness and the declamatory gestures, characteristic of mania. These, as the patient passes from gay to angry excitement, become intensified into destructive and violent motor delirium. The patient vociferates loudly and incoherently, rhymes, yells, and sings for hours, dances about and tears his clothes, smashes windows and furniture, hurls movable articles out on the street, attacks persons (more rarely), and all without a clear motive. The mental processes become more and more confused and incoherent. Hallucinations and delusions, the latter of the unsystemized kind, have a determining influence on many of the maniac's acts; as when, for example, a puerperal maniac with erotic delusions tears all her clothing off, and runs out on the street or climbs on the roof to show her person.

Patients in this condition, which is known as *maniacal fury* (*furor maniacorum*) or *frenzy*, are commonly unclean, urinating and defecating without regard to the surroundings, painting the walls or daubing themselves with their excreta, nay even eating the latter. These disgusting acts indicate that the special sense perceptions, which appeared exaggerated in the lighter and earlier phases of mania, have meanwhile become blunted.

The preliminary expansive phase of mania may be very brief, and angry excitement with pronounced motor activity may be continuous, at the height of the disorder constituting the *Furor* of the Germans. Ordinarily furious and angry excitement is brief and recurs episodically in the same illness, each outbreak being separated by an interval marked by the ordinary characters of mania, or even having the nature of a lucid interval.

In a number of cases the furious stage does not develop, and hallucinations are absent. The patients exhibit merely the maniacal character: they are extravagant and egotistical, undertake the fulfilment of projects which in themselves may be quite feasible; but at the same time they indulge in venereal and alcoholic excesses, visit theatres and balls daily, or undertake extensive travels merely for the sake of travelling. Remonstrance renders them more positive, overbearing, and even brutal to their friends, and not rarely leads the patient into conflicts with the authorities. If brought face to face with his extravagances, by some stronger power, he either denies or endeavors to palliate them by mendacious explanations. To the laity the sufferer from this subdued maniacal condition, the *Hypomania* or mildly-developed mania of Mendel,* appears to be nothing more than a selfish, careless, alternately brutal and good-natured ambitious spendthrift. And this impression is strengthened when the patient, on being sent to an asylum, and appreciating his position, is more guarded in his behavior. Often it is only the writings of such a patient which then exhibit his mental state. Aside from this the latter is expressed in exaggerated ideas of self-importance in the abstract, rather than in formal delusions; in impatience of contradiction and momentary outbreaks of passion without adequate cause; and above all in the fact that the immorality, prodigality, and morbid egotism of the subject were not previously natural to the latter, but involved a *change of character*. The manner of such patients is often the only indication of insanity found at a single examination; and repeated examinations may be necessary to establish the existence of the malady.

From the foregoing it will appear that the essential feature of all varieties of mania is the maniacal character; that is, the peculiarly *mobile* and at the same time *exalted* emotional state which induces in the patient exaggerated ideas (particularly such of self-exaltation) and an undue reaction to trivial external impressions. Here the process may end in the lightest cases. In the typical variety however there are superadded word-delirium, explosions of motor violence, and delusions of a fleeting nature. A predominance of hallucinations with these symptoms characterizes hal-

---

* Who is profoundly in error in attributing to Campagne an intention of describing this disorder as an exaggerated egotism under the term "manie raissonante," which is a form of monomania and not an acute psychosis.

lucinatory mania, and a greater intensity and longer duration of the furious phases is the feature of furious mania. At bottom they are all one and the same disorder, which, however much it varies in the intensity and direction of its symptoms, does not lose the essential characters predicated above.

THE OUTBREAK OF MANIA is rarely if ever sudden or rapid. This is the case in mania occurring in hysterical subjects. In the hallucinatory variety the incubation of the disorder may be apparently restricted to a few days; and it seems as if the hallucinatory state which precedes the outbreak of the characteristic maniacal symptoms takes the place of the ordinarily prolonged prodromal period of typical mania.

In typical mania there is commonly observed, from one to three months prior to the maniacal explosion, a depressed mental and somatic state. Gastric disturbances are early noted, the tongue is furred, the bowels constipated; anorexia, a sense of tension in the head, particularly intensified over the eyes and at the back of the head, and probably like the concomitant insomnia related to the visceral disturbance, are commonly present. With this the patient experiences an inability to concentrate his thoughts; it is with difficulty that he is able to carry out his duties, he feels physically and mentally prostrated, and in his endeavor to account for his condition becomes hypochondriacal, refers his weakened condition to early excesses, or, if of a devout turn of mind, attributes the "punishment" to some early "sin." At the same time he struggles against his frailty, he attends to his duties, however great the effort, thus intensifying the causes of his disorder, and may still pass for nothing more than a dyspeptic. Indeed there is no clinical difference between the condition of many dyspeptics who never become maniacal and of some maniacs in the prodromal stage of their disorder. It is only very exceptionally and then not in uncomplicated cases that the depression preceding the maniacal period reaches the degree of a genuine melancholia; and nothing could be more improper than to designate the initial period of mania its "melancholic stage," as some English and German writers do. The fundamental feature of melancholia (see next chapter) is, with the rare exceptions alluded to, absent.

The transition from the initial stage of depression to the

maniacal culmination is gradual. The appetite returns, the sleep improves, the mental condition appears normal, and to all appearances the patient seems healthy and congratulates himself on his recovery from what he and his relatives have regarded as a simple attack of nervous prostration. But after a few days or weeks his subjective feeling of well-being assumes an exaggerated tinge, and the exalted emotional stage rapidly growing out of it inaugurates that true maniacal condition whose features were detailed at the beginning of this chapter.

The DURATION OF MANIA varies. An average case of typical mania will exhibit an initial stage of depression lasting about six weeks, a maniacal period of about three months, while the period of convalescence will occupy about a fortnight. The average total duration of the illness in a case of moderate severity may be therefore placed at or about the period of five months. But very intense cases may exceptionally last but a few weeks, and very mild ones may similarly last a year or over. Such have been reported in which recovery took place after the disorder had lasted a year and a half.

FREQUENCY AND PROGNOSIS.—Of 5,481 admissions to the Bicêtre, reported by Marcé, 779 were of cases of mania, 300 being males and 479 females. The greater frequency of admissions of females to public asylums is attributable to the large proportion of females in the poorer classes becoming insane in the puerperal state, owing to the combined influence of physical causes: deprivation of food, rapid succession of childbirths, and alcoholic excesses; and unfavorable mental influences: seduction, abandonment, domestic cares, and neglect by the husband. The same disproportion of the sexes is not found among the wealthy. In the consultation practice of the writer mania exhibits an equal percentage in both sexes, and in both plays a less prominent part than in the public asylum statistics. Of 2,297 insane males admitted to the New York Pauper Insane Asylum, and tabulated by the writer, 260 were affected with simple mania, or a little over eleven per cent.*

Mania is more frequent between the 25th and 35th year in males, owing to the greater frequency of alcoholic indulgences and venereal excesses at this period, coupled with the

---

* Race and Insanity, *Journal of Nervous and Mental Diseases* 1880.

fact that at this period of life there is a greater disposition to disorders of an active type than at any other.

The prognosis of mania is very favorable, in fact, more so than that of all other psychoses, with the exception of stuporous, confusional, and transitory insanity. Various authors estimate the recoveries at from sixty to eighty per hundred; the latter figure accords with the writer's experience. Partial recoveries are also noted, where the patient ceases to be maniacal, but exhibits an undue excitability and a permanent enfeeblement of the judgment and memory. Those patients who do not recover either pass into so-called chronic mania with confusion of ideas, and thence to terminal dementia, or directly and gradually into this latter condition. From two to five cases out of the hundred terminate fatally, either through the somatic disorder, such as the fever, or the puerperal state from which the mania originated, or by complicating disorders like pneumonia, developed in consequence of exposure and alcoholic excesses, or finally through *maniacal exhaustion.*

It is rare for mania to terminate by a crisis; cases have been observed where the formation of an abscess, or a diarrhœa, has been instantly followed by amelioration, just as complementarily, in an instance within the writer's observation, the healing of a varicose ulcer was followed by a maniacal condition. Ordinarily convalescence takes place slowly, and is marked by numerous ups and downs. The occurrence of brief lucid periods is the first indication of returning health. The patient either admits that he has acted in an improper manner, apologizes for his misconduct, or more frequently becomes taciturn, and unwilling to make a humiliating confession.* The next hour or the next day may see him exalted and violent again. But the lucid periods recur, they become more and more frequent, and of longer duration, until they run into each other, and the healthy state is definitely re-established. The lucid periods may be marked by a sort of reactionary depression, but this is rather a favorable than an unfavorable sign.

A patient who has completely recovered from an attack of acute mania is not usually liable to a recurrence of the disease. A repetition of the injurious influences which produced the first attack may, however, lead to repetitions of the maniacal attacks, particularly where there is an

---

* Which is a most favorable indication.

hereditary predisposition. Such attacks are as strictly to be considered attacks of simple mania as the first attack, and do not justify the erection of a special form of insanity, designated as "recurrent mania."

CHAPTER II.

MELANCHOLIA.

Melancholia is a form of insanity whose essential and characteristic feature is a depressed, i.e., subjectively arising painful emotional state, which may be associated with a depression of other nervous functions.

At its height the melancholic disorder is the antithesis of the maniacal. Just as every gesture and every thought of the maniac betrays his exalted emotional state, so the attitude and expression of the melancholiac, his thoughts, and his delusions, illusions, and hallucinations—if he have these—announce the dominant emotion of sadness or psychical pain. While the typical maniac shows a tendency to restless mobility, the typical melancholiac exhibits a tendency to motor stagnation. While the maniac is aggressive, communicative, and obtrusive, the melancholiac is glad to be let alone. While the somatic functions appear to be exalted in mania, they share in the universal depression in melancholia: the melancholic patient refuses food, where the maniac is bulimic.* While the maniac indulges in ambitious schemes, or gives himself up to self-satisfied contemplation of his excellent physical, financial, and social condition, the melancholiac is overwhelmed by his physical worthlessness, his moral turpitude, terrible anticipations of the future, and he consequently contemplates, if he does not commit, suicide. While the memory appears more acute, and the conceptions are associated with delirious rapidity in mania, the former is clouded, and the reproduction and association of conceptions are retarded in melancholia. In short, while in mania the typical state is in-

---

* Bulimia is a term applied to excessive and rapacious appetite in the insane. It is ranked by some among the morbid propensities, but may be due to various causes.

augurated by a subjectively arising emotional exaltation, which leads to a suspension of the inhibitions, we have the reverse emotional state in melancholia, leading to an intensification of these very inhibitions.

While the common basis on which the symptoms in all melancholiacs develop is the subjective painful emotion, and all melancholiacs are consequently sad, depressed, dissatisfied, and isolate themselves from, or become reserved and perhaps inimical to, their surroundings; more positive symptoms grow out of this fundamental state in most cases.

The commonest of these symptoms are *unsystematized delusions of a depressive nature.* The patient endeavoring to account to himself for his painful emotional state, unable to understand why his affection for his relatives and friends has ceased or diminished, or seeking in the outer world for some inimical agent responsible for his depression, concludes that he is threatened with bankruptcy, that he has been a bad husband, or that he has been remiss to his Creator in some way not clear to him. From these vague notions there is but a step to positive delusions. The patient may believe himself the subject of persecution by diabolical or human agencies, or by witches. These, he believes, attack either his person or his fortune, and threaten his family, and not unfrequently the melancholiac, to the suicide to which he resorts for the purpose of escaping his personal foes, adds the murder of his family, to save them from the sadder fate which his delusions conjure up for them.

Sometimes an insignificant crime, or an entirely imaginary one, induces the patient to consider his sufferings as a just retribution for his misdeeds. He may even go to court to confess real or imaginary offences committed many years before; and deserving, as he believes himself to be, of capital punishment, he may, in order to accomplish his merited doom, kill another person. Homicides are also resorted to by melancholiacs, in order to gratify their suicidal inclinations, on the one hand, without conflicting with the religious interdiction of that crime, on the other.

The delusions of the melancholiac may be modified by sensory disturbances of central orgin, such as neuralgias, anæsthesias, or disordered smell and taste. In this event hypochondriacal delusions are apt to arise, as such of being poisoned or annoyed by irritant and noisome vapors and gases.

Hallucinations are very frequent in melancholia, and while they are usually secondary to the delusions and partake of their formal character, they serve to fortify and elaborate them. A melancholiac who thinks that he is persecuted by diabolical forces sees the devil and his imps, hears the roaring flames of hell-fire, and smells brimstone. He who imagines himself guilty of a crime sees the corpses of those he has killed, the ruin of those he has cheated, the girl he has betrayed, and the officers of the law who are on his track. He who is pursued by enemies hears taunting and insulting voices, indictments read against his person, attachments against his property, and orders of arrest. Finally he sees the jailer, the hangman, and the scaffold. The melancholiac who suffers from delusions of a hypochondriacal tinge is tortured by a thousand imps, who run needles into his flesh, by powerful electric batteries, or by some animal or person introduced into his body.

With such overwhelming pictures of terror arising in the patient's mind, it is not strange that while the general condition of the melancholiac is abulic and passive, in some cases and at some stages of the disorder the patient should become restless from fear, and wander around seeking for redress and relief in a vague way. In some cases, and then at the height of the disorder, spurious states of fury may be developed, states which by the uninitiated are frequently called "melancholic mania." The consciousness seems to be nearly obliterated during these spells, in vague fright the patient raves that the world is destroyed, that all is lost, a nameless terror seizes on him, and in blind disregard of consequences he destroys or attacks everything within reach. He tears his clothes, murders, commits suicide, or destroys the furniture and resorts to self-mutilation. Cases are on record where patients in this condition have torn out their eyes, cut open the scrotum, disembowelled themselves, have thrown themselves into the fire or against the cages in which wild animals were confined. In one instance a patient who had been sent to an asylum on the recommendation of the writer—but, unfortunately, to an institution conducted as a sort of private side-venture by the authorities of a large pauper institution—after having cut herself with the fragments of a broken pane of glass, and thus sufficiently warned her supervisors of the dangerous tendencies of her frenzy, took a rather blunt stick of wood, and ran it into her abdomen, ripping it up to a terrible extent.

These spells differ from the fury of the maniac in that the delirium if present is never expansive nor as multifarious, and that the violent acts are as apt to be directed against the patient himself as against others. Consciousness is not affected in the same way in the two states. In maniacal fury it is confused rather than obliterated; in the violent outbreaks of melancholia, it is rather obscured or obliterated than confused. In mania the outbreak is the result of an expansive or angry emotion; in melancholia the outbreak, best known as *melancholic frenzy* (*raptus melancholicus*), is the outcome of an anxious terror.

In those cases which are characterized by inter-current attacks of melancholic frenzy the fright immediately preceding it is usually associated with anxious precordial sensations, and these, summarized under the term *precordial fright*, are supposed to depend on some disturbance of the pneumogastric and the sympathetic centres. This view is borne out by the fact that the pulse is frequent, irregular, and small, the bodily surface pale and cold, while the breathing is superficial and the secretions are suppressed.

Frenzy is particularly apt to occur in melancholia developing in alcoholic subjects, owing to the predominance of multitudinous and frightful hallucinations, panphobia, and precordial fright in such patients.*

Melancholic frenzy is of shorter duration than maniacal furor, and unlike the latter it terminates suddenly. The patient appears as if relieved by the explosion, the fright is gone, and he awakes as from a dream of which he has but an obscure if any recollection. The condition is merely an episode of melancholia, as the outbreaks of furor are episodes of mania, and does not justify the erection of a distinct form of melancholia.

There is not as some have claimed a pulse character peculiar to melancholia; that found with frenzy has just been described; a different vaso-motor condition is found in melancholia with stupor. Here the entire muscular system appears to be in a condition of spasm, which is shared by the

---

* "Alcoholic melancholia" is a term applied to melancholia occurring in inebriates. Usually some of the ordinary causes of melancholia are superadded; it is remarkable for its brief duration and favorable prognosis.

muscular coats of the arteries, so that we have vascular spasm as well.

Other of the more common disturbances in melancholia are feelings of tension or of emptiness in the head. Disordered tactile and thermic sensations, so frequently serving as the basis of delusions, as above stated, are found in all the more severe cases. In addition, refusal of food is a characteristic feature with the vast majority of melancholiacs, and may have two different sources. It is either due to the anorexia which is an expression of the generally adynamic state of the system, or it is due to delusions that the food is poisoned, or that the sufferer is not worthy to eat. Anæmia is found in the larger number of patients from the inception of the disorder; in all severe cases it becomes marked as the disease advances. In a single case of mild melancholia in the writer's observation it was absent. There seems to be some obscure source for the general nutritive disturbance in melancholia, for the loss in weight and the rapid wasting, particularly noticeable in the face, and giving the latter a pinched appearance or one of premature age, seem to be out of all proportion to what one would expect as a result of the refusal and insufficient assimilation of food.

As in mania, there are a number of varieties also in melancholia, which, as in the case of the varieties of mania, however great the difference between them as to the intensity and direction of the symptoms, have the fundamental character of melancholia, the depressed emotional state, as a common and as their essential feature.

The ordinary form of melancholia has been just described. When the patient, instead of remaining impassive, becomes restless under the influence of delusions, hallucinations, or anxious feelings, and runs around wringing his hands, weeping and crying out against his persecutors, the wicked world, or a remorseless fate, he is said to be suffering from active melancholia (*melancholia agitata*). It is the episodial intensification of this condition, an intensification which reaches such a degree as to overwhelm the patient's consciousness, that constitutes the melancholic frenzy alluded to above.

While the subjective fears and beliefs of the melancholiac in these cases lead to restlessness, in others the influence of delusions and other factors is of so overwhelming a character as to throw the patient into a state of stupor. The patient sits motionless, and his flexor muscles exert their preponderating influence over the extensors. The head is bent forward, the knees and the elbows flexed, and the tension may involve the facial muscles, giving the face a remarkably frozen and tetanic expression. It has been even claimed that in some cases a true cataleptic waxy flexibility of the extremities may be found. In addition the patient is mute, his breathing is slow and superficial, the temperature sinks, the extremities are cold and blue, and the physiological discharges are suppressed or diminished.

The pulse is frequent in this condition, while it is ordinarily slow in melancholia; but, as in melancholia generally, the arterial tension is greatly increased, the radials feeling almost wiry under the finger.

Constipation, so common in all forms of melancholia, is extreme in the atonic form, and the nutritive disturbance generally is most marked in *melancholia with stupor* (*melancholia attonita*), as this variety is termed. That it is merely a variety of melancholia is shown by the fact that it may alternate with agitated as well as with typical melancholia, and may be interrupted like the other forms by outbreaks of melancholic frenzy. These are usually provoked by disturbing the patient, as when endeavoring to move him, or to compel him to take food. A second series of facts, proving that *melancholia attonita* is a true melancholia, and not a condition of simple neural anenergy, is derived from the recollections of patients recovering from this condition. Under all circumstances the patients have a certain degree of consciousness: immobile and masklike as the features seem, an occasional flicker of an expression passing across the face indicates the persistence of mental life. On recovering they report having had distinct and horrible phantasms, hallucinations far exceeding in their terrible character those found with ordinary melancholia. The most frightful tortures, massacre of their best friends, with whose blood the food given to the patient is supposed to be seasoned, devouring of their flesh by myriads of foul beasts, are witnessed, and fearful imprecations heard by them; so that the frozen attitude and expression appear to be the expression of a fear and dread so intense that the patient is, as it were, struck dumb and paralyzed by them. It is hence not improperly styled thunderstruck melancholia (*angedonnerte Melancholie*) by the Germans.

Just as there is a form of mania in which the maniacal

outbursts may be but feebly marked, and the delirium, incoherency, and hallucinations absent, so there is a form of melancholia in which there are no delusions, no hallucinations, and neither stupor nor restlessness. Just as the form of mania lacking the more violent symptoms is marked exclusively by the development of the maniacal character, so there is a form of melancholia whose sole discoverable morbid feature is the melancholic character; and as the former has been termed a hypo-mania by Mendel, the latter might be termed a hypo-melancholia. It is sometimes designated "reasoning melancholia," and better still as *melancholia sine delirio*.

Much has been recently written about the treatment of patients suffering from this mild melancholia at home; and it is true that a number are there treated and recover. But, in using the adjective "mild," let it be distinctly understood that this form is far more dangerous as to its possible civil and criminal consequences than any other, and that the risk assumed by those who attempt to treat a patient suffering from this form outside of the walls of an institution is a grave one. It is not the insidiousness of the malady—and it is very insidious indeed—that alone makes it so dangerous a one, but the fact that the reasoning powers of the patient are relatively intact, and permit him to carry out his schemes of homicide and suicide with a cunning and a deliberation which could not be exceeded by a person in the full enjoyment of mental health. Another element of danger is the great likelihood of such patients developing morbid impulses. The sight of a weapon, in their subjectively painful emotional state, suggests its use; the sight of a real or supposed foe determines an assault; the reading of accounts of any novel method of suicide or homicide, or the spectacle of an execution, suggests imitation; the sight of the patient's children suggests the desirability of removing them from a world which holds out but a hopeless vista of gloom to the patient's mind. The frequent immolation of a family by one of its members is ordinarily the deed of a melancholiac suffering from this form of mental disorder, and a large proportion of suicides generally are referable to the same cause. A downcast expression of countenance, a frequent recurrence of causeless fits of depression, frequent sighing, or perhaps groaning, possibly dilated pupils, and a contracted condition of the arteries, may be the sole but sufficient indications to the

MELANCHOLIA. 147

experienced physician, warning him of the danger ahead, while to the uninitiated, or the over-confident, the condition appears to be merely a "fit of the blues."

There is not that contrast between the INITIAL STAGE OF MELANCHOLIA and the fully-developed phase of the disorder which we find in mania. The various symptoms detailed appear, as a rule, gradually, after a preliminary period, during which the patient complains of an inability to exert his will, of a mental vacuum, of impairment of the memory, and of the other signs just noticed as comprising the picture of melancholia without delirium. It is only in the melancholia following fevers, exhausting discharges, and sudden emotional shocks that the symptoms rapidly or suddenly attain their maximum.

The DURATION OF MELANCHOLIA may comprise weeks, months, or years. Its average duration is from three to eight months.

The FREQUENCY OF MELANCHOLIA it is difficult to determine, as a large number of the patients never reach the asylum, and recover, die, or commit suicide outside its walls. Of 2,297 admissions to the pauper insane asylum for males, of New York city, 301, or a little over thirteen per cent, were cases of melancholia. Of 1,193 patients of both sexes admitted to the reception wards of Professor Meynert, not quite six per cent were melancholiacs. The greater frequency of melancholia with females, which is in part attributable to the influence of prolonged lactation, pregnancy, exhausting discharges, and chloro-anæmia in provoking this psychosis, is illustrated by the statistics of the Buda-Pesth asylum. Here ten per cent of the male and seventeen per cent of the female patients suffered from melancholia. Of 146 males received at Feldhof by Krafft-Ebing, thirteen, or nearly seven per cent, were melancholiacs; while of 121 females eighteen, or nearly fifteen per cent, suffered from the same disease. In the writer's statistics the striking fact is noticeable, that melancholia is most frequent in the Teutonic peoples, the ones who, according to Morselli, also show the highest suicidal ratio; and among these it was found more frequent with those who had emigrated, than with those who had been born in this country. This illustrates the effect of *nostalgia* in producing melancholia, at least in part, an effect which has been observed among the Bavarian soldiers recruited from the mountain districts of the country.

In private practice melancholia, particularly of the lighter grades, is very common; and is not unfrequently treated as "neuræsthenia"—whatever that may or may not be— and dyspepsia, and, thanks to the self-limiting tendency of the lighter forms of the psychosis, it is frequently cured on either theory.

THE PROGNOSIS OF MELANCHOLIA is less favorable than that of mania, about six out of ten patients recovering completely. The proportion of cured females is much greater than that of cured males; this is due to the fact that the removable causes—those of a purely physical character—are relatively more frequent in the etiology of melancholia in females; while cerebral overstrain and business troubles, which are less amenable to therapeutic measures, are more frequent in males. The prognosis is considered to be more favorable in the young than in the old, and in cases not presenting an hereditary taint ; this is not, however, strictly true: melancholiacs with an hereditary tendency recover as frequently, and sometimes more rapidly, than those whose hereditary history is good; but the danger lies in the more frequent development of *melancholia sine delirio* in hereditary cases — with all its attendant dangers as to suicide and homicide. (Appendix III.)

Aside from the greater likelihood of suicide in *melancholia sine delirio*, as well as the probability of recurrence, the immediate prognosis of this variety is always favorable. It is next most favorable in the typical and the agitated forms, over eighty per cent of the patients suffering from these presenting good prospects as to recovery, under early and proper treatment. With stupid or atonic melancholia the outlook is bad, and although the patients may in exceptional cases recover with apparent suddenness, the larger number recover slowly or not at all, and then die of inter-current affections or pass into terminal dementia. Among the causes of death are mal-nutrition, diarrhœa, and catarrhal as well as tubercular phthisis; the latter form of pulmonary trouble is particularly frequent. Summarizing the prognostic indications, these may be enumerated in the following order of importance : The prognosis is more favorable according as there is less stupor, less nutritive disturbance, more variation in the symptoms from day to day, and with youth and the female sex. It is of specially good import in melancholia when the patient's condition shows a marked improvement from evening to evening. As a rule melancholiacs are worse in the morning than in the later part of the day. It seems as if the irritant effect of the diurnal routine, and perhaps the better nutrition enforced in the waking state, lead to a gradual amelioration of the vaso-motor disturbance, which is at the basis of the precordial fright, and the resulting delusions and frenzy. It has been noted as a confirmation of this view that the administration of cardiac and general stimulants, as well as easily assimilable food, whenever the patient awakes at night, is followed by sleep and an improvement lasting throughout the following day.

When melancholia does not terminate in recovery or death, it passes either into terminal dementia of the apathetic variety—which, as above stated, is a common sequel of *melancholic attonita*—or into a chronic delusional insanity with deterioration—an occasional sequel of the typical, and a more frequent one of the agitated forms.

CHAPTER III.

KATATONIA.

Katatonia is a form of insanity characterized by a pathetical emotional state and verbigeration, combined with a condition of motor tension.

This well-marked though not generally recognized mental disorder was first demarcated by Kahlbaum of Görlitz about eight years ago. In the course of the writer's study of the pauper insane at Ward's Island he became impressed with the genuineness of the grounds on which Kahlbaum based this classification; and the subsequently published and more recently reprinted paper of Kiernan, based on a study of the same patients, was the first confirmation of the important and practical proposition of the Prussian alienist.

The illness begins with an initial stage resembling that of any ordinary melancholia. This is followed by a period in which the patient presents an almost cyclical alternation of atony, excitement of a peculiar type, confusion and depression, which finally merge into a state of mental weakness, approaching if not reaching the degree of a terminal dementia. Any single one of these enumerated phases may be absent. In a large number of cases spasmodic conditions, in the way of muscular *crampi*, chorea-like movements of the facial muscles, epileptiform and hysteroid convulsions, have been observed to accompany the initial period. In not a few of the cases the initial depression was observed to be accompanied by self-reproaches relating to masturbatory excesses, and very frequently disappointment in love determined the morbid ideas of the patient; and, in fact, both factors are exciting causes in many cases of katatonia. On this basis the ordinary melancholic symptoms, fear of poisoning, delusions of persecution, and dread of committing unpardonable crimes, crop up and increase the resemblance of the first stage to genuine melancholia.

The excited stage presents symptoms of a kind different from those of ordinary melancholia, and constitutes a connecting link, as it were, between the symptoms of an agitated melancholiac and those of a lunatic with fixed delusions. Some of the patients present exaggerated, others diminished, self-esteem; and not rarely does the developing delirium assume an expansive tinge. But all katatonics exhibit a peculiar pathos, either in the direction of declamatory gestures and theatrical behavior, or of an ecstatic religious exaltation. Frequently the patients wander about, imitating great actors or preachers, and often express a desire and take steps to become such preachers and actors. In America, as Kiernan remarks, the chronic stump-speaking tendency is more frequently displayed by these patients, and in a negro suffering from katatonia the conversation in excited periods was noted to present the grandiloquent character which has been so aptly rendered in the minutes of the "lime-kiln club." This patient had, without any prodromal symptoms, fallen down suddenly while at work, his face and arms twitching; a lucid period followed, the patient gradually became depressed, and in the asylum his depression gave way to a maniacal condition, which was followed by another fit of depression, marked by numerous hallucinations. With this he refused food and passed into a cataleptoid condition, from which he suddenly emerged one morning, saying that he was "equal to any white man." Apparent recovery took place, but between 1871 and 1875 he was re-admitted three times. On these occasions he presented mainly alternations between atonic stupor with catalepsy and a peculiar condition which was so characteristic that it first called the writer's attention to the distinctness of the disorder. On going up to the patient and loudly addressing him, he lifted his ordinarily bowed head in a very consequential way and prefaced the reply to any question by the words "I do not doubt but what." Asked his name, for example, he said, "I do not doubt but what my name is William Henry G——;" asked his age, "I do not doubt but that I was born in the year 1838, so my mother said;" asked his nativity, "I do not doubt but that I was born in some part of the world;" asked whether he had any desire, "I do not doubt but what I want to *get* out and *go* home to get me *some* work, in order that I may buy me some food, *and* some clothes, and —" here he relapsed into the passive state. Before he answered a most remarkable series of grimaces were gone through, a series of spasms of the oral muscles, which culminated in the explosive enunciation of the first word of his reply. The answers were deliberately and pompously given, something after the fashion of a juvenile actor.*

In a better educated patient, who had made a suicidal attempt in the depressed period, in obedience to hallucinatory suggestions, and in whom an exquisite cataleptoid condition developed, the excited periods were marked by a tendency to contradict everything said by others. On another occasion he said, "I am Arminius, and have swallowed J. E——" (his name); he became very dignified, knocked down a fellow-patient for sitting on the same bench with him, and then followed another cataleptic attack. He showed rhythmical movements of the fingers, and talked incessantly about his noble descent, appealing for and demanding its general recognition. The grandiose ideas were mingled with self-accusatory delirium, and, from intermingling sounds belonging to no known language with his mixed German and English flight of words, he proceeded to talk in such sounds altogether, making the while the most expressive theatrical gestures—risti pili chinko ti ki ti king, ter pilli mimili nono chichotitonifor tikohoforchink—marking his periods well, so that to a casual visitor it would have appeared as if the patient were actually speaking in some Oriental tongue.

It has seemed to the writer as if there were at times a recognition on the part of the patients that their "verbiger-

---

* The italicized words were peculiarly emphasized.

ation" is nonsensical, and that they have a silly enjoyment of the "fun." Although this suspicion can be based only on the facial expression of the subjects as shown when they are sharply cross-questioned, it seemed convincingly strong in the two cases referred to.

Kahlbaum has noted the tendency of katatoniacs to use diminutive expressions, a tendency which can be better gratified by German than by Anglo-American patients, and was observed in the case just cited. One of Kahlbaum's patients would say, for example, " Ach ich bin so klein*chen*, und in zwei Minut*chen* bin ich todt*chen*—Ich bin so schwach-*chen*—jezt bin ich bisweilen so gross*chen*—Ich muss sterb-*chen*, alle menschen todt*chen*, ich muss wein*chen*," etc.

The hallucinations of this form of insanity are commonly of a depressive character. Where the disorder begins in a religious ecstatic state the devil and hell-fire are seen. One patient of Kiernan's saw blood on everything he looked at, another was followed to church by "droves of dogs." But although these hallucinations are accompanied by a depression like that of melancholia, there is rarely the same profound painful emotional state; and the expression of the patient throughout the depressed period, and even in the atonic states, often indicates rather a silly hilarious tendency, which reaches its acme in the excited phases. The acts and ideas of the patient in these latter differ from those of a maniacal subject in being exceedingly monotonous. Instead of being constructive and productive like the maniac, the katatoniac is rather destructive and oppositional in his tendencies: he contradicts where the maniac assents good-humoredly, he refuses where the maniac may be led to consent. It is in harmony with this that refusal of food, so rare in maniacal delirium, is so common in the excited periods of katatonia, and that it is as difficult to get a katatoniac to leave his bed as to induce the maniac to retire to his.

Occipital headache of an occasionally severe character is said to be characteristic of katatonia by Kahlbaum.

The most striking phenomena of the disorder are its *cataleptic* periods. The catalepsy is typical and extreme. For days, weeks, nay months, the patients are immobile, resembling sitting corpses, requiring to be fed by the stomach-pump, to be carried to and from their beds, and betraying neither by look nor word that they have any mental activity left, although on passing out of this state they often recollect something of that which has occurred meanwhile. On raising an extremity in any position it will retain that position. The writer once placed the patient J. E—— in the corridor, with one foot on the ground and the other on the bench behind him, his head extremely flexed, one arm raised out horizontally before, and the other horizontally behind him, and watched him so more than half an hour without observing any material change in position. On making the rounds of several wards and returning the patient was still found in the same position, his arms now showing a tremor and gradually sinking to his side. In another case the patient retained any position in which he was placed for a day at a time, and Kahlbaum reports one of even longer retention of a constrained and uncomfortable position.

The prognosis of katatonia is relatively favorable as regards life, although the danger of pulmonary tuberculosis developing in the depressed and atonic stages of the trouble is not to be lost sight of. As to recovery, Kahlbaum entertains very sanguine views, which the writer's experience does not sustain. It is true that after one or two cycles of the symptomatic series related, the patients can be discharged from the asylum recovered, and that this occurs in the majority of instances; but relapses are exceedingly likely to occur. The progress to dementia is slow, and rarely does the latter reach an extreme degree. Probably this is due to the fact that it is in these very cases that a pulmonary affection early closes the patient's career.

It is not yet possible to make any positive statements as to the frequency of katatonia. The writer found two per cent of over two thousand male lunatics presenting this form of insanity. In one hundred and eighty-seven tabulated cases from private practice but one instance was observed.

CHAPTER IV.

Transitory Frenzy.

Transitory frenzy is a condition of impaired consciousness, characterized by either an intense maniacal fury or a confused hallucinatory delirium, whose duration does not exceed the period of a day or thereabouts.

Numerous instances are recorded where persons, previously of sound mental health, have suddenly broken out in a blind fury or confused delirium, which, passing away in a few minutes or hours, left the subject deprived of a clear, or of any, recollection of the morbid period, and generally concluded with a deep sleep. The superficial resemblance of the delirium and the blind and destructive fury to the delirium and fury of the maniac has led observers to designate this condition as *transitory mania*. Others, noting that the deliria are chiefly of an anxious character, and that the violent outbreak often bears a similar relation to the deliria that the melancholic frenzy bears to the anxious delusions and hallucinations of melancholia, have termed it *transitory melancholia*. Still others, guided by the fact that many of the most acute epileptic mental disorders manifest themselves in similar explosions, and by the yet more suggestive fact that consciousness is entirely or nearly entirely abolished in transitory frenzy, classed it among the epileptic disorders.

But it would have to be considered a remarkable form of epilepsy in which there is but a single epileptic attack (transitory frenzy usually occurring but once in the life of an individual), and that attack manifests itself in the guise of a transitory frenzy rather than in a convulsion, and in which none of the exciting nor the predisposing causes nor any of the somatic signs of epilepsy can ever be determined to exist!

The theory that the disorder is an extremely acute mania in some, and an equally acute attack of melancholic frenzy in other cases, is more reasonable. The superficial resemblances on which this theory is founded have been briefly pointed out. They are strengthened by the fact that a period of depression, either following worry, vexation, or somatic disease, is sometimes discovered to have preceded the outbreak. But, in the absence of more positive signs, and in view of the specific feature of *amnesia* characterizing transitory frenzy, it is best to use this term, as committing us to no doubtful hypothesis, and best expressing the leading symptom of the disorder.

It is a comparatively rare affection, so rare that many asylum physicians have never seen a case of it; the writer has seen only one during the attack. But its existence is too well authenticated by the best observers to be called into question to-day; and, on such grounds as certain writers have advanced. Utilitarian considerations, growing out of the desirability of announcing popular views on the witness-stand as a step to further patronage by the legal fraternity, have had their day, and the eloquent language of Foville* the elder, translated by Kiernan, may be cited as applicable to the rhodomontades indulged in on the subject of transitory insanity. "Here is a fantastic interpretation which we could scarcely have expected, and which is *hardly* calculated to rank as a *scientific* production.† Other than this, it is not to scientific procedures that the author has recourse to combat the existence of moral insanity and mania transitoria; it is only by the aid of appeals thoroughly permeated with religious sentimentality, and drawn from the domain of literature, that the author declares moral insanity and mania transitoria false, absurd, ridiculous, and, above all, unworthy of being received by the courts. To enable the reader to judge of the extra-scientific method adopted by the author, we give the conclusion of his article: 'Lastly, we object to both (mania transitoria and moral insanity) because it is an attempt to set back the clock of the century, and to revert to supernaturalism and superstition in medicine. It is an attempt to curtain the windows (*sic*) ‡ of that science whose religious duty it is to cast light and not mysticism around disease—to treat it not as a personal devil entirely, to be exorcised by *philters* and mummery, but rather as the perversion of a natural state struggling to regain its equilibrium.' Many physicians will be astonished to learn that, according to Dr. Ordronaux, they are deceived in believing themselves in the pathway of modern progress

---

* *Annales Médico-Psychologiques*, 1874.
† Alluding to his statement that the existence of moral insanity was due to Pinel's benevolent attempt to account for the executions of the first French Revolution.
‡ Dr. Foville's interpolation.

and scientific advance, when in reality they are returning to the dark ages. But will the rhetoric of their American colleague induce them to retrace their footsteps?"

Transitory frenzy has been observed to follow an intense emotional strain, as in the case of a little boy reported by Engelhorn, who, after being slightly injured by a gunpowder explosion which killed his brother, was thrown into such a state of excitement by the judicial investigation of the occurrence that he sank back in bed, then suddenly rose and began reciting biblical verses and mortuary songs ecstatically, unmindful of the interruptions of his surroundings. After an ensuing sleep the patient had only the confused recollection of a dream, whose contents he endeavored to recall in vain. Kiernan—who seems to strike at the root of the matter in the following words, written with reference to Cook's claim that transitory mania is a cerebral epilepsy, "You cannot prove the epilepsy; you can the mania, and it is transient; and is it not as easy to accept the theory of transitory mania as it is to go wandering after a far-fetched, forced explanation?"—reports the case of a prisoner who made a sudden violent and unprovoked attack on the other prisoners and their keepers while at supper. Transferred to a cell he became violently maniacal, continuing thus two hours, and then fell into a slumber lasting an hour and a half. In the course of the next fourteen hours he was transferred to an asylum and was found lucid, though slow in speech, and had a perfect recollection of everything that occurred up to the time when he went to get some salt at the table. The next thing he recollected was the finding himself in a cell. Nothing further occurred. Another patient, whose history is reported by the same writer (rendered the more valuable as the reporter was an eyewitness of the attack itself), had had a quarrel with her betrothed, after physical exhaustion following night-watching at her mother's bedside. On going to another appartment after the quarrel alluded to, she found that two live coals had fallen on a dress which she had been occupied in sewing for two days, and which some one had placed near the fire. Hereupon she fell into what was apparently a violent rage, tore the dress to pieces, attempted to smash the furniture, and continued violently excited for an hour when the reporter saw her. She was then in a condition of intense frenzy; said the doctor was so dark he must be the devil, and made two assaults on him and continued destructive. After being treated with restraint and the cold pack she fell into a deep sleep, on awakening from which she was perfectly rational, and recollected nothing that had occurred subsequently to the discovery of the fact that her dress had been spoiled. Kiernan's summary of the cases of Calmeil, Tardieu, Le Grand du Saulle, Marc, Hoffbauer, Krafft-Ebing, Griesinger, Pick, Ray, Rush, and numerous others, is substantially as follows : That transitory mania is an ordinary form of acute mania characterized by the brevity and explosive character of the violence ; that it occurs in persons sane prior and subsequently to the attack, rarely relapses and seldom lasts over six hours ; that there are no very apparent prodromata, and no sequelæ other than the slumber and turgidity of the hands ; that the predisposing causes are heredity and an excitable temperament; that the exciting causes may be alcoholic excesses, physical exhaustion, violent emotion, and mental strain, and that the disorder of itself tends to recovery.

These conclusions seem to be well substantiated by the evidence adduced, though it is clear that the author has formed them under that widely prevalent conception of "mania" which sees in destructive or violent excitement the essential maniacal characteristics. If the word "furor" or better still "frenzy" were substituted no exceptions could be taken to them.

It is a significant fact, that while epilepsy is not found in the ancestry or collateral relations of such patients, acute mental disorders of a furious or maniacal type are occasionally noted to have occurred. A tendency to so-called head-congestion has been noted in the family history of others. In my case there was a history of migraine.

Insolation, prolonged insomnia, exposure to extreme cold, and violent emotional and intellectual strain, have been frequently determined to have been the exciting causes. A classical instance is one related by Reich:* Four boys between the ages of six and ten, who had been out sleighriding on an extremely cold day, on returning home were suddenly ushered into a room which was overheated ; a transitory insanity followed, marked by a maniacal delirium, excitement, and hallucinations, which rapidly subsided giving way to a critical sleep. On the whole it seems that those factors which lead to disturbances in the cerebral cir-

---

* *Berliner Klinische Wochenschrift* 1881. No. 8.

culation, of a probably congestive character, are the ones active in the production of transitory frenzy. In not a few of the recorded cases the fury of the patient and his amnesia for the furious period remind one of the rage of the bull or the male elephant, conditions which are looked upon as transitory nervous disturbances in those animals, and which, like the transitory frenzy of man, are as a rule isolated explosions in the individual's career.

SUB-ORDER B.

Forms not Characterized by a Fundamental Emotional Disturbance.

STUPOROUS INSANITY.
PRIMARY CONFUSIONAL INSANITY.
PRIMARY MENTAL DETERIORATION.
THE SECONDARY AND TERMINAL DETERIORATIONS.

CHAPTER V.

STUPOROUS INSANITY.

Stuporous insanity consists in the simple impairment or suspension of the mental energies, unmarked by any emotional or other perversion.

In some young persons, such who have as a rule not passed the twenty-fifth year, and in whom there is either a congenital or an acquired weakness of the nervous system, there develops a condition of apathy which may reach a degree as intense as the atony which characterizes melancholia attonita. But there is a profound difference between the two conditions: While the atony in melancholia attonita is the result of an overwhelming delusion based on a painful emotional state, it is in stuporous insanity a phenomenon by itself, the direct result of a physical condition, and not translated through an intermediate emotional perversion.

Stupor may be best compared to the condition of indifference and indolence which sometimes follows excesses, or results from prolonged night-watches, and indeed it has a similar causation. It develops in one of two ways: Either the patient, in consequence of masturbation, starvation, or exhausting discharges, develops a gradually deepening apathy and anenergy; or, after a profuse hæmorrhage or a powerful shock to the nervous system, the stupor is brought to its climax in an instant.

At the height of this disorder the patient is in a state of immobility, he does nothing of his own initiative. Sensibility is impaired as much as the mobility, so that even powerful punching, nay, the cautery, may not be perceived by the patient. The reflex acts are sometimes impeded to such an extent that the food which is placed in the mouth of the patient will not excite the act of swallowing, unless it be pushed well backwards into the pharynx. There is also a corresponding anenergy of the involuntary muscles: the pupils are dilated and react poorly, the heart's action is greatly enfeebled, the pulse tardy, small, and frequent, the temperature is slightly lowered, and the extremities are cold, while œdema of the feet is constantly, and that of the hands and face sometimes observed.

The mental activity shares in the depression and abolition of the other nervous and general somatic functions. The stuporous lunatic's recollection of the period of his illness is found to be entirely destroyed on examination after convalescence. The mind throughout the disorder appears to be a blank, and the only indication that a feeble appreciation of the outer world still exists is to be found in the occurrence of tremulous movements of the muscles when the patient is by signs imperatively ordered to do any given thing, and in an acceleration of the pulse when loud noises are made in his immediate neighborhood. Otherwise the stuporous lunatic manifests no reaction, even to the vegetative needs, and his fæces and urine are passed without any knowledge on his part, or any change in his position, while the saliva dribbles from his mouth. The urine is rich in phosphates, and the physiological discharges of the skin and uterus are suppressed as in other atonic states. "Raynaud's Disease" is not uncommon.

At times there are observed changes lasting for a brief period, during which the patient manifests a slight return of mobility, or even speaks a few words, either relating to the subjects which occupied his mind last, or consisting merely in parrot-like repetitions of what is said and done around him. If recovery is to occur, these periods become more frequent, of longer duration, and their lucidity appears more marked.

Stuporous insanity may run its course in a few weeks, but its usual duration is from one to three months. The prognosis is highly favorable, probably ninety per cent of the patients recovering. Recovery is most rapid in very young subjects, and in those in whom the stupor has been produced suddenly, as after a fright or a profuse hæmorrhage. It is least rapid and least likely to occur in cases arising after masturbation. When recovery does not occur the apathetic variety of terminal dementia or pulmonary disease closes the history.

This disorder is known in American and English asylums* as acute or primary dementia. The writer considers the term "dementia" a very unfortunate one; for alienists are accustomed to associate with this term the idea of incurability, or at least of a deterioration following some other psychosis. Again, by employing the adjective "acute," they impress the novice with the idea that a set of symptoms like those of chronic dementia may present themselves in an acute form—which is not the case—or on an analogous basis—which it would be fallacious to maintain! Add to this the fact that there is such a thing as a "primary dementia," a dementia like terminal dementia, but arising independently and developing progressively,† which is as different from stuporous insanity as one thing can well be from another, and the further fact that both "acute" and "primary dementia" have been used as designations for hebephrenia and masturbatory insanity, and it will appear reasonable to recommend the dropping of both terms in this relation.

The important differential diagnostic relations of stuporous insanity will be found detailed in the last part of the work.

---

* And by a few German authorities as "primary curable dementia."
† Described by Voisin as insanity from atheromatous degeneration of the brain-vessels, and in this treatise as Primary Mental Deterioration.

CHAPTER VI.

PRIMARY CONFUSIONAL INSANITY.

Primary confusional insanity is a form of mental derangement characterized by incoherence and confusion of ideas without an essential emotional disturbance or true dementia.

Just as stuporous insanity imitates one of the most frequent phenomena of melancholia, yet differs from it in the absence of the fundamental painful emotional state characterizing the latter, so there is an acute insanity marked by a prevalent confusion in the conceptional sphere resembling maniacal delirium, without having the same pleasurable emotional basis as the latter.

Just as the resemblance between the atony of melancholia and that of stuporous insanity is superficial, so there is only a surface similarity between the confusion of mania and that of acute confusional insanity. The confusion of mania is not the expression of a genuine confusion of ideas, but of a disparity between the number of ideational items and the word-channels through which these seek exit; that of confusional insanity is an expression of a true, essential incoherence in ideation. Whether the patient speak slowly or deliberately, incoherence is equally noticeable. This is not the case in mania.

This disorder is rare and develops rapidly on a basis of cerebral exhaustion. Consciousness is blurred in parallelism with the conceptional disturbance, and the patients on recovering have as a rule but a very crude recollection of their condition. Its duration is variable, comprising weeks or months, and the prognosis is as good as that of stuporous insanity, with which condition it also has a resemblance as to etiology; emotional shock, cerebral overstrain, exhausting diseases, and excesses being the principal factors responsible for confusional insanity.

The patients suffering from this psychosis, after a rapid rise of their symptoms during a period of incubation rarely exceeding a few days, present hallucinations and delusions of a varied and contradictory character. The delusions resemble those of mania and more often those of melancholia, but no emotional state is associated with them. The patients assert in the same breath that their property is being stolen and that they are going to take part in some great state affair. A patient of Fritsch's, after protesting that she was innocent of the disease of a child which she had been nursing through an illness, said that a gold wagon was going to be sent for her; in the asylum the same patient, after rambling incoherently about the medicines the said child received and the accusations made against her, resisted all attempts to transfer her to the ward, saying that she "had to fight for her country, because she was to be a man."

The speech in confusional insanity is characteristic; although there is no richness in ideation as with mania the sentences are left uncompleted, and are entirely irrelevant as well as incoherent. A patient of this kind told the writer the following: "I am, I—, I don't know that—I—is dead—funerals are—how do you do—met you in Boston steamer—this is London—London—I am sure of it—see! I have not forgotten everything—there are not so many now—" Here he was interrupted by the question what he referred to; he replied: "The police of London have their wires to watch—you know, to watch—all the furniture—look at that horse—it is alive—it is not a wooden horse—[it was]—it looks so—I think it is—my poor father—who was the eldest son of my niece—that is a mechanism working on the principle—the wires work up and down—move it off the wall—see them—that furniture is all alive—I know my head is wrong—I can write as good—my wife and my nieces arrested and confined forever—spirits and the headache is intolerable—meat is not for the righteous—I am an orthodox man—the poison causes these fancies—I never had hæmorrhoids—I never had consumption—but how nice to meet you here in the Boston steamer."*

Delusions of identity are very common. The patients believe they are not in the same place, or they recognize as old acquaintances persons to whom these bear no resemblance. It is noteworthy that a large number of the patients are aware that a change has taken place, that they are no longer their former selves, and they may be able to give—by snatches, it is true—a tolerably fair account of the circumstances preceding the outbreak of the disease. But, as the latter develops, the patients cease to recognize their position, or to complain of the "head trouble" whose

---

* Translated from the German.

existence they previously admitted, and at most they speak
of their former selves in the third person, or manifest a
confused variety of double consciousness. When the hallucinations, as is frequently the case, preponderate from the
beginning, the disorder we have here considered is termed
by some *acute hallucinatory confusion*. From the superficial
resemblance of the verbigeration of patients exhibiting this
form of insanity to monomaniacs with episodial excitement,
Westphal was induced to classify it among the monomanias
(Primäre Verrücktheit), a questionable arrangement. It
would be as just to comprise several other forms of insanity under monomania on the same basis. In confusional
insanity there is no method as in monomania, no productiveness as in mania, no origin of the delusions from a process of
reasoning and reflection as in the former, nor a flight of ideas
as in the latter.

Recovery is gradual, the patient becoming progressively
clearer; his somatic complaints, such as headache, then
occupy his attention more than his incoherently recounted
delusive troubles, and finally reason is entirely restored.
In only a small proportion of cases does the insanity remain and the patient become permanently deteriorated,
his disorder then appearing as a form of *chronic* confusional
insanity.

CHAPTER VII.

PRIMARY MENTAL DETERIORATION.

Primary mental deterioration is an uncomplicated enfeeblement of the mind occurring independently of the developmental and involutional periods.

In most persons surviving the sixtieth year a pronounced
and general failure of the mental powers occurs at or after
that period. This is the ordinary senile change, and cannot be considered to be in all cases a pathological one.
But where a similar deterioration anticipates the senile
period it can only be accounted for on a pathological basis.
Such a decay of the mind is observed in paretic, syphilitic,
and organic dementia, and is also found to be a sequel of numerous other forms of insanity. In all these instances the
mental failure is accompanied by active symptoms which in

their association with the dementia characterize the given variety of mental disorder. It is not so, however, with a certain class of cases in which progressive deterioration, chiefly limited to the higher mental faculties, is the only notable indication of a cerebral disturbance.

Crichton Browne described as "chronic brain-wasting" a disorder in which there is confusion and failure of the memory, lack of attention, and general inertia. With this the muscular power is enfeebled, the articulation is affected, the pupils are unequal, and the temperature is subnormal, while the patient generally complains of a sensation of pressure or fulness in the head. Convulsive attacks occurring on one or both sides heighten the resemblance to paretic dementia, and the progress of the disease, with rare exceptions in which recovery occurs, is toward complete extinction of the mental faculties.

The writer has observed a similar condition among business men, particularly among those whose duties were of a varied, exciting, and exhausting character, who, with an expensive domestic establishment on the one hand and a tottering firm on the other, resorted to Wall Street to make good the difference. It is also not uncommon with members of the legal and other professions, to the practice of which excitement and strain are incidental. In short, the etiology of this affection is very similar to that of paretic dementia, and it may not be improper to consider it an, as it were, functional analogue of that organic malady. The paralytic and convulsive symptoms noted by Crichton Browne have not been observed in the writer's cases, and, judging by the serious prognosis given by that author, it is probable that he has considered genuine cases of paretic, syphilitic, and "organic" dementia in conjunction with those cases to which the writer would limit the designation "primary mental deterioration."

The first signs noticed are generally recognized by the patient himself. He experiences a lack of energy both mental and physical. The warning being disregarded, and the strain kept up, the abused nervous system replies with insomnia. The patient finds it difficult to go to sleep, and when he finally drops off into a brief and fitful slumber it fails to refresh him, and the irritable condition of his brain manifests itself in dreams, whose subjects are generally taken from his daily occupations and cares. The patient now becomes dyspeptic, and signs of functional or organic heart disorder, or of the prodromal period of Bright's disease may be noted by the examining physician. Often the patient becomes prematurely gray. There can be little doubt that continuous mental worry and emotional strain are competent to provoke all these disorders, particularly in predisposed individuals. At this stage the warning may be heeded, and a comparatively healthful mental state resumed under treatment; but if the exciting causes are kept in operation actual dementia may be the result. At first the subject is noted to be absent-minded: the lawyer finds that he is unable to fix his attention on his opponent's argumentation; the physician discovers that he is at a sudden loss in writing prescriptions and forgets to add important directions, not in single instances, but repeatedly; the stenographer finds that his hand fails him; and the literary man omits words, or misspells where he was previously methodical and accurate. Important engagements are broken, articles of value mislaid, addresses forgotten, expenditures unrecorded, and, with the intensification of all these symptoms, complete fatuity may be developed. Yet it is noteworthy that, while the memory fails, attention becomes difficult, and the power of acquiring new impressions is impaired, the patient may in fits and starts show his old brilliancy in reasoning. Let him, however, attempt to keep up the effort any considerable length of time and he will break down.

On the basis of the condition just described any of the primary simple psychoses may develop, and it may prove to be the preliminary phase of a paretic dementia. But it may also continue to exist by itself, and terminate in a relative recovery or in death without further complication. As a rule complete rest and proper tonic and moral treatment are capable of checking the disorder at any but its later periods, and while a complete *restitutio ad integrum* has never been observed by the writer, and even the most favorable cases reveal some permanent damage, however slight and however unnoticeable, to those who have not known the individual before his illness, some of the patients remain free from a renewed attack, and may even return to business of a less exciting character, and successfully fill a responsible position in life.

This disorder rarely comes under the notice of the asylum physician. The absence of delusions, of morbid propensities, and of excitement account for this fact. Occasionally a suicidal tendency may render sequestration necessary, and the mistake is apt to be made of confounding such a case with paretic dementia of a melancholic or hypochondriacal invasion-type. Add to this the fact that a laxity of the facial and a weakness of other muscles, as well as forgetfulness of words and facts, are common accompaniments, and the possibility and probability of this error being committed will be understood. The future history of the case exposes its true nature, and a careful analysis will show that the suicidal attempt was not the outcome of emotional depression, delusion, or hallucination, but the result of a process of reasoning, often correctly based on correct premises, by a patient fully appreciating his sad position. The misery of the sufferer is often aggravated by his recognition of the fact that his affection for his dearest friends and relatives, like his more strictly intellectual faculty, has become blunted, and that he is unable to recall these feelings in that intensity which characterized them in his healthy state.

CHAPTER VIII.

The Secondary and Terminal Deteriorations.

In the foregoing chapters on mania, melancholia, stuporous and other primary forms of insanity, reference has been repeatedly made to the fact, that in a certain series of cases, while death does not ensue, recovery is not effected; and a secondary and chronic psychosis develops from the primary disorder.

A thorough consideration of dementia is nearly tantamount to a study of all that which the older authorities designated as "secondary forms." As the term is generally used, however, it refers to *terminal dementia*, which is the ordinary conclusion of most chronic, and the uncured acute insanities. Inasmuch as terminal dementia develops from primary forms differing greatly among themselves, and the transition from the primary insanity to dementia is gradual and progressive, it will be perceived that numerous grades and varieties of this affection must exist. It is customary in order to fully characterize their varieties, to state what primary form preceded the dementia. Thus we say: dementia follows mania, melancholia, or stuporous insanity. Sometimes we are enabled to determine from the demented patient's symptoms what the primary form of his insanity was; in one case we may find residua of the delusions of marital infidelity with physical symptoms indicating the previous existence of alcoholic mania, in another the delusions of persecution, and incoherent ideas growing out of such, which point to the previous existence of melancholia.

Dementia must not be confounded with imbecility; while both dementia and imbecility imply a profound general defect in the mental sphere, the former term should be always limited to *acquired enfeeblement*, the latter to the *original feeblemindedness* due to fœtal or infantile arrest of development. Much confusion has also arisen from the unfortunate use of the terms "acute" and "primary dementia." Acute dementia is applied to a primary insanity more properly designated as acute stupor, while "primary dementia" is indifferently applied to stupor, insanity of pubescence, and primary deterioration. The designation dementia should be limited to permanent mental deteriorations, and a discrimination should be made between the dementia from gross organic disease of the brain, the paretic dementia to be considered in a later chapter, and that senile dementia which is a natural manifestation of brain-involution on the one hand, and the trouble we are here considering, which implies the previous existence of some well-marked primary form or the other.

The course of the development of this secondary insanity is twofold: either the primary disorder passes directly into dementia, or it does so indirectly, through an intermediate stage of chronic secondary mania, with confusion of ideas and mental enfeeblement as prominent features. When dementia follows the latter affection it is "tertiary." But, as it does not in this case materially differ from the dementia which is a more direct sequence of and secondary to the primary forms of mental disturbance, it is best to devise some common term for both. The framing of such a term may be based on the fact that, whether secondary or tertiary, these varieties of dementia, in contradistinction to the primary dementia from coarse brain disease and senescence, are the *terminal* epochs in the history of prior psychosis. They may hence be termed TERMINAL DEMENTIAS.

In its widest sense this designation might apply also to the dementia which closes the history of epilepsy, as well as of epileptic and alcoholic insanity. But, as the dementia in these cases is customarily designated as epileptic, alcoholic, etc., according to its etiology, and these adjectives indicate also the clinical characters of dementias which are different from those of the dementia following the ordinary forms of insanity, the group of terminal dementia may be advantageously limited to the latter.

For stuporous insanity and melancholia attonita it requires nothing further than for the patient to remain in the atonic and stupid condition a longer period than in favorable cases to constitute a terminal dementia. Occasionally the exhaustion following violent outbreaks of maniacal furor passes into dementia as directly. In all these cases the mental deterioration is of a *passive* variety, one whose characters are simply negative; the mental processes generally are nullified, the countenance is devoid of expression, the extensors are not innervated, the flexors consequently predominate, and the patients in their inactivity resemble cowering statues or animals whose cerebral hemispheres have been partly removed. What was a merely functional and temporary clouding of the mental sphere in atonic melancholiacs and the stuporous insane now becomes an organic, progressive, and permanent condition, which finds an anatomical expression in the accompanying cerebral atrophy.

These unhappy creatures constitute a considerable proportion of the pauper asylum * or poorhouse population, and they largely people the " unclean" wards of all asylums. Here they may be seen on the benches, mute, expressionless, devoid of any spontaneity, requiring to be fed, conducted to the water-closet, dressed and brought to bed like children. As deterioration proceeds even the few words retained in their limited vocabulary become lost, and complete mental annihilation precedes physical death, which occurs either through the extension of central paralysis to the centres of vegetative life, or by inter-current diarrhœas and pulmonary affections. As a rule these patients do not live more than a few years.

Other demented patients appear docile, willing to assist

---

* Of the 2,297 pauper lunatics, referred to elsewhere as tabulated according to their form of insanity by the author, 334, or over fourteen per cent, were terminal dements.

the attendants in the performance of routine duties, are employed in copying records, in nursing debilitated comrades, and attending to the cattle on the farm, while the great majority are lounging listlessly around the corridor, or stand or sit in one place all day, indulging in some rythmical movements, or vociferating the same set phrases. Some dements pass their evacuations without any regard to time and place, and even delight in doing so in the most unusual localities, and in the most unseemly manner; others do so because they are simply oblivious to the calls of nature. Certain of these patients require to be fed by force, others will eat as soon as the automatic processes are started by seating them at table, and putting eating utensils in their hands, still others are ravenous eaters, and their ideations revolve within the limits of the daily bill of fare, whose items perhaps they will recite or chant all day. The fundamental feature of terminal dementia is *an acquired mental defect*, and this may vary from a mere loss of memory, usually of recent events, or of the reasoning power, to the nearly complete extinction of mind. The loss of memory may be of every grade; in some it involves a special period of life; in others the period of the primary disease; in all more or less the memory of recent events. Old recollections may exist with normal intensity, but there is a failure of the receptive sphere to register new impressions, or at least to register them perfectly; indeed it is to be presumed that it is the struggle between the old healthily-established mental combinations, and the imperfect and hampered products of the newer ones, which accounts for the phenomenon of double consciousness, and other disturbances of the sense of personal identity sometimes found in the early phase of terminal, as in other varieties of dementia.

There is another form of terminal dementia in which apathy is not so pronounced, and inactivity cannot be said to exist. On the contrary, the patients are restless, talkative, and even obtrusive or destructive. But their violence is without purpose, and even without that emotional basis which, the maniac's violence always has. Their speech is verbose but the sentences are without connection and sense: in fact the logical and associating bonds are altogether wanting, and, under the confused medley of disconnected acts and words, the progressive dementia is apparent to the experienced observer. Even to the inexperienced observer the expression of the patients, as well as their random and confused talk, seem the outcome of a silly and childish condition. This variety is known as *active* or *agitated dementia*; it is a sequel of mania and of agitated melancholia, and is progressive, though of longer duration than passive dementia.

This dementia, in which fragments of delusions and delusive ideas are still retained, constitutes a transition to that secondary form of chronic insanity which some have called "chronic mania," others, "secondary partial insanity," and which still others have unfortunately classed among the "monomanias." It is observed that some maniacs as well as melancholiacs lose the dominant emotional character of their insanity, without regaining mental health. The mind, in other words, is no longer stimulated by an emotional state to construct expansive or depressive delusions; but, on the other hand, it loses the logical power to correct the delusions formed during the previous period, *i.e.*, the primary insanity. These consequently remain integral parts of the patient's *psyche*, and become fixed delusions. Unlike the delusions of the monomaniac—which are also fixed—the delusions of *secondary insanity with confusion of ideas*, or "chronic confusional insanity," as the writer proposes to designate this disorder, are not elaborate, not defended with skill and a show of judgment; in short they are not truly systematized. The delusions resemble ruins left over from the destruction of the more elaborate and multitudinous if less fixed delusions of mania and melancholia, around which the gathering tide of a slowly progressing dementia rises, till the assertion of the delusions becomes a mere parrot-like repetition, and is finally buried under that same levelling sea of dementia which closes the history of all those primary psychoses entering the domain of the secondary deteriorations.

The weakening of the logical power and the memory accounts for the frequent observation in these patients of a change in their sense of identity.

In marked contrast with the primary insanities, the chronic deterioration last mentioned shows few if any anomalies of vegetative life. The appetite and assimilation as well as the sleep become normal or nearly so, and not unfrequently the patients become very stout. A rapid increase of the adipose tissues of a patient who is becoming calmer than he was during his primary period of mental disorder is hence, not without justice, looked upon as a sign of evil augury. All recovering maniacs and melancholiacs increase in weight, it is true; but that increase is usually compensatory for the loss of weight occurring at the onset of the disease, and does not as a rule go further.

While the general nutrition does not always suffer in the terminal deteriorations, certain trophic disturbances are quite common. Hæmatoma auris, cutaneous eruptions, premature grayness, and fatty and fibrous changes of the blood-vessels are frequent accompaniments. These, like the deep structural changes in the nerve-centres, are collateral phenomena, and do not stand in a direct causal relation to the insanity.

II. Division of the First Class. Attacking the individuals in Essential Connection with the Developmental and Involutional Periods.

SENILE DEMENTIA.
INSANITY OF PUBESCENCE.

CHAPTER IX.

SENILE DEMENTIA.

Senile dementia is a progressive and primary deterioration of the mind connected with the period of involution, but exceeding the ordinary extent of such involution to a pathological degree.

As stated in the seventh chapter, a certain degree of mental enfeeblement is an ordinary accompaniment of old age, and cannot be considered pathological. Simple diminution of the mental powers, and the intensified conservatism, lethargy, and habits of economy incidental to senility, do not constitute a true insanity, and therefore should not be called dementia. But when the ordinary limits of these conditions are exceeded, when lethargy becomes fatuity, when conservatism becomes suspicion, and penuriousness provokes delusions of attacks on property, the senile subject is the victim of an insanity which is only found with the aged, and is therefore called senile dementia.

Senile dementia is to be considered as an entirely distinct conception from "senile insanity." Senile insanity, so called, includes senile dementia, but senile dementia does not include all of senile insanity. Any form of ordinary insanity, such as mania, melancholia, dementia from active organic disease, and monomania, may be found in the aged, and present at least in the main the features which characterize these affections at other periods of life. There is, indeed, no need for discriminating between senile and other periods of life as far as the ordinary forms of insanity are concerned. There is no senile mania any more than there is a middle-age melancholia or an adolescent stupor, and it is best to speak of the ordinary forms of insanity as mania, melancholia, or monomania in a senile subject. The only characteristic form of senile insanity is the one now about to be considered. Senile dementia should not be confounded with other conditions occurring in old age, of which mental enfeeblement may be a symptom. There are certain gross organic diseases, affecting the brain in advanced life, which produce a set of symptoms often and improperly classed as senile dementia. Thus, an old person, after a paralytic attack due to hæmorrhagic or necrotic brain lesion, may become feeble-minded, forgetful of the proprieties, morbidly irritable, and filthy in his habits. Such a condition is, however, a complication of what is commonly recognized as an ordinary brain disorder, which may produce similar results at any period of life; there is nothing essentially senile in its character, although it is more frequent in the senile state, because the conditions causing it are more common in the aged than in the young. Paralytic and epileptiform seizures may be accompaniments of senile dementia, but in that case they are epiphenomena, and not essential features of that psychosis; this disorder begins as a senile dementia, and is not secondary to other affections, as are the forms of dementia from coarse organic disease just alluded to. Senile dementia is to be attributed to a slowly progressing marasmus of the nervous tissues transcending the ordinary degree of intensity; to which more active nutritive changes, in the way of encephalomalacia or hæmorrhage may or may not be added. The complicating dementia occurring in old age after coarse disease referred to above, and which is distinct from it, is constantly and characteristically associated with such coarse disease.

Senile dementia is manifested by an increased egotism, or by penuriousness, which sometimes reaches such a degree that the millionaire may starve in the midst of his or her millions, and, though residing in a palace, grovel in filth. The memory becomes enfeebled, particularly with regard to recent events, while those of an earlier period of life may be well remembered. It thus happens that senile dements frequently lose their way in the streets, do not recognize their own houses and apartments, and confound the property of others with their own. Prejudices are formed on trivial grounds, or on no grounds at all; and wealthy senile dements have in all ages been made the subjects of speculative and designing persons, to the detriment of their real interests and of those who were the subjects of their natural affection in the healthy period of these patients' lives. As the disorder advances the memory continues to decrease; the incidents of whole years seem to be blotted from the mind; and patients have been known to forget the names and number of their children, or even that they had been married when such was the case. A profound moral deterioration is frequently a marked accompanying feature. Coarse and vulgar expressions are used by persons previously accustomed to select language, or the patient becomes filthy or intemperate in his habits, and assaults or scolds his children, treating them and the servants like dogs. To this there may be added—particularly in male persons—a pathological sexual desire, a senile satyriasis, which with some manifests itself in indecent assaults on young girls or even on infants, and with others in absurd and ridiculous marriage plans.

While some senile dements exhibit delusions of an ambitious character—always unsystematized, however—the majority have depressive delusions, and rare instances are on record where senile dements have committed suicide, either in consequence of such delusions, or because they recognized their deteriorating mental condition.

The chief and most common delusions of senile dements relate to their property. They suspect that they are being defrauded or robbed; in consequence they take what they think are the best measures to prevent defraudation and robbery. If their property is in the charge of an agent, they will discharge him and employ another, and another, till they find one who possesses the undesirable qualifications necessary to the management of a senile dement—for experience teaches that intrigants and time-servers have had more success in this direction than straightforward and independent business men or the honest friends of the patient.

The anxiety as to the security of their earthly possessions, and their delusions of robbery, produce a lachrymose disposition and a restless and purposeless activity in these patients. Some of them roam about at night continuously, watching for thieves, while other patients do so without being able to give any reasons for their acts whatever. Others are continually engaged in devising new fastenings for their doors and windows, and new hiding-places for their treasures. Hallucinations and illusions may complicate this phase of senile dementia, and the patients then cry out that they are being murdered, robbed, burned up, cut to pieces, or poisoned.

Should no other inter-current illness cut short the course of the psychosis, bed-sores and colliquative diarrhœas close its history. Sometimes affections of the bladder are very troublesome toward the end of the patient's life, and may lead to fatal cystitis or pyelitis; indeed, incontinence of urine is one of the most constant physical accompaniments of senile dementia from its inception. If the patient lives long enough complete fatuity sets in; he may then become voracious and filthy, to finally die with apoplectiform symptoms, or with those of a gradual and general paralysis. Aside from a temporary improvement, which is exceptionally observed in those cases where the delusions have a melancholy tinge, the progress of senile dementia is chronic, and consistently in a downward direction.

The physical indications of extreme age are always found in senile dementia. The most important of these signs, because it is related to the cerebral condition, is arterial sclerosis; the radials are hard, giving the impression to the finger of a tendinous cord, instead of the normal arterial resilience; with this the temporals are tortuous. Often there are observable a marked *arcus senilis*, opacities of the vitreous body, and sometimes cataract, as expressions of a vitiated state of nutrition. An invariable symptom is tremor, but this does not differ in degree from that which is commonly found in very old persons. In certain patients marked hyperæsthesia has been observed by Güntz,[*] and vertigo, anorexia, paraparesis, hemiparesis, disturbances of speech, and epileptiform attacks have been recorded in others. It is cases presenting these symptoms, which are associated with more considerable cerebral atrophy and

---

[*] "Allgemeine Zeitschrift fur Psychiatrie," xxx.

nutritive as well as membranous lesions than the ordinary ones, which have suggested the view that paretic dementia is a pre-senile involution of an active type.

Of 2,297 patients whose form of insanity was made the subject of a statistical study by the writer, 82, or a little over three and a half per cent, were classed as cases of senile insanity. With three exceptions these were all senile dements. It may be assumed that the proportion of senile dementia is much greater than the one shown by asylum statistics, as only the agitated and troublesome patients suffering from this malady are sent to asylums.

CHAPTER X.

Insanity of Pubescence.

Insanity of pubescence is characterized by mental enfeeblement, marked by a silly disposition, following a preliminary period of depression, which has the same tinge as, without the depth of, that characterizing melancholia, and which coincides with or follows the period of puberty.

Probably few persons pass the period of puberty without manifesting some indications of the profound change which the mental organism undergoes at this important physiological crisis; a change which in not a few cases is a real change of character, without being for that reason—as those who define insanity as essentially consisting in a "change of character," might be compelled to admit—an indication of mental disorder. Particularly in the male sex is the transition between the childish and boyish period preceding, and the adult period following puberty, marked by many comical, ridiculous, and even disgusting conflicts of the boy's nature with that of the coming man. The carelessness, lack of judgment, natural egotism, and sportive tendencies of youth are out of harmony with the aspirations and feelings which now develop and which are destined to characterize the man. The result of their union is a silly ambition, a mawkish sentimentality, and an obtrusive self-assertion, which, in a more or less pronounced degree, are manifested by most youths; to control which is one of the main objects of every sound educational system,

and which in healthy subjects with or without such system
are soon corrected by experience through its incidental
and beneficial hard knocks.

In certain rare cases this correction does not occur; the
patients retain the absurd notions, the silly propensities,
and the obtrusive egotism of adolescence. Whether it be
the existence of a hereditary taint, or masturbation, which
weakens the nervous centres, it is certain that the transformation of the childish into the adult character is arrested.
This is the essential feature of the hebephrenia of Hecker,
the "insanity of pubescence" of Skae and Maudsley.

This psychosis begins with a period of sadness; the patients are depressed without being able to assign any reason for their sadness, and suicidal attempts are not unfrequent. There is, however, no depth to the depressive
emotion as in melancholia, and in the midst of the depressed period the patients appear rather obtrusive in communicating their sufferings to others, and will not hesitate
to simulate in order to awaken, not sympathy—which they
care little for—but interest! In the midst of these periods
they may suddenly burst out in causeless laughter or even
joke in a silly manner. In short, the contrast in the character of the changing emotions is great, but the emotions
are in no case as deep as in the mania and melancholia of
the adult.

After this preliminary period the patients exhibit vague
or blind propensities; they enter a business, to leave it the
next day, wander about aimlessly, or display a stupid
malice toward their surroundings. While there is no
incoherence the patients manifest a peculiar tendency to
adopt verbose language. They will use long words, or
such of an odd sound, or ride certain grammatical hobbies.
Others will use slang or foreign expressions and quotations
by preference. Gradually the condition changes; the intellect weakens progressively, and the patient, who is usually
a confirmed masturbator, will pass into a terminal dementia
marked by occasional furious outbreaks, determined in their
occurrence by his unnatural excesses or by powerful external impressions.

Everything connected with the mental state of these patients appears shallow and even unreal. They have sham
emotions, sham regrets, sham anger, and sham complaints.
Even their hypocrisy, which is a common characteristic, is
shallow. In the same breath in which they affect religious

aspirations they will indulge in slangy vituperation, and then break out in causeless laughter. The expression of the countenance is an indication of the condition within; it expresses the leading character of lack of emotional depth, silliness, and insincerity.

The course of this form of insanity is protracted. Enfeeblement of the mental faculties is noted from the very beginning, and the process may be arrested and remain stationary for years without material progress toward terminal dementia. In one case in the writer's experience a relative cure was effected, the disorder early arriving at a standstill, and the positive characters of the illness disappearing. But on the whole the prognosis is exceedingly unfavorable. Imperfectly developed cases, such in which the disturbance is limited to a slightly strained emotional condition, with a tendency to writing silly and extravagant poetry, and which appear to be merely instances of a pathological intensification or undue prolongation of the ordinary pubescent state, present better prospects.

Pubescent insanity has been observed in but three out of one hundred and eighty-seven private patients by the writer. A computation from the statistics of a pauper asylum yielded the high figure of nearly five per cent of cases of this psychosis.

As indicated by its name pubescent insanity is found in subjects between the fifteenth and twenty-second years. Many of the cases are still classed as "primary dementia," particularly when the deterioration is very rapid. Where masturbation is a pronounced feature some writers use the designation "insanity of masturbation." In reality the masturbation, although a frequent accompaniment and perhaps a result of hebephrenia, is not its cause, however much this habit may ultimately modify the character of the psychosis.

SECOND CLASS: SIMPLE PSYCHOSES ASSOCIATED WITH DEMONSTRABLE
ACTIVE ORGANIC CHANGES OF THE BRAIN.

DEMENTIA PARETICA.
DEMENTIA SYPHILITICA.
DEMENTIA ORGANICA.
DELIRIUM GRAVE.

CHAPTER XI.

PARETIC DEMENTIA—PRELIMINARY CONSIDERATIONS.

There is a form of insanity which, from its constant assoication with the classical symptoms of ordinary organic disease of the brain and spinal cord, merits most attentive consideration. There have been thus far discussed mental affections whose essential characteristics are the mental symptoms proper. We have found that with most of these forms of insanity disturbances of the bodily functions are indeed present; but these are rather attendant and subsidiary phenomena, of importance to the speculative somatic psychologist, than striking features of the insanity. In short, the psychoses thus far considered could be defined and recognized in a crude way without taking into account the coarser bodily conditions; while, with the psychosis we are now about to treat of, this is different.

Here the mental symptoms generally present the picture of unsystematized ambitious delusions, combined with progressive paresis and dementia; they may range, however, from atonic depression to the most furious delirium, from the construction of fanciful projects to extreme incoherency, and from slight and almost undemonstrable mental impairment to the absolute extinction of higher mental life. In like manner the physical signs, whose combination with these varied mental disturbances is essential to the picture of the disease, may vary from slight disturbances of speech to gross paralysis, or may present themselves under the mask of a posterior spinal sclerosis (locomotor ataxia), of a disseminated organic disease, or of apoplectiform and epileptiform seizures. Among the individual signs there may be found almost any and every focal and general symptom known to the neurologist: paresis of various voluntary and involuntary muscles, anæsthesias, paræsthesias, and hyperæsthesias, pains and trophic disturbances, changes in the vascular tone, amblyopia, hemiopia, color-blindness, and aphasia; not to mention choreiform and athetoid movements, progressive muscular atrophy, pseudo-hypertrophic and bulbar paralysis: all these may be found co-existent with the mental disorder, and indeed depending on the same morbid process as the latter.

In the case of no other form of insanity are the pathological findings so constant and satisfactory. It may be safely asserted that in all advanced cases of this disease a diffuse lesion of the brain, sometimes involving other parts of the central nervous system, is to be looked for. That a disease whose pathological basis is so extensive, affecting numerous centres in varying degree, should present almost every conceivable variation within the outline of symptoms just drawn, is not surprising. That the mental symptoms predominate in one case, the motor in another, and the sensorial in a third; that their order of appearance differs: in some instances, disturbances of vision; in others of speech; in others, absent-mindedness, fits of fury, or hypochondrical *tædium vitæ* opening the history, is perfectly natural in view of the complexity of the functional rôle of the nervous structures, any one of which may be the first to weaken and break down under the diffuse morbid process of this disease. With the recognition of these facts we may waive any consideration of the mooted question, whether paretic dementia is a simple insanity or a complication of insanity by the features of ordinary central nervous disease. All observers are now agreed in considering this disorder as a primary form of insanity, existing by itself, and they attribute to its physical signs the same value that is assigned to its mental signs.

This disorder is known by a number of names. Some of these are too obsolete to call for mention here, and the designation employed in this treatise is one which is now gaining ground, particularly in England. In Germany it is known as *dementia paralytica* and "*progressive Paralyse.*" These terms are ambiguous, for the dementia sometimes accompanying hemiplegia has been earlier known under the former term; and while the objection is a finical one, yet it has been raised, that the affection is only in very rare instances evenly progressive, being generally marked like that similarly progressive affection, locomotor ataxia, by exacerbations and latent periods.*

About two facts there can be no dispute or quibble, that the essential and constant feature of the disease is dementia, and that it is associated with paresis of certain muscles. These features are hence incorporated in one term, which is as little ambiguous as it can be made. The only other affections to which the same designation might apply are syphilitic dementia † and dementia from coarse organic disease.

Paretic dementia is a very common affection, more frequent in communities whose members are subjected to great mental strain than in those whose members are engaged in mechanical pursuits or are able to indulge in a *dolce far niente*.

Of 2,297 male patients at the pauper insane asylum of New York city, 284,‡ or a little over twelve per cent, were paretic dements, or dements with organic diseases. In the same statistics the writer found that the nationalities represented by a small quota of the asylum population, such as the Scandinavians, Dutch, Scotch, Italians, and Sclaves, had a larger proportion of their insane among the paretic dements than the nationalities represented by larger numbers. This excess is attributable to the facts, that the members of wandering professions, such as agents, sailors

---

* Paretic dementia also passes under the names "general paresis," and "general paralysis." The alienist's position with regard to these terms is similar to that above stated as the one held with regard to the use of the adjective "progressive." No scientific alienist will misunderstand the term "general paresis," he will recognize it as a legitimate label for a well-marked affection. But he will abandon its use after being met on the witness-stand, as the writer was in the Gosling case, by an opponent who states, amid the tumultuous applause of the court-room crowd, and the commendatory glances of the "intelligent" jury, that there is no general paralysis except in death. Popularly the disease has been called "softening of the brain," and is diagnosticated and treated as such, and is hence a fruitful source of diagnostic errors to this very day.

† Syphilis found as an accessory etiological factor in paretic dementia of the typical kind does not justify the ranging of the case under the head of syphilitic dementia; the latter is a clinically and pathologically distinct affection.

‡ The source of error involved in the confounding of syphilitic and organic dementia is not sufficiently great to affect the proportionate values of these figures.

or firemen on board of steamers, are proportionately numerous among those of these nationalities arriving at the port of New York; and that exposure to caloric and to syphilis, two potent causes in the etiology of paretic dementia, are very common with these professions. The wandering tendency (*mania errabunda*) of paretics may also account in part for the accumulation of paretics of foreign extraction in this metropolis.

Among the five nationalities or races represented in large numbers in the asylum mentioned the proportion of paretic dements was as follows:

```
Anglo-Saxons............................ 13.29 in 100
Celts................................... 11.58 in 100
Germans................................. 11.13 in 100
Hebrews................................. 10.29 in 100
Negroes.................................  8.82 in 100
```

It is here seen that the Anglo-Saxon race, the race of the greatest speculative business tendencies, and of a high, if not the highest intellectual development among the races inhabiting the United States, has the largest percentage of paretics. That mere business exertion is not the essential and most fertile cause of the disease is shown by the fact that the Hebrew race, equally as active, and equally if not more successful in the mercantile world, occupies one of the lowest places in the list. That intellectual exertion *per se* is not a cause, is shown by the lesser percentage among the Germanic races, who have always stood foremost in the abstract and speculative sciences. Either the high proportion must be directly due to a race predisposition or to some inherent tendency of the race. England and America are the lands of the most active and feverish progress in civilization, of great facilities for rapid travel, of large mercantile and manufacturing establishments, of hurry, bustle, and restlessness generally, and all these features seem to be implanted in the Anglo-Saxon people. The German, on the other hand, still retains in this land the so-called phlegmatic disposition of his forefathers; the Celt preserves, as a rule, that quality for which he was noted in his native island, of "taking things easier" than the Saxon; and the negro is, as a rule, indifferent and lethargic in those matters which call for the interest and action of the higher races. The claim that there is a constant relation of sexual excesses to the development of paretic dementia as primary causes, contradicted as it has been by high authority, is not supported by these figures. No one will claim that the Anglo-Saxon is more libidinous or less able to endure indulgences than the other races. If a reflection were to be cast on any race in this respect, it would be the negro race—which shows the least percentage of paretic dements—to which a libidinous character might be assigned. When it is borne in mind, too, that where the negro lives under conditions natural to him, and where he is not compelled to enter into competition with a higher race, paretic dementia is almost unknown, the conclusion will seem reasonable that paretic dementia is more frequent with races of a high than of a low cerebral organization, because their higher civilization induces a restless mental activity with its attendant emotional strains, and that the disease is hence attributable to the excessive wear and tear of the brain induced by such civilization. A confirmation of this view is the fact that, while paretic dementia is much less common in females than in males, it is most common in those females who have entered into competition with the male sex in occupations ordinarily carried on by males.* Paretic dementia, therefore, is not as some have thought a penalty of high cerebral development, but the expression of a discrepancy between the instrument and its purposes; in other words, of the inadequacy of *some* brains to support the strain to which the race, *as a whole*, is subjected. It is one of the methods by which the contest for existence is continually being decided in favor of the strong and against the weak; and its greater frequency at the present day is in harmony with the fact, that the contest for existence, which in earlier epochs was decided on battle-fields and in the arena, is now carried on more largely in parliamentary halls, in the *bourse*, or on "change," and with the pen instead of the sword.

That the disease did not exist, however rarely, in ancient times is not demonstrated by the fact that no descriptions recognizable as those of cases of paretic dementia have been handed down to us by the earlier masters of medicine, as is so frequently urged by modern writers. If this reason is to stand, then we must assume that a number of disorders, not only of the nervous system, like "spastic paralysis" and "bulbar paralysis," have first appeared within

---

* Excluding the influence of alcohol and syphilis.

a few decades, but that many diseases of a general nature, which pathology has recognized the existence of only within the last few years, have not existed prior to their discovery. The same kind of argumentation would support the view that the planet Uranus was a new creation. It should be borne in mind, too, that many of the statements made to the effect that paretic dementia is on the increase, are exaggerated, however true the general tenor of this claim undoubtedly is. In many of the asylums in the west of this country the disorder was not recognized in the tables for other reasons than its non-existence; and the widely-circulated error has thus gained ground that the disease, first recognized in the Bloomingdale asylum, is "travelling" from the East to the West. In the asylum at Kankakee, in the east of Illinois, which is one of the institutions where a scientific classification is carried out, the proportion of paretics from the surrounding agricultural districts, and before the admission of patients from Chicago was not quite one and a half per cent. (6 in 424 patients admitted from 1881–83) is not quite one and a half per cent; but in Chicago, the metropolis of the same State, the writer is assured by competent correspondents, that it is as frequent in private practice as in New York. Indeed the writer found a tolerably large number of patients in the pauper asylum of New York city to have acquired their disorder, as they had been born and brought up, in the large cities of the West. It is not safe to venture too far in speculation on the apparent fact of a rapid increase of paretic dementia from year to year, made manifest in some statistics.* Where the diagnostic acumen of the medical officers is unquestioned, as in the large German and French asylums, while an increase is noticeable, it is but a slight one; in some cities, as in Hamburg, it is actually at a standstill, and in one year at least there appeared to be rather a decrease in this place. But, whatever inferences may be drawn from the imperfect records now at our disposal, there can be no question that

---

* A student of the writer's, himself an asylum physician, visited an asylum in one of the Middle States in which about a fifth of the male and a large proportion of the female inmates were exhibited to him as paretic dements. In many of these cases nothing beyond an emotional tremor of the hands could be advanced to justify the diagnosis, and it seemed that wherever this symptom was discovered, particularly if there were present expansive delusions, no matter whether these were systematized or not, the diagnosis of paretic dementia was made.

one of the great problems with which the preventive medicine of the future will have to deal is the grappling successfully with this scourge of the civilized portion of mankind.

CHAPTER XII.

PARETIC DEMENTIA, ITS COURSE AND SYMPTOMS.

As indicated in the last chapter, paretic dementia in its full development is characterized by a combination of mental and somatic deteriorations. But these constitute merely the permanent and constant background of the disorder, on which we may find developed at various periods of the disease and in bold relief almost all of the main positive symptoms of insanity. It is customary for purposes of convenience to divide the malady into stages demarcated by these accessory symptoms, and prominent writers on the subject have established three such: a first stage, marked by moral deterioration and other changes of character; a second, characterized by exalted delusions; and a third, in which these exalted delusions disappearing, a progressing mental and physical failure closes the history of the disease. Others speak of successive stages of depression, of exaltation, and of dementia. But while there are a number of cases in which these stages undoubtedly exist, there are a larger number in which they are not sufficiently well marked to justify the discrimination. In a few cases the progress of the disease is even and unmarked by exacerbations, in others the only noticeable symptoms are a progressing dementia and ataxia with paresis; in some the physical symptoms are prominent from the beginning, in others not; in a few mental deterioration is rapid, in most slow, and in still others it is checked and retrogrades, to advance again. But, however much the disorder may vary with regard to the existence of separate stages, all typical cases have a well-marked prodromal period whose recognition is most important to the general practitioner of medicine; for here medical treatment, which is practically powerless in the fully-developed disease, may accomplish a great deal.

THE PRODROMAL PERIOD is marked by so insidious a development of the symptoms that it is difficult to say anything positive as to its duration. The writer has never seen a typical case in which these symptoms did not cover a period of at least a year. In the majority of patients observed in private and consultation practice, and where the relatives had been observant of the approach of the disease, it was determined to have lasted between two and four years. In some it was even of longer duration, and in one case the first *outbreak* of the illness occurred in 1877, while a preliminary change in character, occasional amnesia, purposeless and unprovoked fits of fury, and hesitation in speech could be traced back to 1856. Morel speaks of patients in whom this incubatory stage may be said to have extended throughout a lifetime. Undoubtedly there are persons whose career is marked by a behavior very similar to that characterizing the paretic dement. Such individuals are full of extravagant projects, they are considered "hail fellows well met," being generous to a fault with strangers, though tyrannical and breaking out in causeless fits of anger at home. The most prominent feature of their characters is a silly boastfulness, manifesting itself in boyish claims of superior qualifications for almost every and any position in life. In the case of an intelligent merchant in good social standing, whose generally excellent mental training was manifest even in the deteriorating period of his disease, this tendency showed itself in an ambition to acquire physical prowess and to become known as a pugilist. He frequented taverns and other low places of amusement, became an intimate friend of the prize-fighters Heenan and Sayers and, as he paid his way very liberally, was allowed to gain easy victories in the various encounters which he boastfully provoked. Such a "paretic disposition" must not be confounded with another condition, namely, with primary expansive monomania complicated by paretic dementia, to which detailed reference will be made later on. (See chapter on Diagnosis.)

In that group of paretic dements whose disorder may be designated as being of the "spinal" or ascending type, the symptoms of the prodromal period are such as precede organic disease of the spinal cord, or suggest the existence of an insidious affection of the entire axial portion of the central nervous system. Pains in the lower extremities, usually described as being of a rheumatoid character, though sometimes of the dolorous kind found in locomotor ataxia, double sciatica, early color-blindness, belt-like sensations in various parts of the body, particularly the head, photopsia, tinnitus aurium, and temporary double vision are the chief of these symptoms. In the "cerebral" or "descending form," while these symptoms may coexist in a less prominent degree, a change of character is the most notable sign. The careful business man becomes negligent, and the good father or husband indifferent to his family. Frequently the patient himself notices this, and becoming morbidly emotional, he may weep, or show genuine melancholic depression, because he feels his brain power failing and cannot call up his natural affections as of yore. There is a great similarity between this condition and primary mental deterioration. (Chapter VII.)

All the mental symptoms of this period are attributable to simple brain failure. The attention is not as readily aroused, and the patient engages in conversation, and after a prolonged harangue of the one speaking to him, interrupts him with the exclamation, "What did you say? I was not listening." This inattention is not the inattention of an abstracted normal mind which is able to recall the subject of its abstracted reverie; for the paretic, while not hearing what his friend said, cannot tell what he was thinking of that occupied his attention. That faculty has become entirely dormant for the time being, and the paretic's abstraction is only a lesser degree of another symptom commonly observed at this period, namely, a tendency to fall asleep in the middle of the day, in the counting-house, and particularly after meals, or at lectures and entertainments.

Amnesia is noted from the beginning. At first the failure of the memory relates to trifles; the patient does not recollect whether he has wound his watch, and may wind it half a dozen times one day and not at all on the next. He may forget to button his clothes, to pay for his meals at restaurants, or to take his purse with him on leaving the house. More serious omissions and errors are made as the prodromal period progresses : the business man makes wrong entries, or omits to record important items; and the cases of two physicians are related by medical jurists, one having in this period of paretic dementia prescribed 16 gr. of tartar-emetic instead of $\frac{1}{16}$ gr., while the other, a Russian doctor, was sent to Siberia for having caused the death of a colleague by a similar error. In his "empty abstraction" the patient may take the wrong train, or a car going in the opposite direction to the one he should take; and is particularly apt to neglect appointments. These acts may be committed by persons normally abstracted, and some of them are habitually committed by those who have failed to cultivate systematic habits; their diagnostic importance in the case of a paretic dement, therefore, lies solely in the fact that they constitute *a persistent change* from the patient's previous and normal condition.

In this stage thefts are very apt to occur, in some cases with a *quasi*-criminal intent; in others, merely from forgetfulness. Thus a patient will pick up an article to look at it, and then pocket it in his abstraction. Sometimes forgeries are committed with considerable skill by previously upright business men, owing to their loss of moral tone; at other times useless as well as valuable articles are stolen in a stupid and random way. Brierre de Boismont relates the case of an old government officer who for eight years prior to his reception in an asylum had been guilty of repeated abstractions of articles at public sales which he attended officially. His insanity was not suspected until several months before his interdiction. On the occasion of the last theft he was arrested, and Brierre de Boismont examined him. On entering the room this physician immediately saw what kind of a patient he had to deal with; he had the embarrassed pronunciation, "petrified" face, heavy walk, in short, the characteristic signs of paretic dementia. On being interrogated as to the circumstances of his arrest he answered, without the slightest appearance of remorse or shame, "the people who put me in prison are imbeciles, who know nothing of our professional usages; it is the custom among us, a custom known as the '*cote G*,' to choose some object of slight value and retain it when taking the inventory, and see, here are two which I thus appropriated." With this he drew from his pockets a handsome meerschaum pipe and a gold-mounted tobacco pouch. The distinguished physician mentioned pronounced him to be suffering from paretic dementia, and a few months later the patient died of this affection, verifying the opinion.* Simon relates a case, presenting a similar tendency to the manu-

---

* "Études Médico-légales Sur la Perversion des Facultés Morales et Affectives dans la Période Prodromique de la Paralysie Générale." Paris, 1860.

facturing of stupid excuses,* at a later period of the disease. A fisherman who, it was subsequently ascertained, had presented signs of paretic dementia for half a year, was detected emptying the nets of others, and appropriating their contents. He was first beaten by the owners, and then taken before court. Here he declared that his oars had become entangled in the nets, and that he had taken the fish out in order to rearrange the nets, intending to replace the former. This explanation was rejected as a "cunning evasion," and the physician called in by court pronounced him of sound mind, notwithstanding the fact that the prisoner had been suspected to be—and the suspicion was confirmed by witnesses—insane by the police authorities for several months previous. Incidentally to his other declaration, the prisoner announced the characteristic project of running a net across the Elbe River, to be dragged by two steamers, thus intending to catch all the fish at one swoop.

Indecent exposures of the person may also be made, in some cases from satyrical motives; in others they are due to the forgetfulness of the patient, who neglects to button his trousers, or fails to bear in mind that he is exposed to the public gaze. It must also be borne in mind that the free determination of the will is gravely impaired in paretic dements. Chorinski, one of the Austrian nobility, was prevailed upon by the Baroness Ebergenyi, his paramour, to poison his wife. A few years later he died a paretic dement. Several instances are on record in which such patients have been induced to marry courtesans or other speculating women, in some instances thereby committing bigamy. The undue influence in such cases has the way prepared by the patient's forgetfulness of the fact of a previous marriage, or his moral deterioration.

In this period there is also developed a morbid irritability. The previously sedate and calm head of the family will fly into a furious passion on hearing of a trifling loss, a slight expense, the breaking of crockery at table, or on finding his meat overdone. The simplest contradiction will cause a fierce denunciation, or even a violent assault. One of the first observed manifestations in one patient under the writer's care was the throwing of a large bottle filled with ink at his brother and business partner, on the latter's asking

---

* "Die Gehirnerweichung der Irren" (dementia paralytica). Hamburg, 1871.

him the meaning of a certain entry in the ledger. Impatient as the patient is of contradiction, he is impatient in regard to other little matters. One paretic dement, who was turned out of a theatre because he was unable to show his ticket (having in his amnesia forgotten where he put it, or thoughtlessly thrown it away), broke a large pane of glass to climb in by another entry. Another, because the atmosphere of a carriage seemed too close for him, and finding some difficulty in opening the window, took his cane and broke out the glass.*

With all this irritability in regard to the little affairs of life, the patient is singularly apathetic with reference to more important matters. A patient who threw a knife at the servant, because she took his plate away before he had, as he alleged, finished dining, heard unmoved a few hours later of the collapse of a large business undertaking, which involved a loss to him of over a hundred thousand dollars. It is remarkable how frequently a patient, who has perhaps brutally abused his wife and children for calling in a physician to prescribe for him, on finding that a carriage drive, undertaken at the suggestion of some friend, terminates at the asylum in which he is to be confined, hears the news without manifesting the slightest feeling or making any protest whatever. In one case, the day before the commitment, a patient of the writer's had had a physical encounter with an expressman, for leaving one of his trunks on the street instead of immediately carrying it in; on finding himself within the walls of Bloomingdale he walked up to the scales to be weighed with an air of bravado, and said to his companions that they should also avail themselves of the chance of being weighed *gratis*.

Irritability which breaks out on slight provocation is a sign of a weakening of inhibitory power and of a general loss of nerve tone. The reverse of the condition, described by Wundt as the normal one, is hence found in paretic dementia. A healthy person displays a sanguine temperament in regard to the lesser affairs of life, the melancholy temperament in the serious phases of his career, is cholerical in connection with events which most deeply affect his interests, and should be immovably phlegmatic in carrying

---

* Both of these patients were brought before juries on a *habeas corpus*, and these acts were successfully paraded before the laity as rational ones.

out his intentions after these are deliberately formed.*
The paretic dement, on the contrary, is cholerical with regard to the petty affairs of life, phlegmatic at important turning-points in his career, and sanguine with regard to, as well as easily diverted from, the carrying out of his purposes.

It is in harmony with the readiness with which paretic dements may be controlled by their surroundings that, under the influence of "jolly companions," they become spendthrifts, while they may be penurious misers at home. And it is an evidence of the frailty of their purposes, and the readiness with which they may be diverted from them, that although many paretic dements develop suicidal intentions in their depressive moods, they very rarely carry them out.

Simultaneously with the memory, will, and emotional balance, the morals begin to totter. Often moral alienation is the most prominent of the earlier symptoms of the disorder. Just as the paretic becomes irregular in his habits generally, unpunctual in business hours, and forgets his appointments, he loses sight of the proprieties and of his moral obligations to his family and to society. Just as he becomes careless in the spelling of words, he begins to use improper ones in conversation, employs lewd language before females, and oaths as expletives in ordinary conversation. Sexual and alcoholic excesses are now indulged in. The previously prudent and temperate business man orders cases of wine sent to his office, in order to have the means for indulgence close at hand. The once faithful husband begins amours with the serving-maids before his wife and children, or goes to theatres, to balls, and shows himself in public with notorious courtesans. The accompanying excesses, like the similar ones indulged in by the maniac, precipitate the development of the disease, particularly as an intolerance to alcohol is one of its early and marked features. A well-meant remonstrance leads to an outbreak of furious violence, the intervention of the police or other authorities to conflicts with the latter, and, the patient's disorder being consequently recognized as a mental trouble, he is perhaps sent to an asylum.

This EXPLOSION OF THE ACTIVE PHASE of paretic dementia is commonly marked by exacerbations of the physical

---

* "Psychologische Physiologie."

signs. The slight defect in the movements of the tongue and lips, hitherto noted only at times, now becomes more permanent. The patient finds it difficult to pronounce particularly the explosive and hissing sounds, and the longer a word containing such sounds is, the more manifest does this difficulty become. In addition the voice changes, becoming hoarser, and, as Marcé claimed, those patients whose disease is due to alcoholic excesses exhibit a more tremulous intonation than is ordinary in paretic dements. The speech defect is aggravated by the increasing amnesia of the patient, and he often employs the wrong consonants— " b" for " p," or "t" for "d," and "m" for "n." Later on whole syllables are suppressed, and it is difficult to decide how much of the speech disturbance is really ataxic and how much is amnesic in origin. A most characteristic feature is the association of other and normally unnecessary movements with those of the lips and tongue. The patient, when about to speak, moves the lips as if to fix them more firmly; there is a tremor at the angles of the mouth, an exaggerated and spasmodic movement of the zygomatici, alternate dilatation and contraction of the nostrils, the usually habitual corrugation of the brow increases, and, after all these preparations, the word is thrown out precipitately, as if it had had to force its way through some impediment.

With the speech innervations all the finer motor co-ordinations seem to suffer.* The patient's walk becomes less steady and regular. His legs are thrown wider apart to increase the basis of support, and such motions as dancing and skating, et cetera, if among the previous accomplishments of the patient, can no longer be executed. The musician forgets his notes and loses the mechanical skill necessary in wielding the bow of the violin, in executing rapid tremolos on the piano, and can no longer regulate the inflation of the cheeks necessary in playing on the brass instruments. The stenographer becomes unable to follow the speaker whose words he is to report. Watch-makers, engravers, or other mechanical artisans, who depend on the use of their hands, find their occupation more laborious, their attention tiring easily, and their fingers failing them, so that their work is spoiled or clumsily performed. It is

---

* One patient, a ventriloquist, lost his art in the early period of his illness.

at this period that the handwriting of the patient may first present the characteristic features to be referred to.

The hypochondriacal ideas, depressive moods of the patient, and complaints about head symptoms, if they existed, disappear about this time in the majority of cases. A subjective sense of power and general well-being takes their place. The so-called delusions of grandeur then manifest themselves, and are often coupled with morbid projects and extravagant expenditures.* Both the delusions and the resulting projects are unsystematized, and in this respect widely different from those of a monomaniac; they resemble the corresponding symptoms of acute maniacal delirium in many features. But it is usually easy for the skilled observer to detect the lacunæ in the intelligence behind the veil of delirium in paretic dementia; whereas in the maniac such lacunæ do not exist, except temporarily in maniacal frenzy. The latter condition is, therefore, not always distinguishable from the similar phases of paretic dementia.

---

* The distinguished French alienist Brierre de Boismont was called in consultation about the nervous condition of a wealthy man, whose mental disposition had been recognized by the family physician, although the patient dissimulated his infirmity pretty well, under that show of reasoning specious argumentation, and habitual decorum which is so apt to mislead the laity. The alienist speedily unearthed his prodigious vanity and egotism, and recognizing that the patient was suffering from the prodromal stage of paretic dementia, called the attention of the family to the fact, and advised them to be on their guard as to the disposal which he made of his fortune. A year passed without his hearing of the case, when one day the gentleman in question was brought to the asylum, after a scene of violence which nearly cost one of his family her life, and after he had squandered about two hundred thousand francs in absurd speculations.

Hammond (General Paralysis of the Insane, with Special Reference to the Case of Abraham Gosling: an address delivered before the Medico-Legal Society, April, 1880) describes the following characteristic case: "Another undertook the task of buying nearly all the jewelry at Tiffany's, and only stopped when the proprietors, becoming alarmed, refused to sell him any more. This man took the jewelry he purchased home, and bedecking his wife until she glittered with gems from head to foot, compelled her to walk up and down before him. Then he drew a check for $5,000, and gave it to his servant who returned with a glass of water which he had called for. When I was sent for, the patient told me he was going to Europe. He intended to make the voyage over in the Great Eastern, and would charter the Scotia as a tender. He would pay me $1,000,000 a month, and he would have a corps of physicians on the vessels, the members of which should be attired in a uniform of blue velvet with diamond buttons."

After one of these explosions the patient may become comparatively calm and rational, his physical signs retrograde, with slight exceptions, and the only remaining mental defect may be a feebleness of judgment and a difficulty experienced in sustaining a prolonged mental effort. The relatives of the patient, with whom the wish is father to the thought, and the inexperienced medical adviser may regard him as entirely recovered. But, after a more or less prolonged lucid or rather para-lucid interval, the patient breaks out in another fit of excitement or depression, and this may recur at irregular intervals; so that the history of many paretic dements is a series of asylum sojourns, separated by intervals, in which they have been able to attend to their business, or have travelled under the advice of their friends or attendants. But with each attack the patient is left in a more crippled condition bodily and mentally, the resisting power of the brain is gradually weakened, and the patient sinks lower and lower on the down grade to absolute dementia. The loss of the finer motor co-ordinations is succeeded by the abolition of the coarser ones; gross speech defects, or absolute aphasia, ataxia of movement and inability to write mark the decline; and when the latter is far advanced, the slight paresis of the earlier period becomes so much intensified that the patient may be unable to leave his bed. The trophic disturbances, which were but faintly indicated at first, as a herpes zoster, for example, also become prominent, and frequently terminate the patient's life; malignant bed-sores, furuncles, hæmatoma of the lower bowel, diarrhœa, gastric hæmorrhage, or pulmonary gangrene may then supervene. Finally, if the patient escape or survive these dangers, while the night of utter mental darkness is settling on him, so that he is unable perhaps to utter even the infantile delusions of grandeur entertained in the earlier period, he succumbs to apoplectiform or epileptiform seizures of a kind peculiar to paretic dementia, and which may sometimes mark the course of this disorder from the beginning.

Throughout the latter phases of paretic dementia, and aside from the maniacal exacerbations, the patient's demeanor is marked by good-humored self-satisfaction in the majority of cases. He is consequently generous with his imaginary riches. At the second interview the writer had with a paretic dement, whose disease was a complication of a pre-existing monomania, the patient offered him three (actual) patent-rights as presents. An almost characteristic feature of these patients in their quiet intervals is their enthusiastic and demonstrative greeting of strangers, to whom they will almost invariably state that they are in excellent spirits, and in the best possible condition of bodily health. "Fat and saucy" responded a paretic dement, as he half-stumbled and half-swaggered into the lecture-room of the college, when asked how he was getting along; "all right" and "first-rate" are the usual responses in the paretic wards of an asylum. These patients are enthusiastic admirers of anything novel, or which they are unable to understand. When Obersteiner tested a number of lunatics with his "psychodonometer,"* while the melancholiacs developed ideas of persecution based on the formidable appearance of the instrument, and the terminal dements remained indifferent, the paretic dements were unable to find words extravagant enough to express their unintelligent admiration for the new device. The writer has often found the exhibition and application of the sphygmograph a most useful means for securing an examination of a refractory paretic. A brief explanation of the mechanism of some medical appliance, such as the ophthalmoscope, will elicit from such patients the declaration that their doctor is the greatest man in the world—next to themselves of course—he having looked "right into their brains." Quackery which treats these patients as sufferers from "brain-softening" consequently finds an occasional votary here, and the writer has heard no more enthusiastic praises of "static electricity" and similar therapeutical impositions on the credulity of the profession and laity, than from a paretic dement who had been treated by these means.

But not all paretic dements are habitually good-humored, self-satisfied, and "hail fellow well met" at this stage of their illness. Some remain ill-natured and distrustful, when they were so before their illness, and many are more demonstrative with the closed fist than the open hand.

Any one of the symptoms hurriedly related in the foregoing may be prominent at one or other stage of the disease, absent in a few cases, and appear earlier or later in the histories of different patients. They therefore merit

---

* An instrument for measuring the rapidity of the mental reactions.

detailed consideration before we proceed to consider the varieties of the disease.

The unsystematized DELUSIONS OF PARETIC DEMENTIA have from the time this disease was first recognized been assumed to be always of the expansive kind, and its constant and unvarying features. But this view is erroneous. It is true that such delusions are present at some period of the illness in most patients, but they are of a depressive kind in a few, while in some very exceptional cases no delusions whatever are observed.* When present they are almost pathognomonic. The patient claims to be the most powerful, the richest and ablest man in his community. He can raise the asylum with his little finger, he has trunks filled with gold in every city in the Union, he is married to all the handsome women in the world, can speak all the living and dead languages, has the best-developed sexual organs extant, and is the intimate friend of every contemporary great man, sometimes himself Napoleon, Cæsar, Shakespeare, Grant, Buffalo Bill, and every other celebrity in one person, and the fortunate owner of numerous patents. The following is a partial list of the "possessions" of a paretic dement, who had at one time been a stock-broker in Chicago:

Six trunks of gold in Chicago at $30,000 each	$ 180,000
Patent watch per year	50,000
Patent knife per year	75,000
Four trunks of gold at Governor's Island at $16,000	64,000
Stock in Chicago	1,200,000
Patent billiard cue per year	15,000
Real estate in Chicago	184,000
Real estate in Washington	90,000
Interest in Chicago	8,000
Interest in Washington	19,000

This patient made at the time few or no errors in his arithmetic and spelling, and was perfectly competent to compute interest; his alleging a larger amount of interest in Washington where he had less property, and his assigning different values to his items in different papers show how little reflection and system enter into such delusions

---

* One patient in the writer's experience had advanced far in dementia, paresis, and ataxia, without manifesting a single delusion up to the time when it was deemed advisable to commit him to an asylum. This is the only case observed by the writer which corresponds to the descriptions of paretic dementia without delusion given by recent English writers.

as elements. The lack of real originality in the delusions and projects of paretic dements is illustrated by the fact that in this case the "patent knife" had "four blades, one to saw with," the "patent watch" "could go two days without being wound up," and the "patent billiard cue" had a "rubber tip." A common day-laborer who attended the writer's clinic, alleging that there was some kind of an animal in his stomach, claimed that the female patients had all remarked the peculiar expression of his eyes, and that he was generally fancied on their account. With an air of greater secrecy he added that his virile member was two feet long and nine inches in diameter, and that he had forty-four houses in New York. In a remission which followed, the size of the organ in question gradually "diminished;" he admitted that he did not own the houses, but had a lease on them, and later still he claimed no extra allowance, either of real estate or anatomical property, beyond the ordinary male citizen. Extravagant ideas relating to sexual matters are exceedingly common in male paretics. In females they are less common, and when present usually devoid of the lewd tinge characterizing the sexual ideas of the male; females may claim, for example, that they are pregnant, and delivered every week of a beautiful child, or a child with gold teeth, or some other valuable addenda.

These delusions are as manifold as the number of paretic dements is great, but they have the common characters of extravagance and lack of system. In the later periods they are also exceedingly unstable, and vary greatly from day to day, so that a patient who had ten thousand dollars yesterday claims to have a hundredfold that amount to-day, and to-morrow may find no figures adequate to express his wealth. The general of to-day is the president to-morrow, and "God above all other gods" the day thereafter. One patient lives in a marble palace in the morning, which becomes transformed into a golden one by noon, and if a conversation with him is kept up long enough, his residence will be transformed into diamonds before he gets through. The patient wishes to be whatever he believes to be great and powerful, and his wish is speedily gratified by the enfeebled brain.*

---

* It is an interesting fact, illustrating that the aspirations of the paretic determine his delusions, that the wealthier paretic dements in the writer's private practice have not displayed as extravagant monetary delusions as

Simon aptly says that the position in life of the patient should be borne in mind in estimating the signification of these delusions. Thus the claim of possessing a thousand dollars by a pauper, or of an income of a hundred dollars a week by a common day-laborer, are as grave delusions of grandeur as the belief of having millions, or a daily income of thousands would be on the part of a well-to-do patient. Most patients having such delusions are given to delusive boasting of their past achievements. A Wall Street broker claimed that he had beaten Jay Gould, Vanderbilt, and Russell Sage at every point, time and again. Another boasted of his adventures with wild animals, which he had torn limb from limb in single encounters. The anecdotes of paretics relating to their physical strength are generally embellished by the most brutal and offensive details. Thus one whom the writer took to an asylum, because his relatives did not venture to assume the responsibility of accompanying the powerful and excited patient, related how at a previous asylum sojourn he had seized an attendant, hurled him through the air down eight flights of stairs, and at the bottom of each landing had jumped on his victim, till at the last one the viscera of the latter "squirted" out of the mangled body, and covered the walls and vestibule. The story was of course entirely fictitious; one of the first persons at the asylum whom the patient greeted with customary paretic hilarity was the alleged victim, and while the account of the "massacre" started with locating the incident at the top floor of the asylum, where the patient had once been, and which was only three stories high, the scene shifted, even while the victim's body was on its way down eight flights of stairs, to the vestibule of his residence, which was four stories high; finally he confounded the past with the present, and made a vigorous pass at the writer, for which he as vigorously apologized. The boasts of a military paretic dement, narrated by Mickle in his instructive treatise were yet more extravagant and as cruel. "He said that he was commander-in-chief; that the queen was his mother; that he went with her in a yacht to Russia to see his sister, who was married to the czar; that he with forty comrades killed 10,000 Russians at the Malakoff tower, and

---

the pauper patients at the city asylum. It is in insanity as in health: what the subject has he cares little for; to get what he has not, seems most necessary to his happiness.

on the same day stripped the corpses, dug a hole, buried them, and sold their clothing for £20."

In one case, that of a patient whose disorder dated from an injury by a shell fragment which struck his head at Bull Run, and who boasted of the "boys of Company K," and the gold and guns they had buried at Fort Hamilton, the writer drew out the admission that the patient knew he was boasting; and it seems that sometimes the delusions are merely vague assertions, which have become settled beliefs, owing to the enfeeblement of the logical power, and consequent inability to recognize the absurdity of the boast.

The delusions of grandeur are usually associated with EXTRAVAGANT PROJECTS. One patient whom the writer saw at Meynert's clinic, a Hungarian hairdresser, had invested his entire fortune in buying up all the hair in the Austrian empire having a certain rare shade of gray. He imagined that this was the color of the empress's hair (which at the time was raven black); that by having the monopoly of the hair required for her artificial curls, etc., he would necessarily become hairdresser to the court; and that his way to future preferment was thus opened. A keeper of a small-beer saloon proposed to build an enormous concert-hall in New York, in an out-of-the-way part of the city, and to engage a celebrated prima-donna at a salary of $250 a night to sing there; at the same time he advertised the "Jumbo glass of beer," and scattered five-dollar gold pieces among the street boys. A Cuban patient, who was discharged as "recovered" from the pauper asylum of the same city, during a partial remission of his illness bought several gross of red and blue pencils to distribute among his fellow patients. Frequently these patients propose to marry all the fine looking women in a given city, country, or the entire world; one intended to marry a specimen of each race, and another all the women with eyes of a hazel color. Occasionally the patients claim that they are acting under commands from God. They may then order the instant execution of all persons having red hair on account of their antipathy to such, or in more favorable moods and without a cent to their names donate several million dollars to charitable institutions. A demented paretic physician* proposed giving a lecture on the

---

* He resumed his practice and continued it for five years, though when seen in the street by the writer showing the characteristic gait of his disease. Since then track of him has been lost.

"Diamond Cross, for the benefit of the Little Sisters of the Poor." For this purpose he was going to hire Steinway Hall and the Academy of Music (distant from each other about two hundred feet), having sold, as he said, ten thousand tickets.

The incongruity of the paretic's schemes is very well illustrated in one of Mickle's cases: the patient ordered "$25\frac{1}{2}$ pounds of tobacco, half a dozen of eau de cologne, four concertinas, a paper shirt and a paper cravat, $60\frac{1}{2}$ dozen pocket-handkerchiefs, a field-marshal's uniform and baton, 1009 boxes of hams, 26,000 pounds of currants, a stage and a carpenter." To pay for this he gave an order for £150,000, or "more if necessary." One of the writer's patients, who had been prevented from accomplishing a suicidal purpose by his wife, drew diagrams of a tombstone to be erected to his memory, whose inscription recited all his achievements, and sang the praises of his wife for saving the life of so valuable a citizen.

Patients having such delusions and entertaining such projects are usually as vain and obtrusive in their demeanor as in their speech and writings. Sometimes the delusions are absent, but the inflated ideas of self-importance are just as prominent: a subaltern government officer believes the land will go to ruin unless he remains at his post, or a cashier threatens to resign unless his dignity and importance are properly recognized by the directors of a bank. It is characteristic of these patients that there is nothing small about them. Their wives are the handsomest wives, their children the "smartest" children, their friends are all great men, and they have no disposition to bother about such "trifles" as the ordinary daily occupation to which they owe their bread.

As already stated, the delusions of paretic dements are not always expansive. Even at the height of the typical disease a sudden change in their character may occur. The patient who was the emperor of the whole world yesterday is the poorest beggar to-morrow; the God of yesterday is thrown into the deepest pit of hell to-day; he who was a giant of more than mountain height suddenly shrinks to an invisible dwarf; and another who had the best brains, a stomach that could accommodate tons of the rarest delicacies, and boasted of having a most powerful animal frame the night previous, wakes up one morning to the discovery that his brains are running out at his meatus urinarius; that his stomach is gone—the seat of cancer—or gnawed by some wild animal; that his bowels are impacted, and that he is as physically weak as he is mentally annihilated. This condition was by the French termed "micromania," so called in contradistinction from the ordinary state of delusive grandeur or "megalomania."*

It is noteworthy that the delusions of belittlement in paretic dementia are as absurd, extravagant, unstable, and unsystematized as those of aggrandizement.

While depressive delusions are among the rarer episodes of the fully-developed disease, they are common enough at its earlier stages; indeed, they characterize the "hypochondriacal" and depressive forms of the malady, and to some extent are developed in the earlier phases of the disease in most patients. As a consequence of his inability to collect his thoughts and regulate his business affairs such a person alleges that he has ruined or beggared his family, when this is not true; another claims that he is all burned up, that his abdominal cavity is being scraped out inside by some mysterious agency, or is the seat of the exploits of "something alive." Hallucinations of smell of a disgusting character, and "magnetic" or gustatory illusions modify and determine these ideas.

Just as suddenly as the extravagant delusion of grandeur may undergo a transition into a depressive one, so the reverse may occur here. The patient who has been bemoaning the ruin of his family all along, some morning may scatter money among the boys on the street; he who was a "worthless wretch, physically and mentally," yesterday, and suspicious of everybody else, fearing that he would be arrested and imprisoned, to-day addresses all persons he meets as his best friends, invites them to gorgeous banquets, or offers them shares in his extensive mining and railroad undertakings. A patient who could neither eat nor digest, and who had not a penny in the world, according to his statements made during the hypochondriacal period, awoke one morning with the project to get up a monopoly of the entire sardine and Bermuda onion trade in the world; and having, as he alleged, secured it, proposed to eat all the sardines and onions himself.

HALLUCINATIONS and ILLUSIONS are much more common in paretic dementia than is ordinarily supposed. Many of

---

* Also applied to expansive monomania.

the delusions are, as has been already hinted, based on faulty sense perceptions, and we consequently find that the hallucinations of an unpleasant character are found in the "micromaniacal" and melancholy phases of the disorder, and those of an exhilarating nature with the ambitious deliria. They are overlooked in paretic dementia as in simple mania, owing to the greater prominence of other symptoms. Mickle,* the first who has systematized the study of these symptoms, believes that at some period of the disease hallucinations occur in about one-half the cases. He finds that visual and auditory hallucinations are most, and olfactory hallucinations least common. In the writer's experience auditory hallucinations and illusions were most frequently found, next those of the tactile and visceral sensations, then those of sight, and lastly those of smell, the latter in rapidly deteriorating cases. As Mickle says, they "are often variable, unstable, inconsistent, being usually less fixed and systematized than the hallucinations of many of the insane of other groups." It would be still better to say that they are more like the hallucinations of mania, alcoholism, and melancholia than those of monomania. The patient hears his name whispered or the sounds of approaching footsteps, and sees people with ugly countenances making faces at him. In one case in the writer's experience "countless frogs," whose intestines had bulged out from the vent and been "stuffed into their mouths," "hopped around" the patient. Very frequent are the visions of heaps of putrefying corpses, which are often associated with corresponding hallucinations of smell. Sometimes voices are heard commanding the patient to do a certain deed; in an impure case, one of traumatic paretic dementia complicating an undetermined pre-existing mental disorder, the patient heard voices commanding him to kill some one, in order that he might himself be compelled to commit suicide.

According to Mickle, hallucinations and illusions of the sense of touch are manifested as "fæcal lumps adhering to the skin, or dirty fluids thrown upon it," and the illusions of the sense of touch generally are apt to be of a disagreeable nature. In one case of the writer's, that of an aged paretic dement who had advanced spinal lesions, vermin were complained of, which he alleged the superintendent

---

* *Journal of Mental Science*, October, 1881, and April, 1882.

had bred to annoy him. On admission several scabs were found in his vest pocket, which he carried with him as an antidote to similar inflictions, which he said were imposed on him before his admission. More frequent are the sense disturbances of a pleasurable kind. The wall-paper of the patient's room is changed to gold, his furniture to diamonds, worthless rags and scraps of paper are hundred-dollar bills, and pebbles and fragments of glass are diamonds. Often the patient sits in a corner, in rapt ecstacy over the kaleidoscopic visions of heavenly and military pageants doing him honor.

In the writer's experience unpleasant hallucinations involving a multiplicity and sameness of objects are indicative of a rapid progress of the disease. In one of the most acute cases observed, that of a young man aged twenty-two, lights were seen everywhere: torch-light processions across the island on which the asylum was situated; boats on the river covered with torch-bearing soldiers; and whole regiments carrying lampions, and intending to march against the patient, were awaiting transportation across on the opposite shore.

DISTURBANCES OF THE SPECIAL SENSES AND VISCERAL INNERVATIONS are found aside from those just mentioned. Many paretics become amblyopic,* in consequence of central processes or affections of the optic nerve. Others suffer from permanent or temporary hemianopsia, due to cortical disease or dropsical distension of the third ventricle. Loss of smell (anosmia) is occasionally observed early in the disease, and becomes usually marked as the morbid condition progresses. Voisin is decidedly in error in claiming that it is a constant symptom in the early stages, and not a single writer agrees with him on this head. An insatiable craving for food (bulimia), which is noted particularly in patients with considerable deterioration, is attributable to a disturbance of the vagus nerve; while anæsthesia, paræsthesia, hyperæsthesia, analgesia, or hyperalgesia, are noted in those cases in which the spinal disease is prominent, and they sometimes serve as the basis of illusional delusions. Thus one of Mickle's patients believed that his skin was tucked in,

---

* In a will-contest, the Perrin case, tried in a Western State, sclerotic degeneration of an entire occipital lobe was reported by the pathologist witness, and it was singularly enough omitted to bring this fact into connection with the amblyopia noted during the decedent's life.

another that it was hung up to dry, and a patient of the writer's was continually picking off "gold leaf" from his bodily surface.

The EPISODICAL ATTACKS of paretic dementia are among its most important signs. They are of three kinds, which, from their resemblance to maniacal delirium, epileptic fits, and apoplectic seizures, are called respectively the maniacal, epileptiform, and apoplectiform attacks of paretic dementia.

The nature of the MANIACAL ATTACKS of paretic dementia varies with the period at which they appear; for, like the other episodial outbreaks, they may mark the disease at an earlier or a later period, in rare cases recur from the beginning to the end, and in still rarer cases be absent altogether. It is evident that in the earlier periods before the mind has undergone deep decay, the deliria must be more creative, the flight of ideas more extensive, and a chain of reasoning occasionally visible in the patient's words; while in later periods the fancies of the sufferer will be hampered by the dementia, their expression checked by aphasia, and reasoning impossible, because its essential foundation, the memory, is grossly impaired, and the association of ideas interrupted. Meschede reports a case of a paretic who, brought to the asylum early in the disease, suffered from a maniacal attack of three hours' duration, in which the flight of ideas and rapidity of speech were actually delirious. He did not interrupt the torrent of sentences which issued from him but once or twice, to moisten his parched lips with a little water, and all this time announced his scheme to measure the orbits of the planets, thought he was determining the distance of the dog-star, undertook to square the circle, and finally gave a feast to the whole world of truly Arabian Night's profusion. There are a few cases on record where this maniacal condition continued for a long period, or even marked the entire course of the disease. Such cases terminate rapidly through maniacal exhaustion or other complications, and have been designated *galloping paretic dementia*. In others there is a subacute maniacal condition, analogous to hypomania (p. 136); the patients then are not actually delirious, but display a restless activity, often leading them to the performance of boyish and silly acts, such as dressing themselves up as women. A noted pantomime actor examined by Hammond first manifested his disease by hurling loaves of bread, turnips, cabbages, and other objects employed in the pantomime among the audience, and later by uniforming a number of children in "Humpty Dumpty" costume, intending to teach them his art, and thus to perpetuate it. Just as this analogue of the milder attacks of mania is found in this protean disease, so the severe attacks of maniacal furor are also and more faithfully copied in it. The furor of the paretic dement is one of the most fearful of all the occurrences of the asylum ward. Day and night these patients rave, tearing and breaking everything within reach, besmear themselves with their excrement, or even devour it, and shout at the top of their voices. They yell alternately that they are being murdered, that they wish to get out, announce the most extravagant delusions, claiming that they have millions on millions of palaces, all the wealth of the world, can lift the solar system with a finger, or threaten their attendants with the vilest and most cruel punishments. The brutality of these patients is something remarkable. It is the possibility of the maniacal attacks occurring, and the great likelihood of their leading to violent and fatal assaults, that should be borne in mind by those who let loose such patients on society in the remissions of the disease.

The attacks of *paretic furor* last a few hours, days, and occasionally weeks, and may cause the death of the patient by exhaustion; particularly when they recur in rapid succession. It is remarkable to what extent the dementia and certain physical signs of this disease may be masked by these attacks. The furious paretic dement has a more extensive vocabulary, more expansive ideation, less ataxia and aphasia than he had during the previous period; and one who previously was bed-ridden, or tottered about the wards with the characteristic paretic stagger, now steps more firmly and destroys heavy doors and furniture. One night the writer was suddenly called to the residence of a demented paretic patient, and in the absence of conveyances and assistance was compelled to stay up with him until the morning. A heavy blow was the first greeting, but a little art elicited a characteristically profuse apology from the patient. An hour before, he had broken the panels and driven a heavy door from its hinges with the intention of murdering his wife, in whose behalf he subsequently employed the writer's medical services, and, to satisfy a simulated hysterical desire of hers, tasted the medicine she was to receive, thus taking what was really intended for him. He had attempted to set fire to his house three times that night. Subsequently his delirium, which toward morning became modified and diminished by the conium and hyoscyamus given, assumed a less destructive and more expansive character. The patient went over his school attainments and almost every boyhood reminiscence and event of his life, in an incredibly rapid speech which lasted eight hours, and was occasionally marked by quite poetic flights.

In some cases the maniacal explosions are followed by stupor or aphasia, and complicated by the attacks to be next considered.

The EPILEPTIFORM SEIZURES may, as already indicated, take place at any period of the disease; though usually observed only near its termination. In the following case they were the first symptoms noted, and had—what is very rarely the case—the true epileptic character: A porter in a down-town warehouse had been promoted to a higher position, greater responsibilities and labors of a mental character were thrown on him; in the midst of apparent health, having been slightly "worried," he was seized with a convulsion lasting several hours, with partial consciousness, and later on these convulsions occurred in status-like succession, at intervals of a week, for some months. Eighteen months after, the convulsions having been absent for a year, he died with the "quiet type" of paretic dementia. In another now under observation a remission of over eight months followed a series of such attacks.

In ordinary cases these attacks occur after motor paresis is already indicated, and begin as imperfect fits of the clonic kind, affecting the muscles of the face, or of both the face and arm, on one, or more rarely on both sides. The spasms are not usually as violent nor as excursive as those of epilepsy, and in many instances, particularly in those onsets which last for whole days, resemble a convulsive tremor rather than an epileptic fit; consciousness is impaired or not notably affected; at times, however, an initial spasm of a tonic kind is observed, and then there may be well-marked convulsive action of all the muscles of one half of the body and conjugated deviation of the eyes and head with abolition of consciousness. The appearance of a patient lying for many days in a continued convulsion involving all the muscles of one half of the body with conjugated deviation is one of the most surprising ones experienced by the novice in an asylum. And still more surprising is the frequent recovery of the patient from so formidably appearing an attack.

APOPLECTIFORM SEIZURES may, like the epileptiform ones, inaugurate the disease in exceptional cases. Ordinarily their appearance is heralded by "congestive spells." The patient having for some weeks observed that his head feels heavy and dull, or as if a tight band encompassed it, after an unusually liberal meal or a slight indulgence in alcoholic beverages, experiences a sudden rush of blood to the head; his face becomes crimson or purple, the temporals throb violently, and for a moment there is an inability to speak or to collect the thoughts.* These attacks may occur in the midst of conversation, and while the continuity of ideas is interrupted by them, the thread of thought is resumed when the normal or approximately normal condition of the circulation is re-established. They are but momentary; the more severe ones resemble the apoplexies † due to extensive cerebral hæmorrhages, and are of longer duration. Here the patient may suddenly fall down as if struck by lightning, and the entire half of the body may then be found limp and removed from the influence of the will, or but imperfectly controlled by it. These apoplectiform attacks may be complicated by convulsions, by tetantic spasms, *mouvements en manége*, and, as Kiernan was the first to notice (1876) in a case shown the writer, by athetoid motions. It is not the least remarkable feature of the strange disease we are considering that these attacks, like the epileptiform seizures, are often and rapidly recovered from, as far as the life of the patient and gross motility are concerned.

The effect of the epileptiform and apoplectiform episodes on the patient's general condition is disastrous. Occurring as they often do during the remissions, just at the moment when the patient has been apparently improving, and leaving him more or less enfeebled or aphasic and paralyzed, they destroy what little hope as to a delay in the progress of the disease may have existed.

Although no invariable rule can be framed it may be as-

---

* A condition which is a pathological imitation of the action of nitrite of amyl.

† True apoplectic attacks dependent on hæmorrhage do occur in the terminal stages. Such were determined at the *post mortem* of three of the writer's cases.

sumed that those paretic patients who experienced numerous syncopal or vertiginous attacks in the prodromal period of their disease, as many do, will be more likely to suffer from epileptiform, and those who had chiefly congestive spells and "word-stoppages" will have apoplectiform seizures toward the end of their lives. Both classes of attacks may however be, and frequently are, associated in the same patient.

Among the continuous motor disturbances of paretic dementia those of the PUPIL merit special consideration. Its most characteristic condition in this disease is inequality, due to paresis of the circular fibres of the iris of one eye, or to a greater degree of paresis of the iris of the side where the pupil is relatively dilated. Although Lasègue could only find such a difference in one third and Simon in one half of his cases, the writer is inclined to believe that, on comparing—not a large number of patients simultaneously—but their records extending over the entire history of the disease, this inequality will be found to have been present in the majority of paretic dements at some time or other. The inequality is usually not constant; one week the pupil of the right, the next that of the left eye may be the narrowert, and in exceptional cases bilateral dilatation may alternate with bilateral pin-hole contraction. In some cases extreme pin-hole contraction is noted from the beginning; these run a rapid course,* but not, according to the writer's observations, because pachymeningitis is apt to be present under these circumstances, as Simon claims.

In one patient, whose commital to an asylum was made the subject of litigation, maximal and symmetrical dilatation, with normal contraction under the influence of light and efforts at accommodation, were found. His symptoms were of the typical kind; while Simon, who observed a similar condition, found his patient to have symptoms resembling those of apathetic melancholia.

An extreme dilatation of both pupils following pin-hole contraction is an indication of rapid decline, and is accompanied by œdema or other trophic disturbance.

It would be hardly necessary to refer here to the opinion of Austin—that the left pupil is more frequently dilated in

---

* Particularly if the contour of the pupil is not round but irregular—a condition which must be distinguished from the residual irregularity following syphilitic iritis, a not uncommon condition in paretic dements.

paretic dements having exuberant ideas, and the right in those with depressive ideas—if it were not for the fortunately isolated attempt which a recent writer on the disease made to resuscitate this exploded—not to say *a priori* absurd and extravagant view—by certain statistics which happen to answer themselves.*

The pupil is found to have the characteristic features known as the "Argyll Robertson pupil" in those cases where the spinal symptoms are well marked and ataxia or abolished tendon reflex and other evidences of posterior sclerosis are early signs. This symptom is not as frequent in paretic dementia as has been claimed by some recent writers.

It is not necessary to refer here in detail to the signs accompanying the organic affections of the cord sometimes found in paretic dements, or to the signs of focal hemispheral lesion which are its frequent accompaniments. The

---

* Austin ("On General Paralysis," London, 1859) says: "When the right pupil has been the more affected the *general* tone of the delusions has been more melancholic, and with a more implicated left pupil, their *usual* complexion has been elated, and their coloring gorgeous." (Italics Austin's.) Pelman and Nasse took the trouble to demonstrate the untenability of this view, but no one succeeds better than the writer referred to in the text, who, in the belief that he is supporting Austin, says: " From an examination of eighty cases in the asylum in which there was a perceptible tendency in one or the other direction, it would appear as if there was something in the theory, for in the melancholic cases the left pupil was the more dilated in thirty and the less in only eight; while of the maniacal the right was the larger in thirty-three, and the left in but nine." ("General Paresis," by A. E. Macdonald, M.D., *Am. Jour .of Insanity*, April, 1877.) A comparison of the respective statistics furnishes the most sinister disproval which any theory has ever experienced, and also constitutes a significant commentary on the reliability of certain pamphlets. Austin's figures are cited from his table on p. 36.

	Number of paretics with pronounced elation or depression.	Number having elation with right pupil larger.	Number having elation with left pupil larger.	Number having elation with pupils equal.	Number having depression with right pupil larger.	Number having depression with left pupil larger.	Number having depression with equal pupils.
Austin's figures........	64	1	15	5	39	1	3
His supporter's figures.	80	33	9	..	8	30	..

attempt to do justice to these themes would necessitate the extending this chapter to the dimensions of a volume.

Among the motor disturbances those of the facial muscles, the hands, and tongue are of the greatest diagnostic importance; in fact, the expressions of the paretic's face, like his speech, are the most prominent, constant, and characteristic physical indications of his disease.

The prodromal period is not always marked by a permanent disturbance of the facial and lingual innervations. There is less of the normal play of the features, or it is exaggerated owing to a slight ataxia: usually fibrillary tremors accompany the more pronounced changes of expression, particularly when the patient is excited.* With this the explosive opening of spoken sentences referred to in the earlier part of the chapter may be occasionally observed, and it increases after the various exacerbations and paralytic episodes of the disease, being particularly marked with the consonants requiring labial apposition and lingual firmness. The patient, like an intoxicated person, finds it difficult or impossible to say "truly rural," or "Peregrine Pickle," and will instead say t-t-t-tooly roodal—t-t-t-trural roo–roo-roodial. "Emotional tremor," as it is called, is also frequently noted. The patient, when about to speak or when suddenly accosted, is seen to have a fine tremor of the lips, particularly marked about the angles of the mouth, as if he were about to break out in sobs. But there is no real emotional state; the patient may be extravagantly hilarious at the time, and the designation is a misnomer,† for this tremor is an ataxic associated movement and should be designated "paretic" or "ataxic tremor."

As the disease progresses all these symptoms become intensified; a variable degree of ptosis, or drooping of the upper eyelid, is noted, and the features generally are coarser and finally become obliterated altogether. A characteristic element of the facial expression in advanced paretic dementia is a tonic contraction of the corrugator supercilii

---

* The "Nachbewegungen" and "Mitbewegungen" of the Germans are very commonly observed.

† There is sometimes observed very early in the disease a morbid pseudo-emotional condition. The patient experiences the *expression of emotions* without a corresponding *emotional state;* reminiscences of a pleasant character, for example, are accompanied by choking sensations in the throat and a flow of tears, while those of an opposite kind may be associated with a vacant smile.

and the occipito frontalis. Probably this action is at first a sort of automatic equipoise for the paralysis of the levator palpebræ, and then becomes habitual; for it is most marked in those cases where the ptosis is extreme, and on that side of the face where the latter is most pronounced. The fine tremor of the lips becomes coarser, and fibrillary twitches of the zygomatici, the levator labii superioris, and particularly of the muscles of the tongue, which may have been only occasional occurrences in the earlier periods, now become constant. A pronounced coarse tremor is observed in the hand on ordering the patient to stretch it out while spreading the fingers; and the handwriting, which shows at first only a similar tremor, degenerates to scrawling, and the deviations from straight lines are more considerable. The patient frequently erases or blots his words, and a most constant feature is a gradual deterioration of the handwriting in lengthy documents; the patient begins a letter very fairly, but, as he goes on, the words are formed more irregularly, and finally he is unable to keep on the line, writes above or below it, but usually runs obliquely down across the page. While the opening of the letter may be, aside from the tremor, written in a good business hand, the signature may be illegible, a mere scrawl, or a blot. The omission of words, the meaningless repetition of whole sentences, the doubling of single and the reduction of double consonants are among the features of more strictly psychical origin which serve to characterize the documents of paretic dements. Of these such patients usually carry a quantity in their pockets, and many of them exhibit a stupid letter-writing tendency which in less educated ones is replaced by as empty a word diarrhœa.

The other co-ordinated movements of the hands suffer with the writing. The patient who, if a mechanic, first noticed an inability to carry out his finer manual work, now becomes unable to button his clothes, or, in extreme cases, to carry a spoon to his mouth without spilling its contents.

In the lower extremities the motor disorder manifests itself, as a rule, in a combination of paraparesis and ataxia, whose characters will vary according as the lesion of the cord or that of the brain preponderates; for symptoms having a superficial resemblance may be due to lesions in either locality, contrary to the general belief.

It is exceptional to find typical locomotor ataxia in paretic dementia; there is usually less of the throwing out of the leg and bringing down of the heel, and much less swaying on the patient shutting his eyes,* and equally less uncertainty on walking in the dark. In advanced cases the legs are dragged along the floor, often unequally so, giving the impression of a halt or limp, the chief movement of the extremity is at the hip joint, the knees being stiff and the patient consequently sways to and fro in walking. Finally —scarcely able to lift his leg from the ground, stumbling over his feet—while announcing the project to walk around the world in twenty days, or "to take the belt from Rowell," the patient is compelled to take to his bed. It is remarkable how, in testing the resisting power of the muscles when the patient is in bed, these may be found quite powerful, albeit the patient is unable to walk. This discrepancy is to be explained on different grounds from that observed in locomotor ataxia.

A very important motor disturbance in paretic dementia is that of the muscles of deglutition and phonation. How much of this is really an ataxic disorder the writer is unable to determine; though it is to be supposed, in view of the anæsthesia of the larynx and pharynx observed in several cases, that a sensory disturbance may enter as an element into the dysphagia so frequently noted. The patient is very often suffocated by a bolus of food, and more than once has the tube of the stomach-pump been passed into the larynx and trachea without any of the usual indications of this accident on the part of the patient.

It is difficult to determine the precise extent and character of the numerous sensory disorders in paretic dementia; owing to the inattention and dementia of the patients. These signs will therefore never have the diagnostic importance which the other symptoms have, and we may therefore pass by them with this reference.

Disturbances of the bladder and the renal excretion are frequently observed in paretic dements. Aside from the episodial albuminuria, reported by various observers as a sequel of the apoplectiform attacks, and such rare phenomena as hæmaturia (noted in one case within the writer's experience), cystitis, and pyo-nephritis are common occurrences toward the close of the disease, and often end the patient's life. The more marked the spinal lesion the

---

* Commonly there is a decided unsteadiness in standing with the eyes open, which is not greatly increased by closing them.

earlier will paresis of the bladder and its attendant phenomena appear. There are cases, however, in which the urinary secretion does not present any anomalies whatever, nearly to the last moment, and then they may be the distant result of over-distension of the bladder through the amnesic neglect of the patient.

The VASO-MOTOR and TROPHIC DISTURBANCES OF PARETIC DEMENTIA are among the most interesting and striking symptoms of its later stages. Early disturbances of a vaso-motor nature are the flushings of the head after meals, and the similar spells which herald the apoplectiform and epileptiform attacks already described.* Anomalies of the body temperature have also been claimed to exist at various periods, but the evidence on this point is still very contradictory, and the writer's own observations, made some years ago in conjunction with Dr. Kiernan, were not sufficiently systematized to be of value except with regard to two points: In the first place, those patients who, comparatively bright and active in the morning, deteriorated through the day, becoming listless, stupid, and having to be taken to bed in the afternoon, were found to have no rise in temperature, and in a few cases a fall of nearly a degree (Fahrenheit) toward evening. In the second place, a rise in temperature amounting to between one and five degrees,† more marked in severe than in mild cases, was noted to occur after the apoplectiform attacks, and to gradually decrease with recovery or death. On the whole, however, and particularly in the earlier periods of the disease, the revelations of the thermometer are not constant nor pathognomonic.

The PULSE in the early stages reveals very high tension in the active forms of the disease; in a large number of patients it is normal, and in the depressive forms the writer has found unusually low tension in several cases. The "plateau" at the summit, claimed by Voisin even for the early period, is found only in these cases and toward the end of the disease, when there is marked cardiac enfeeblement; it then does not differ from the flattening of the percussion apex found in other forms of dementia. The revelations of the

---

* In the prodromal period of the disease the writer has found remarkable and undoubtedly pathological variations of the *surface temperature* particularly of the hands and forehead, it being very high after meals and mental strain.
† Fatal termination.

sphygmograph, like those of the thermometer in paretic dementia are of high scientific but not of any great diagnostic value, except indirectly in this way: there is often— and in advanced cases constantly—found an irregular and coarsely wavy character of the line of descent, which is the expression of the irregular muscular tremor of this disease.

Among the more important vaso-motor disorders are the changes in the bones, gangrene of the lung, and the malignant bed-sore. The former,* like the othæmatomata sometimes found in advanced paretic dementia, have been referred to in Chapter X.

The gangrene of the lungs found in paretic dementia may be due to septic absorption from bed-sores, to the passing of food into the trachea and bronchi, and, finally, it may result from central processes, developing in numerous foci with the apoplectiform attacks, and probably in a manner analogous to the multilocular pulmonary lesions found after ordinary cerebral hemorrhage.

Decubitus is common in paretic dements who are bedridden; but there is a kind of bed-sore which is not due to pressure or to maceration by urine like the ordinary variety, and which develops particularly after the apoplectiform attacks. It begins as an erythematous spot of a purplish color, on which vesicles appear, after whose bursting the livid surface of a deep tissue infiltration becomes visible. This latter rapidly undergoes necrosis, and the destructive process may extend so deep as to involve the sacrum and reach the spinal canal.† This is one of the most furibund of the complications of paretic dementia; and the *malignant bedsore*, as it is properly called, may develop in a few days and

---

* In paretic dementia with pronounced posterior sclerosis the joint and bone changes found with that disease may be observed.

†At the thirty-eighth meeting of the "Berliner Psychiatrischer Verein," Dr. C. Reinhard reported a case of a female paretic, in whom numerous microparasites (microsporon septicum) were found in the nerve-centres, with septic cerebro spinal leptomeningitis. These lesions were due to septic invasion by way of the intervertebral openings and the cerebrospinal fluid from a decubitus.

It is not necessary to refer here in detail to certain pathological curiosities, such as Addison's disease (observed in one case), mottling of the skin, exophthalmus (in one dispensary and in one asylum patient), pemphigus, purpura, unilateral sweats, spontaneous gangrene, hæmorrhage in the stomach, rhinhæmatoma and hæmatoma of the lower bowel, which, with other trophic disturbances too numerous to mention, are occasionally found, though not characteristic of the disease.

run a fatal course in a week. Sometimes several of these sores appear simultaneously in large numbers, at the trochanters, heels, occiput, and elbows; and their rapid development, the absence of the ordinary causes, and the fact that they chiefly appear after the apoplectiform attacks, justify us in considering them to be of trophic origin.

VARIETIES OF PARETIC DEMENTIA.—Most modern authors make a number of subdivisions of this disease. Some years ago the writer* differentiated from the typical variety, in which the prodromal symptoms are mental and are followed by disturbances of the eyeball, face, tongue, and pharynx movements, and which appears to be a "descending" affection, that form in which the mental symptoms appear after serious evidences of a spinal or axial affection of the nervous system have been observed, and which may be classed as an ascending affection.† There is no necessity for making any further subdivisions. Some cases, as already mentioned, run a "galloping" course, others are evenly progressive (the so-called quiet cases), and most are marked by remissions.

The REMISSIONS of paretic dementia merit our special attention. Countless have been the errors made by those who have looked on these *hiati* in the disease as recoveries. There is no more remarkable and deceptive observation in neuropathology than the abatement of a dementia with delusions of grandeur, which permits the patient to return to his vocation, and the simultaneous disappearance of a paralysis and ataxia, which may have been complicated by episodes of an almost fatal character. Although residua of the symptoms may mark the period of remission, yet there are exceptional cases where even the expert may be

---

* Psychological Pathology of Progressive Paresis, *Journal of Nervous and Mental Diseases*, 1877.

† At the time the writer was unable to separate a series of cases, in which there was a quiet progressive dementia, with progressive paresis, ataxia, and epileptiform episodes, from typical paretic dementia, although aware that the lesion which produced this combination was a peculiarly distributed multiple sclerosis. The opinion of authorities generally is to the effect that these cases, like certain clinically similar ones of cerebral syphilis, should not be included in paretic dementia. The question, however, is still *sub judice* whether there is not every connecting link between these various affections, and it is greatly complicated by the fact that undoubted paretic dementia is found in numerous syphilitic subjects on the one hand, while there is a special form of syphilitic mental disorder on the other.

unable to detect any deviation from the standard of mental and physical health. In the vast majority of cases, however, tremor of the hands, inequality of the pupils—if it previously existed—and a slight speech defect and clumsy walk are found more or less prominent even in the remissions. An anomaly of the moral or mental character, or of both, is also quite common. The patient is given to purposeless lying, is irritable and extravagant; to the expert the continuing dementia is but imperfectly masked by the superficial signs of recovery;* while to the laity the occurrence of an assault, the expenditure of a fortune, or an apoplectiform attack, may prove tragical or costly comments on their ready credulity.

Oddities of behavior not previously noted in the patient may characterize the remission. One of the writer's patients would stop before every looking-glass, manipulating his side-whiskers whenever he thought he was not observed. Another became an active politician and controversialist, although in his sane period he had a great contempt for the political career and the general complexion of politics.

These remissions may last from weeks to years; their average duration is from two to four months. Lionet and Taliet agree in believing that they are more perfect in the congestive variety. In one case in the writer's observation, which has since been rapidly running a downward course, the remission lasted three years, during which time the patient attended to extensive commercial undertakings with fair success, and took charge of several assignments. Such remissions may be regarded as constituting a transition to a genuine recovery, and are particularly frequent with "alcoholic" paretics.

Although the PROGNOSIS of paretic dementia is almost unqualifiedly bad, and most of the cases reported as recovered have subsequently relapsed, yet there are a few well-authenticated instances where the history of the patient has been traced for five and six years after his last asylum discharge, and he has not given the slightest indication of a relapse in that time. The writer met such a patient, in whose case there had been a rheumatic etiology, and who had had a typical

---

* Morel aptly says, that when the patient, however well he may carry on certain routine duties, retains the stolid expression, the stony stare, and the corrugated forehead of paretic dementia, he is not cured, but that his disease is progressing under the "mask of a remission."

outbreak, five years after his discharge, and was unable to find any indication of paretic dementia in him. A remarkable and rather comical instance occurred in Austria: a paretic dement *escaped* from the asylum, and five years later paid a visit to the authorities to demonstrate his recovery. Gauster* and Flemming† have also reported several undoubted cases of recovery; and, in a case of Schüle's,‡ *restitutio ad integrum* even occurred after an apoplectiform attack.

An observation is cited by Simon from Ferrus of a patient recovering and remaining free from the disease for twenty-five years, while Baillarger, Bayle, Calmeil, and Sutherland report others where the patients' histories were followed up for from six to ten years after their discharge, and no relapse occurred. Baillarger himself questioned whether these were genuine cases of paretic dementia; and more recent observers believe that they were of the syphilitic, alcoholic, or rheumatic varieties in which cases the prognosis is relatively better.

The duration of the disease as a whole is variable. It has been already stated that the prodromal period may last only a few months, usually a few years, and in rare cases nearly a lifetime. Dating from the explosion of the malady the lethal termination may occur in six months, more commonly in three years; and, in not a small number of instances, in six or ten or even more years.§

Paretic dementia usually develops in patients between the thirtieth and fortieth years; it has been exceptionally observed in very aged individuals, and occurred in patients over sixty, in five out of three hundred and forty-six cases observed by the writer. The youngest paretic dement observed in this series was aged eighteen. Turnbull reports a case of this disease in a boy of twelve; ‖ it is, however, rare before the twenty-fifth year, and in young subjects generally runs a more rapid course than in older ones.

---

* *Psychiatrisches Centralblatt*, Oct. 12, 1876.   †"Irrenfreund," 1876.
‡ *Allgemeine Zeitschrift fuer Psychiatrie*, xxxi.
§ There was in 1882 a "show patient," frequently brought out before visitors and reporters at the Ward's Island Asylum, if a report in a daily paper is to be credited, who had been ascertained by the writer to have for three years prior to 1877 manifested the characteristic signs of paretic dementia, and who had an epileptiform attack in that year.

‖ *Journal of Mental Science*, 1882. The father of this patient died of paretic dementia *after* his son. A Continental alienist, the reference to whose paper the writer has lost, reports the case of an imbecile infant whose mental deficiency was complicated by a paretic trouble.

Paretic dementia is much less frequent among females than among males; the writer has seen but one female paretic among fifty-eight instances of this disorder in private and dispensary practice.* In various European countries the proportion of female to male paretic dements is found to fluctuate considerably. Schuele gives the highest figure, finding four paretic females to ten paretic males. Most observers give the proportion as being between 1 : 5 and 1 : 8. Neumann, an experienced Prussian alienist, did not see a single case of paretic dementia in females, and hence was led to deny its existence in that sex. The observations of no single alienist can, however, be taken as gauges of the true relation of the sexes to this disease. The unusual experience of the writer, who found but three female paretics in ninety-two cases,† chiefly observed in private practice, is probably due to the fact, that females among the wealthier classes are not exposed to the emotional strain and worry to which females in the lower walks of life are exposed. With this it is in accord that paretic dementia is more frequent among females who enter into competition with the male sex, and among prostitutes who like males are given to alcoholic excesses and exposed to syphilis. According to an old report of the Prussian statistical office, there were in the year 1878 20 female and 106 male paretic dements in private asylums in France, while at the same time there were 454 female and 826 male paretic dements in the public institutions of the same country. The proportion of females to males in the wealthier classes was, therefore, as 18 : 100; in the poorer classes, as 54 : 100, in the latter instance exceeding even the figures of Schuele.

Paretic dementia in females runs a more even, less explosive, and slower course than in males. The maniacal attacks in the former are not as expansive, the delusions of grandeur not as pronounced, and neither the episodical exacerbations nor the remissions are as abrupt as in the male patients. It is the persistent physical signs and the progressing dementia that chiefly serve to characterize the disease in females, as these are the same signs which in the "quiet" male cases suffice to demonstrate its existence,

---

* The 284 remaining cases are excluded, because they were observed in an asylum for males, and the corresponding statistics of the asylum for females were unreliable.

† On the occasion of a visit to the Bloomingdale Asylum (1880), one female and nineteen males were found paretic.

even when delusions, hallucinations, morbid projects, and other perversions are absent. It is dementia, motor paralysis, and incoordination of the character described that are the necessary clinical expressions of the progressive brain wasting resulting from the morbid processes to be considered in the following chapter.

CHAPTER XIII.

The Morbid Anatomy and Nature of Paretic Dementia.

The organs of those dying with paretic dementia examined by pathologists are usually obtained from subjects who have reached the last stages of that affection. The brain and spinal cord, in such cases, show the results of a long-continued and often intense degenerative process, which not a few authorities have regarded as of an inflammatory character; and, in the sense in which the term "inflammation" is applied to the chronic interstitial changes dependent on an altered blood-supply in other organs, such as the liver and kidneys, the analogous changes in the paretic dement's brain may properly be considered to be the results of a similar inflammatory process of a low grade.

The brain itself is found to be wasted; the wasting, however, is not generally even as in simple dementia, being usually more marked in some and less marked in other districts. Thus the convolutions near the base of the brain may be full, and show the normal contours, while those of the paracentral lobule, of the infra-parietal lobule, or of the entire convexity of the frontal lobe, may be atrophied, and separated by widely gaping sulci. Usually the basilar parts show no gross wasting; in two out of fifteen˙subjects examined by the writer there was marked reduction in the depth (dorso-ventral diameter) of the pons, which on a closer examination was found to be due to the wasting of the transverse fasciculi of that segment; in a third case there was a general diminution in all dimensions of the medulla oblongata as well as of other portions of the isthmus. This exceptional observation is of but little value, however, as the patient was over seventy

THE MORBID ANATOMY OF PARETIC DEMENTIA. 219

years of age. At this period of life wasting of the axial parts of the nervous system is not uncommon.

The characteristic feature of the structural cortical changes in paretic dementia is (in harmony with the gross appearances) the fact that they are rarely general, but that they affect certain cortical provinces more than others, and leave some of them nearly or entirely intact. Usually the cortex is discolored, sometimes being preternaturally pale, at others presenting a marked rosy tint, due to an injection of the minute blood-vessels. In one case a rusty color was noted in several spots of the deep cortical layers, extending deeply into the white substance, in large and occasionally confluent patches. The consistency varies in two ways: in some cases the cortex is less firm than normal—without, however, reaching the degree of a necrotic softening, as some have claimed,—in others it is firmer than normal, and in a few the induration approaches the degree of sclerosis. Occasionally these different conditions are associated in the same case, different parts of the brain being differently affected. The writer has found that commonly the white substance immediately subjacent to the cortex is firmer than the cortex itself, having frequently a faint bluish or grayish tinge. As the pia is more intimately adherent to the cerebral surface than in health the result is, that in some examinations, on removing the membranes, the entire cortex follows the latter, separating at the point where the softest gray tissues adjoin the firmest white layer, and thus leaving the white substance behind in a shape repeating all the anfractuosities of the surface.*

A very frequent appearance, in advanced paretic dementia, is *cystic degeneration* of the cortex. The gray and sometimes both the gray and white substances of some one or other area are found to be the seat of numerous cavities, varying in size from a pin's point to a millet seed." When these are very closely crowded the so-called *gruyère* cheese appearance, described by Lockhart Clarke, results. (Fig. 2†.) In several cases the writer has found these

---

* Baillarger (Note sur une Altération du Cerveau Caractérisée par la Séparation de la Substance Grise et de la Substance Blanche des Circonvolutions. *Annales Médico-Psychologiques*, January, 1882) first accurately described this lesion.

† Two convolutions from the mesal face of the right cerebral hemisphere of an aged paretic dement. The cavities in this case opened on

cavities to be branched; in one case the remnant of an obliterated blood-vessel protruded into it; and from these and a number of other observations there can be no doubt that the larger cavities at least are of perivascular origin, and the result of a retardation of the lymph out-flow, with a consequent dilatation of the spaces of His and Robin. It is possible that the smaller cavities are the result of an analogous expansion of the periganglionic (pericellular) spaces. The view that all these gaps are analogous in their origin to retention cysts, is supported by the fact that sclerosis, thickening. infiltration, and adhesion of the pia, all factors which are apt to prove obstructive to the lymph out-flow, if not to the venous return circulation, are

FIG. 2. FIG. 3.†

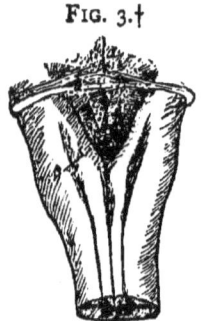

found most marked over those areas exhibiting cystic degeneration. Probably the dilatations of the lymph space in the posterior fissure of the spinal cord (Fig. 7), found in one case by the writer, are susceptible of a similar interpretation.

The ventricles may be enlarged in advanced cases; often they exhibit no change in dimensions, and a more characteristic pathological feature of the disease is the granular change of their endyma or lining membrane.* This con-

---

the surface, which is rare; h, the cortical surface; $c\ w$, cross section; c being the cortical, and w the medullary portion of the section.

* Wilder suggests this term as preferable to ependyma, which latter term may be restricted to the barren layer of the cortex to which Rokitansky applied it.

† Dorsal view of medulla oblongata: o, striæ acustici; c, coarse granulations near the apex and over the alæ cinereæ (nuclei of the glossopharyngeal and pneumogastric nerves); a, finer granulations approaching those producing the ground.glass appearance.

sists in a connective tissue growth of the ventricular lining, which takes place in numerous and closely-crowded areas, raising the latter in little hillocks. As long as these remain minute they manifest their presence to the naked eye by the dulness of the normally smooth and polished lining of the ventricle; in other words, by a *ground-glass appearance*. When they increase they assume the shape of warty excrescences, and these are found particularly well marked in the posterior half of the fourth ventricle (Fig. 3), at the foramen of Monro and over the striæ corneæ. Quite odd forms are sometimes seen among these bodies. Of the larger ones, which usually have a constricted pedicle (Fig. 5), two occasionally join leaving a tunnel between them. In one case the writer found the aqueduct of Sylvius divided into two channels by a series of them.

The ganglionic bodies of the cortex generally exhibit marked degeneration. But, side by side with areas in which it is difficult to find a single healthy ganglionic element of fair dimensions, there may be found regions in which no change, or but very slight changes, can be found; and in one recent case the writer was unable to find any pathological condition in these elements, although sections from every district of the cerebral surface were carefully examined. Mendel found that the changes in the ganglionic bodies are not always evenly developed with the ordinary signs of that interstitial encephalitis, which is ordinarily supposed to characterize the affection. In three cases in which he failed to find any indications of this process he found the peri-ganglionic spaces filled with yellowish flocculi, the nuclei of the ganglionic bodies being indistinct or invisible, with other indications of necrobiosis.* The writer has found the following varieties of degeneration in the cortical elements: 1st. An even shrinking of the pyramidal bodies, without protoplasmic deterioration; the protoplasm is merely condensed, and the bodies stain more deeply than normal ones, while their processes are fragile. 2d. A diffuse yellowish discoloration of the entire ganglionic body, extending into its processes, with a disappearance of the finely granular structure of the protoplasm; the body appears hyaline, does not take carmine staining, and its processes cannot be traced any considerable distance,

---

* Report of the meeting of the Berlin Medical Society of February 14th, 1883. *Deutsche Medizinal-Zeitung*, iv. 8.

the nucleus may stain faintly, or not at all, but is usually visible. 3d. A coarse pigmentary change; a part of or the entire cell is filled with coarse granules of a brownish or yellowish green, and rarely of a decidedly black color; the pyramidal cells usually maintain their contour and exhibit the origin of their processes, but the prolongations of the latter are destroyed, and the nucleus is rarely visible in extreme degrees of this change. 4th. A "granular wasting" at the periphery of the ganglionic body, leaving a part of the latter apparently intact; usually a large number of free bodies are found in the periganglionic spaces in this condition, and it seems that their presence is in some way

FIG. 4.*

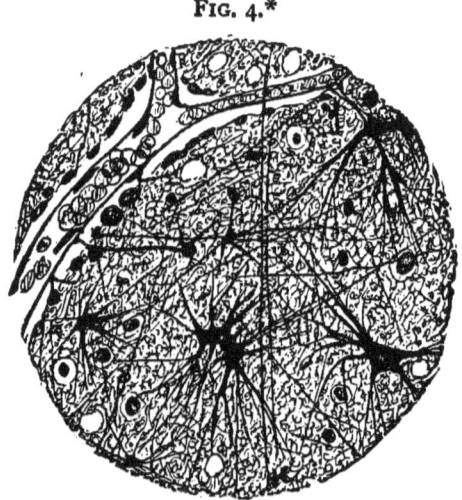

associated with this change. 5th. A progressive deterioration in the protoplasmic composition; the ganglionic body is found to have its normal outline, but does not stain at all, or very imperfectly, the nucleus is shrunken, and the nerve processes appear to have broken off sharply. This is probably a sclerotic condition. All these changes are better marked in the larger elements, with the exception of the

---

* Section from the white substance, adjoining the gray matter of the lower frontal convolution, and showing four large and several small spider-shaped cells. At *y* one of these cells has contracted a union with a small capillary, and one of its processes is becoming transformed into a capillary process; at *n* two coarse axis cylinders are seen.

second variety, which is found mostly in pyramidal cells of the second and third layers.

Beside the ganglionic bodies, which show various degrees of the enumerated changes, patches of loose pigment or irregular masses of no decided histological characters are found, marking the former sites of destroyed ones.

Little of the positive is known about the fibres of the white substance. A striking appearance in advanced cases is the distinctness and coarseness of the axis cylinder, which often appears as if dusted over with a fine powder. The course of the fibres in the white substance is much more clearly demonstrable in the brains of paretic dements than in those of healthy persons, owing to the condition alluded to, which is probably only a preliminary step to the disintegration of the fibre.

Disease of the neuroglia is almost constant in paretic dementia. It may be of every degree and present every connecting link between a general indurating and rarefying change of slight intensity, and the process known as disseminated sclerosis. A common feature is the presence of cells, staining deeply in carmine and provided with numerous brush-like processes: the *spider-shaped cells* of Meynert and Lubimoff. (Fig. 4.) These bodies are often found in large numbers, and the writer has never failed to discover them in advanced cases. In the shape and size in which they are found in paretic dementia they can be confounded with no normal structure, as some writers have suspected to be the case. The basis substance of the neuroglia exhibits the degenerative changes enumerated on page 105, and in addition considerable nuclear prolifieration.

The most intense, certainly the most constant, changes of the neuroglia are found in the pons and medulla oblongata. Even in cases where the spinal cord is not involved, and where the cortex exhibits only a diffuse change of the kind just described, changes in color, consistency, and texture are found in these parts. Sections taken from them generally stain diffusely in carmine ; a coarse molecular material scattered between the fibres, along the septa, and the raphe is found to absorb the staining fluid in a higher degree than the ganglionic elements,* and in advanced cases this material becomes the seat of a true sclerotic fibrous transformation. (Fig. 5.)

---

* In properly hardened preparations these should always stain earliest and most intensely.

Special attention has been given by observers to the changes of the blood-vessels. Nearly everything said with regard to the changes of the vascular channels on page 106 applies to their condition in paretic dementia. In the earliest stages nuclear proliferation of the walls is observed; this

FIG. 5.*

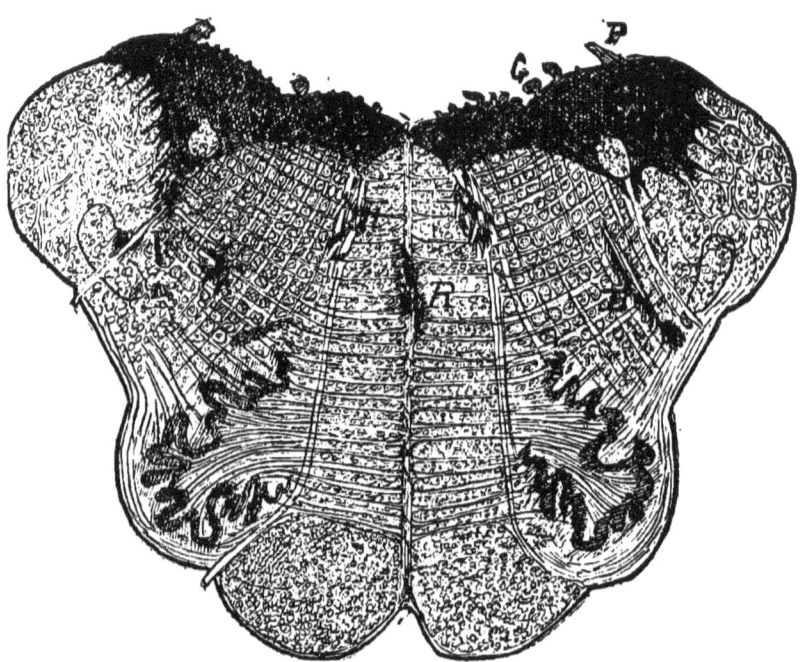

is particularly noticeable in the muscular coat of the arterioles and in the adventitia of the smaller vessels. In the former case the proliferated cells can be readily distinguished from the normal nuclei of the muscular tunic by the roundish or irregular shape and irregular disposition of

---

* Fig. 5. Transverse section of the oblongata, at the level of the tenth pair. *G*, endymal granulations of the gray and white floor (the alæ albæ have lost their white color, through the connective tissue hypertrophy); *R*, sclerotic patch in the raphe; *H*, same around the roots of the twelfth pair; *V*, same at the ascending root of the fifth pair where crossed by the tenth; *B*, same around vascular gap; *P*, "ponticulus." The left pyramid is darker than the right.

the former. In addition, the spots where there is the greatest amount of nuclear proliferation (usually at the bifurcations of, or sudden bends in, the vessels) are also marked by the presence of granular hæmatoidin and other products of the retrogressive metamorphosis of blood pigment. Formed elements of the blood, both red and white corpuscles, are always found in the adventitial space and beyond it, usually in very large numbers; a very characteristic picture of the microscopic sections of the cortex in paretic dementia is the presence of a series of "nuclear" bodies along the borders of the perivascular space, many of which appear to be undergoing a transformation into spider-shaped cells (Fig. 4). With this there is apt to be found an accumulation of similar bodies in the periganglionic spaces.

An amorphous yellowish substance is sometimes noted in the adventitial space, and appears to be taken up by the neuroglia nuclei in some instances.* As a rule the latter lie in clear roundish spaces, which may contain a little coarsely-granular material.

In later stages of the disease the nuclear proliferation increases, and the "free bodies" undergo a transformation into branched cells whose processes are connected with the adventitia on the one hand, and with the neuroglia surrounding the vessel on the other. When a blood-vessel in this condition is isolated from the cerebral substance it presents a villous appearance, due to the fine processes of the branched cells attached to it. This change is most noticeable in blood-vessels of moderate dimensions, and not well marked in the capillaries. The latter are sometimes observed to establish a communication with a process of some large spider-shaped cell, which subsequently becomes hollow, thus leading to the formation of a new vascular channel.†

Still later the infiltration of the adventitia and muscularis with new elements becomes less marked than the passive phenomena of degeneration. The muscular tunic becomes

---

*This appearance has been observed not only in specimens hardened in chromic acid or its salts, but also in alcoholic and fresh preparations. It is found in other conditions.

† If any analogous process occurs in the healthy state it does not occur in the same prominent manner, and certainly not as extensively as in paretic dementia. The new formation in question was first described by Lubimoff and confirmed by the writer in 1877.

granular, its nuclei decrease in number and distinctness, and the adventitia either exhibits sclerotic meshes or wasting. The resisting power of the vessels is evidently decreased and fusiform dilatations are common. Sometimes extravasation of blood into the adventitial space occurs, but, on the whole, this accident is rarer than is ordinarily claimed. Very frequently the vessels are kinked and contorted, doubled on themselves and almost thrown into coils by the excessive strain on their weakened walls (page 107);

FIG. 6.*

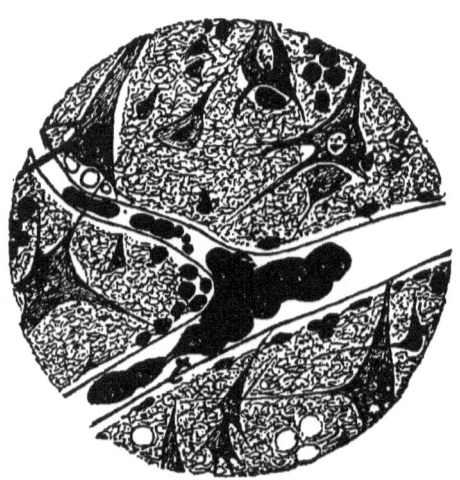

and while Sankey, who first called special attention to this condition, undoubtedly employed methods which might have led to similar and therefore false appearances in the healthy brain, and erroneously believed the adventitial sheath to be a morbid product, yet his observations as to vascular kinking are fully sustained by hardened preparations. The writer has found the lumen of one vessel and ectasies of that vessel almost approaching in degree aneurismal dilatations divided five times by the knife in a single section.

A most important field for study in this branch of morbid

---

* Thrombic cylinder undergoing separation and checked at a bifurcation of a cortical capillary.

anatomy is the condition of injection of the blood-vessels. Usually these are more or less injected, but the most characteristic condition found is one of thrombic stasis (Fig. 6). Where a patient had died in consequence of a maniacal outbreak, an apoplectiform or epileptiform attack, or shortly after any of these episodes of paretic dementia (thirteen out of fifteen of the cases examined), the writer always found a high degree of engorgement in the cerebral capillaries, which in places reached the degree of a stasis more intense than any observed by the general pathologist. The bloodcorpuscles in this state are so closely crowded that their outlines are no longer distinguishable, and they appear fused into a hyaline and opalescent cylinder which stains deeply in carmine and hæmatoxylin.* When resolution takes place—a condition which is frequently observable—this cylinder breaks up into spherical and oval fragments, which are carried onward in the vascular current, becoming further subdivided at each bifurcation, and finally are represented by a number of fine granules having each the same optical appearance as the larger masses.† Thrombic stasis appears to be most persistent in the white substance subjacent to the cortex and the deepest layers of the latter.

Much discussion has grown out of the claim that there is a new formation of spaces around the cerebral blood-vessels. It will be recollected that His claimed the existence of an extravascular space separating the blood-vessel from the surrounding parenchyma. In health such a space certainly does not exist, in paretic dementia it is undoubtedly found; we must therefore look upon it as a morbid product: the result of the distension of the true adventitial lymph space, which, subsequently retracting or wasting, leaves spaces behind. These are the beginnings of the cortical and medullary cysts previously referred to.

Before leaving the brain proper and passing to the consideration of its appendages, it may be stated, that the most

---

*This appearance has been correctly interpreted by Meynert and Lubimoff. Earlier observers have described a similar appearance as exudations, fatty emboli, etc.

† Should it be shown that any of these thrombi on resolution pass the cortical vessels—which, owing to the tenuity of the latter (they being the narrowest in the body), is not likely—it might be possible to trace some of the multilocular pulmonary lesions observed after the apoplectiform attacks to emboli of the pulmonary vessels. At present it is safer to attribute these lesions to trophic influences, although in some cases t... y have been shown to be due to the inhalation of foreign bodies.

intense changes are—in a crude way—symmetrical. While the left cerebral hemisphere is usually most involved, yet the difference between the two hemispheres in respect to the anatomical changes is not striking.

There is scarcely a ganglion or fibre tract that may not be affected in paretic dementia, just as there is scarcely a symptom studied by neurologists which may not have been observed during the life of the sufferer from this affection. The skull may be thickened or the seat of exostoses, as in other forms of insanity. In rare instances it may be thinned and softened in consequence of a trophic change. But a more characteristic condition is an intense injection of the cranial diploe, the vessels of whose Haversian canals are filled to repletion in those patients who reach the autopsy room early in their disease. With the nutrition of the skull that of the *dura* is also affected. As a rule the periosteal or outer layer of this membrane is much more adherent to the cranium than in health. In one case within the author's experience the adhesion of the dura at the convexity was so firm that a novice pathologist neglecting to divide the dura with the bone (as should be done in all cases where an unusual degree of adhesion is found) pulled the entire and intact brain out of the cranial cavity, by tugging at the calvarium; the sac of the dura came away almost entirely, tearing off at various points at the base. Such and related changes in the dura are indications of an over-nutrition, as is particularly well shown in the production of genuine bony plates. Such plates are usually found in the great falx, and sometimes in the tentorium.* They are more frequently found in paretic dementia than in other forms of insanity, with the exception of the traumatic varieties. A typical specimen of the kind in the author's possession is about an inch long, half an inch high, and a third of an inch thick; it shows a median slit, into which the falx enters, so that it really consists of two halves, each having developed on one side of the falx, and the two communicating through a hiatus in the membrane. These bodies have the appearance and structure of true bone. They cannot be confounded with the calcareous plates sometimes noted in the arachnoid. These are of a creamy white color

---

* They must not be confounded with the small spiculæ of bone which are normally found, particularly in negroes, in the neighborhood of its basilar insertion.

and translucent, in no connection with the dura, with an irregular thin margin, and gaps, and are probably the result of the calcareous transformation of lymph exudations. The writer has never found them to have the true bony structure of the osseous plates found on the dura, although some authors claim that bone corpuscles occur in them.

A most important feature of a number of cases of paretic dementia is so-called hæmorrhagic pachymeningitis. It was found in three out of fifteen cases of paretic dementia examined *post mortem* by the writer, and in one case combined with a corresponding condition of the spinal dura. Baillarger observed this lesion in one eighth, and Mendel in nearly a third of his cases. In two of the three cases recorded by the writer an extensive meningeal hæmorrhage accompanied this lesion, and in one of them the patient, whose spinal dura showed intense pachymeningitis, this hæmorrhage extended from the olfactory lobes down to the lumbar enlargement of the cord.

In the earliest stages of hæmorrhagic pachymeningitis the inner layer of the dura is said to exhibit a rosy tinge, due to a vascular hyperæmia. This, according to Virchow and Kremiansky, is followed by an exudation of formed elements of the blood which, in their union on the inner face of the membrane, constitute a delicate neomembrane. It is a question whether the exudation is a genuine diapedesis or a hæmorrhage, but the writer's observations support the probability of the latter occurrence. He believes that, in a considerable number of cases, a hæmorrhage, not from the dura but from the leptomeninges, is the primary lesion; that the greater portion of the exuded blood is resorbed, but that the portion nearest the dura becomes organized and attached to it in the manner described by Huguenin. The leucocytes become transformed into spider-shaped connective tissue cells, whose processes uniting form the ground net-work for a new connective tissue in which blood-vessels with very fragile walls are developed. These grow from the dura and extend into the new formation, and their development indicates a secondary irritation of the dura proper. A renewed hyperæmia may lead to the rupture of these feeble vessels, and a second hæmorrhage then occurs between the neomembrane and the dura, or between the different layers of the neomembrane. The same histological metamorphosis then takes place in the new clot, and hæmorrhage after hæmorrhage may recur until the enormous blood cysts known as *hæmatomata* * of the dura mater are formed.

A strong support for the view of Huguenin, that the hæmorrhage is the primary factor, is derived from the following considerations: It is well known that hæmatomata and other signs of pachymeninigitis interna are not only found with paretic dementia and traumatism, but also with alcoholism, apathetic and senile dementia, epilepsy, and phthisis. In several of these disorders positive signs of lesion of the dura proper are absent; indeed, in that stage of the neomembranous formation, when it represents a sort of rust-colored lining of the dura, the inner epithelium of the latter membrane may be found intact; while there is at first, as Huguenin has shown, no tissue connection between the two. Then again, large hæmatomata are found as sequelæ of brain atrophy, and in that event we may be surprised to find enormous blood cysts, whose existence no symptom observed during the life of the patient could have induced us to suspect. It would be remarkable—if the hæmatoma were always the result of an inflammatory process of the dura—that pain, ordinarily so prominent a symptom when that membrane is affected, should be entirely absent in some cases. Mendel thinks it inconsistent that if, as Huguenin claims, the neomembrane were a metamorphosis from a hæmorrhage, no traces of blood pigment should be discoverable in some cases. Against this objection the writer has a remarkable observation to advance. In a case of katatonia, on the verge of terminal deterioration and complicated by a pulmonary affection, which proved fatal, a gelatinous material was found in the meshes of the arachnoid which had a very pale rusty tinge, but was quite transparent, and at some places attached loosely to the dura as a thin film. Microscopically this material was found to be almost entirely composed of red blood corpuscles—at least, bodies resembling in every way red blood corpuscles deprived of their color, after exposure to the action of aqueous solutions. A very few black pigment granules, some streaks of fibrin, and a large number of white corpuscles were the only other elements discoverable. It required no extravagant speculation to imagine that the material in the meshes of the arachnoid was destined to undergo a liquefaction and ab-

---

* Durhæmatomata.

sorption, while that near the dura was preparing to contract a permanent union with that membrane, in which event it would, had the patient lived longer and the membrane had time to become organized, have presented itself as one of those rusty-colored pseudo-membranous patches of the dura, in which the pathologist finds but few if any indications of pigment. Another objection of Mendel almost answers itself. That writer finds it difficult to understand how, if Huguenin's theory is true, a fresh hæmorrhage could be included in a hæmatoma, and entirely separated from the dura by newly-formed tissue. On referring to his excellent plate illustrating the lesion in question, the inner dural epithelium is found to be nearly intact, and the transition between the " fresh hæmorrhage" and the peripheral parts of the neomembrane as delineated is so gradual that it is a matter of doubt whether the two were not coeval in origin. At any rate, the statement of Huguenin, that newly-formed and fragile blood-vessels penetrate from the dura into the neoplasm, covers the possible occurrence of an early hæmorrhage.

The changes of the arachnoid and pia may be considered together. A chronic form of leptomeningitis (inflammation involving both these membranes), with connective tissue new formation, and milky opacity, rather than purulent infiltration as a result, is one of the more common gross findings in paretic dementia. The laminæ of the arachnoid bridging the convolutions are with this unusually firm, and the pia is found abnormally adherent to the cortex, sometimes over the entire surface, but usually only in insulated places, particularly at the apices of the convolutions. The opposite condition, an abnormal looseness of the pia, is sometimes found, and authors have described cases where this membrane was raised in blebs from the surface by fluid accumulations. This separation is probably due to the formation of a pathological sub-pial space, bearing the same relation to the true lymph meshes of the brain envelopes that the pathological perivascular space of His bears to the true adventitial space.

Among the organs outside of the cranial cavity which are found diseased in paretic dementia, the spinal cord deserves the first place, for its morbid processes are often anatomically continuous with those of the encephalon. Thus arachnoid hæmorrhage, changes in the dura, and inflammation of the surface are found involving the envelopes of the brain and cord simultaneously; while sclerotic changes are sometimes traceable from one to the other. Again, there are cases where clinical observation demonstrates that the coarser anatomical changes must have begun in the cord, and involved the brain secondarily.

While sclerosis of the posterior columns of the cord, in the distribution which is typical of locomotor ataxia, is sometimes found in paretic dements, a less fascicular and symmetrical form of sclerosis is more common. Sometimes

FIG. 7.*

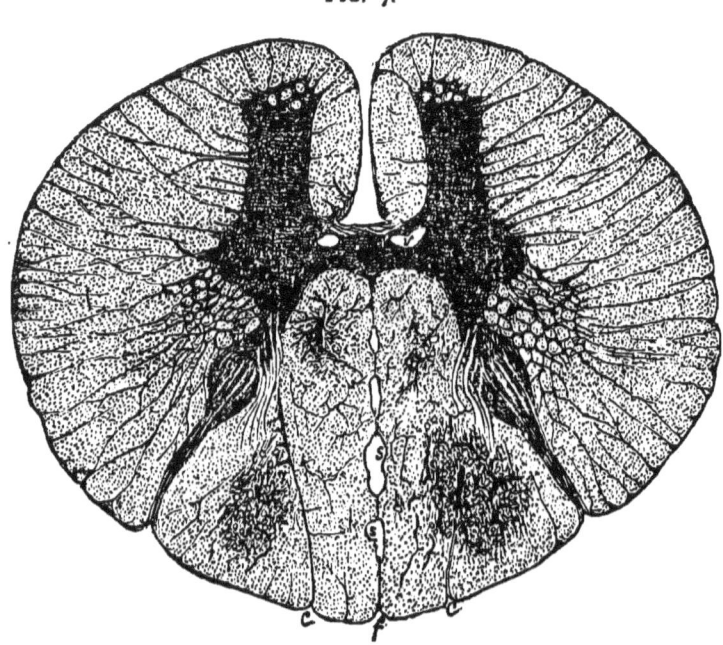

it is peripheral and the result of a meningo-myelitis, more commonly it is distributed as shown in the figure, and seems to originate in a sclerosis of the blood-vessels, and a subsequent development of a formless connective tissue, with an occasional Deiter's or spider-shaped cell. To the

---

* Transverse section of the spinal cord from an advanced paretic dement. k, sclerotic patch in the posterior column apparently concentrated around a sclerotic blood-vessel; c, collateral sulci, with sclerotic patches in their depths; f, posterior fissure, with abnormal ectasies (lymphatic) at s, s; v, vascular spaces (normal) of gray substance.

THE MORBID ANATOMY OF PARETIC DEMENTIA. 233

naked eye the gray matter of such a cord appears normal, while a reddish-gray discoloration of the region bordering the collateral sulci and the deeper portions of the posterior columns indicates their diseased condition. The discolored patches are firm. Under the microscope it is found that slight changes similar to those in the posterior columns exist in the lateral columns, particularly near the reticular processes. An exquisite example of the vesicular degeneration described by Leyden was found in two cases by the writer, who believes the vesicular spaces to be tubular in character, and to contain fluid or semi-fluid contents of

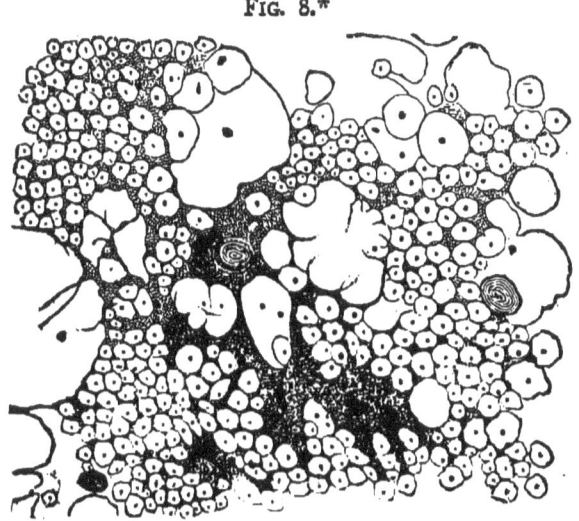

FIG. 8.*

the nature of myelin or some product of the degeneration of myelin. Often they contain one or more axis cylinders, and a glance at the specimen suggests the probability of these tubular spaces being the product of the fusion of several hypertrophying myelin tubes, accompanied by obliteration of the neurilemma, and subsequent degeneration of the axis cylinders. Granular cells are sometimes found

---

* Section magnified 500 diameters from one of the sclerotic patches of Fig. 7. Two sclerotic vessels are clearly recognizable, their lumina being nearly obliterated. The neurilemma is hypertrophied (rendered too coarsely granular in the cut), and large nuclei are found scattered in the formless connective tissue which constitutes this hypertrophy.

in and between these spaces; these bodies seem to be leucocytes, which have taken up the products of myelin disintegration. A strict line of division is to be drawn between these bodies and the so-called "granule cells" of several authors, which can be conclusively shown to be not cells, but fragments of disintegrated nerve tubules, whose carmine-absorbing centre (so-called nucleus) is an axis cylinder fragment, and whose supposed protoplasm is the granularly degenerating myelin. Frommann's cells are sometimes observed in large numbers in the sclerotic areas.

In one case—that of a negro paretic with cerebro-spinal pachymeningitis—the writer found a fascicular softening involving the left lateral column, and only a millimetre and a half in diameter, though it extended for the entire length of the dorsal and cervical cord. Associated with this there were wedge-shaped sclerotic patches from the level of the first to that of the eighth dorsal nerve exits. The whole area of the transverse section of the cord was of a dirty yellowish tint.*

The ganglionic groups of the anterior horn frequently show marked changes in advanced paretic dementia, of a kind similar to that found in locomotor ataxia of long standing. In such cases the coexistence of trophic disturbances has been noted by Kiernan. The nerve bodies appear to be sclerotic, their processes break off easily, and their protoplasm contains immense accumulations of a yellow or brown coarsely-granular material, the precise chemical nature of which it is difficult to determine. In some bodies no nucleus can be seen, it being destroyed or obscured by this morbid deposit which, as shown in the accompanying cut (Fig. 9), may almost fill the protoplasmic area. At the same time the bodies are fewer in number than in the normal cord, and residua of destroyed ganglionic bodies are found, though much less frequently than in the cortex. In one case the writer found the nerve bodies of the anterior horn, yellow in color, not taking up carmine, and with few or no traces of a nucleus—a condition which, as stated, is also found in the cortex of the hemispheres.

The nerve roots may show all the changes found in posterior sclerosis in those cases where the spinal lesion is marked. In the majority of cases they exhibit no apprecia-

---

* A similar discoloration was noted in nearly all the medullary districts of the brain of the same patient,

THE MORBID ANATOMY OF PARETIC DEMENTIA. 235

ble diminution in size, nor any change in color or consistency.

Much interest has been awakened by the claim of Poincaré and Bonnet, that the sympathetic ganglia show the most constant lesions in this disease. While these observers based too sweeping a generalization on their findings, it is worthy of note that in one case, where there were pronounced trophic disturbances, the writer found a larger number of cells than usual with double nuclei, thickening

FIG. 9.*

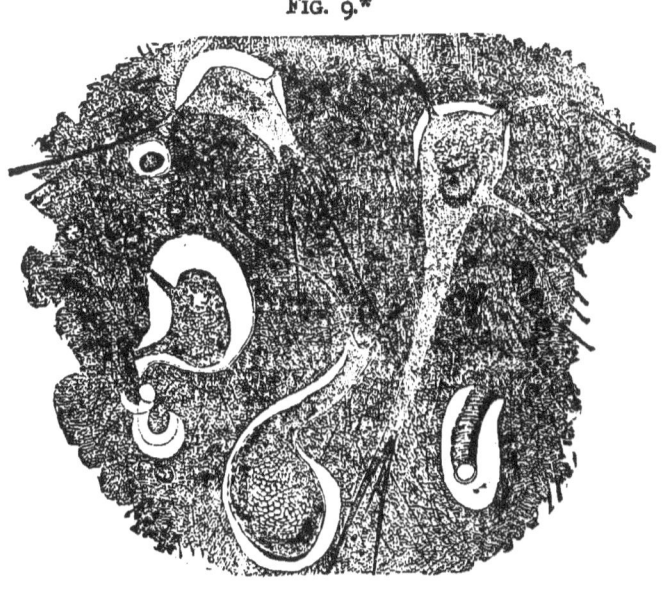

of the cell capsules, and multiplication of the free nuclei between them in the interspinal ganglia, as well as pigmentary degeneration of the inferior and superior cervical ganglia.

The optic papilla is most constantly affected in that form of paretic dementia which complicates locomotor ataxia. In one case of this kind, noteworthy for the youth of the patient who was only twenty-three, the writer found white

---

* Changes in the large nerve bodies of the anterior horn of the same cord represented in Fig. 7. In the upper part there is an apparently healthy body; the long ganglionic body shows the nucleus and granular clump side by side; in the lowermost one, the clump occupies nearly the whole protoplasmic area.

atrophy of both papillæ. In an old-standing case, clinically a typical paretic dementia, but with which a luetic nature was probable, there was sclerosis of the temporal side of the papillæ, and a corresponding defect of vision on the right side, while the left eye appeared healthy. A third case of this disease, in which there were photopsia and extensive hallucinations of fiery visions (page 202), showed marked hyperæmia of both papillæ. These were the only cases of pronounced retinal disease or disturbance observed in eighteen asylum patients examined. In thirty-nine cases in private practice in which ophthalmoscopic examinations could be made, there was found optic nerve atrophy in three cases, choked disk in one, and a pronounced hyperæmia in four; the single female case had the most pronounced atrophy found in this disease by the writer.

Such assertions as those of Clifford Albutt, who claimed that the papilla was atrophied in forty-one out of fifty-three, less severely affected in seven, and healthy in only five cases, are, to say the least, startling..

The changes of other organs and structures, such as those of the lungs, heart, stomach, and bowels, the bones and skin, inasmuch as they are, as far as an intrinsic relation to paretic dementia is concerned, secondary to the cerebral disorder, and have a similar pathology and etiology as the same changes, studied by the general neurologist, require no detailed consideration here.

THE RELATION BETWEEN THE LESIONS OF THE NERVOUS SYSTEM AND THE SYMPTOMS OF PARETIC DEMENTIA.

As previously indicated the majority of those pathologists engaged in the study of morbid changes which are most frequently found in paretic dementia have, through the frequent discovery of material and destructive lesions of the cerebral cortex, been led to consider this disease as an essentially inflammatory, or at least as a degenerative, process. While this is true with regard to the terminal and active phases of the disorder, it is a question whether the condition of the brain in the earlier periods of paretic dementia justifies us in assuming an inflammatory or degenerative change to exist from the beginning.

If the intrinsic factor of the morbid process underlying this disease were an inflammation, or a primary progressive histological deterioration, the evidences of these conditions, which are readily discovered and unmistakable, would be found in the brain of every paretic who reaches the autopsy table. This, however, is not the case. Well-authenticated instances are on record where no marked lesion was discoverable, and one of the most reliable investigators* in this field takes special occasion to mention this fact. In the writer's experience, one case has occurred in which nothing was found beyond the ordinary appearances presented by the brains of sane persons, excluding the stasis-like condition above described, and which can be interpreted neither as a degenerative nor as a strictly inflammatory process in the ordinary acceptation of the term.

Then again, if a degenerative or inflammatory process, or, in short, any profound tissue change were the fundamental lesion of this interesting disease, it should, in its extensity and intensity, stand in a rather constant relation to the amount and character of the mental disturbance. This has been found to be so in all cases which had run a long course; but, in several others which had terminated within two years from the date of the first manifestations, the writer failed to find any harmony between the extent and severity of the tissue lesions and the symptoms.

It was the consideration of this fact that led Poincaré and Bonnet to search in the sympathetic ganglionic system for those disturbances which the brain refused to reveal. That they mistook appearances occasionally occurring in health for disease, and secondary processes for primary lesions, does not detract from the value of the principle they were endeavoring to establish, though their observations as observations cannot be utilized to support it. As we have seen, changes in the sympathetic ganglia are sometimes found, but they are much less pronounced than those of the central nervous system.

It is more than probable that the vaso-motor system controlling the cerebral circulation is, at least in part, in the brain itself, and that its initial disturbances are as little tangible to the microscope as they are with other disturbances of the same system—for example, in epilepsy.

The most furibund of the symptoms ever manifested by paretics are: First, high maniacal explosions, with great destructive tendencies, and often with rapid flight of ideas; second, apoplectiform attacks; third, epileptiform attacks.

---

* Theo. Simon. "Die Gehirnerweichung der Irren." Hamburg, 1871.

The first of these symptoms, which is not at all as constant as is generally supposed, is found in its most perfect development in early periods of the disease. It appears as if the profound lesions ensuing in the later stages of the affection were inimical to its full development. The second is found variably, more frequently and fatally in the last stages. The third is also found variably and more frequently in the last stages of the disease, but exceptionally convulsions are the very first evidences of the latter. Each maniacal, apoplectic, or epileptiform exacerbation leaves the patient permanently more crippled than before, either in his motor or psychical field; rarely, however, the contrary, a relatively complete restitutio ad integrum, takes place.

A patho-physiological theory of the disease must take into account and harmonize all these facts. It must offer an explanation for the facts that the maniacal explosions are less perfect in the last than in earlier stages of the disease; that the most violent symptoms may occasionally open the clinical picture; that, as a rule, each exacerbation leaves the patient worse, though a remission and temporary improvement are not out of the question.

The supposition of a strictly inflammatory process is incompatible with the occasional appearance, as the first evidences of the disease, of epileptiform spasms which are not followed by those immediate sequelæ ordinarily following such an inflammation. It is incompatible with the very rapid and relatively complete remission of the symptoms. It is also known that certain of the injurious physical influences provoking the disease, such as violence directly or indirectly affecting the cranium, and insolation as well as other forms of overheating, do not always act on the brain through the channel of a meningitis or other inflammatory process.*

All these facts can be harmonized by assuming the essential and primitive anomaly in the paretic's brain to be a vaso-motor disturbance. This assumption is in strict accordance with the observations made on the brain after death.

In every case where the patient died in a maniacal attack,

---

* In Arndt's cases (soldiers dying from overheating after a forced march) the majority showed a pale brain, without visible morbid appearances.

or shortly after such a one, or in an epileptiform state, or with apoplectiform symptoms, the writer constantly found the capillary thrombi, which with Lubimoff he considers to be the expression of a blood stasis. That stasis is to be regarded as the result of a paralysis of the muscular coat of the blood-vessels, over-distended by the efferent blood column, in its turn an indication of hyperæmia.

A cortical hyperæmia would explain the expansive ideation, and the motor excitation; the arrest of the blood current through stasis, the subsequent congestive and comatose states. A sudden stasis, causing sudden arrest of the cortical functions, would satisfactorily account for the epileptic manifestations. A cortical hyperæmia, as a factor that may, on the one hand, vanish with the most violent storm sweeping over the mental plain, without leaving a permanent defect; and, on the other hand, in its repeated recurrence, determine those structural changes which account for the permanent symptoms of the disease, would also, in its necessarily progressive severity, account for the progressive greater gravity of each exacerbation, and the final preponderance of symptoms of subtraction such as paralysis, lacunæ in the memory, aphasia and coma, over those of functional excitation, such as the destructive tendencies, constructive schemes, ambitious delusions, and flight of ideas of the earlier periods.

As the disease progresses, and the resisting tone of the vessels decreases more and more, stases are found to occur not only in the exacerbations of the disease, but also in the intervals; here more restricted in extent and less pronounced, so that with proper methods of preparing histological specimens the writer believes that no lesion will be found so constantly in the terminal periods of the disease as the capillary thrombi resulting from such stases.

Of course, with this explanation we are as much in the dark as ever as to the organic basis of the vaso-motor difficulty. As above stated, this consists in a probably impalpable morbid state of the encephalic vaso-motor centre. Such a morbid state it requires no stretch of theory to consider inducible by mental overstrain, by the repeated hyperæmias of alcoholism, rheumatism, and certain forms of syphilis, or by typhus fever, insolation, and the molecular disturbances determined by concussion, directly or indirectly involving the skull contents. While the essential factor in the development of paralytic dementia would, in this light, be constituted by the vaso-motor difficulty, a direct influence of many of the causes supposed and known to provoke disease of the cerebral tissues need not be excluded. In fact, for some of the syphilitic forms and those cases in which the writer believes a multiple sclerosis to be the essential lesion this seems to be the case, and we may some day have to draw a line between those cases with which the vaso-motor anomalies are in the foreground, and those with which they are in the background.

A careful study of certain cases of cerebro-spinal sclerosis, constituting connecting links between the typical disease and "chronic" varieties of dementia, may yet show that vaso-motor derangement plays a far more important rôle also in that disease than is generally imagined. Apoplectiform and congestive attacks are also encountered here, and they present many points of resemblance to those of genuine paretic dementia. When it is recollected that a large percentage of the cases of multiple sclerosis is ascribed to concussion as a cause, by Erb and other authors, the attributing of certain cases of paralytic dementia to the same cause by prominent alienists merits renewed attention, as bearing on the vaso-motor etiology of the disease. In this connection it may be well to refer to some remarkable analogies of paretic dementia, with typical posterior spinal sclerosis. Both diseases are more common in the male than in the female, and under certain circumstances occur nearly in the same proportion in the sexes. Both have been attributed to sexual excess as a cause. With regard to both the syphilitic dyscrasia plays the same rôle. Cranial or spinal symptoms may open the history of either, and these symptoms show a striking analogy. Thus amblyopia and diplopia may be the first indications of posterior sclerosis, so may color-blindness, optic nerve atrophy, and diplopia be the first signs of paretic dementia. The essential cerebral signs of paretic dementia may be complicated by the typical picture of posterior sclerosis; so the latter may be complicated by paretic dementia. Paretic dementia is pathologically a progressive brain-tissue deterioration, which is the basis of the progressive and constant symptoms of subtraction found in this disease; while certain episodes—maniacal, apoplectiform, epileptiform, and trophic accidents—mark its progress, which cannot be explained by that lesion alone. Similarly, posterior sclerosis is pathologically a tissue degeneration of a considerable part of the cord, and certain phenomena are constant and progressive in its clinical history, and can be explained by its lesions; but there are episodes such as gastric, nephritic, and other crises which cannot be accounted for by the lesion alone; and some of these episodes, like the acute maniacal outbreaks exceptionally observed, show that they may even assume a psychical character. In short paretic dementia and posterior spinal sclerosis appear to be similar if not pathologically identical processes, only differing in their location, the former being concentrated in the brain, the latter in the cord.

There are a number of facts which point to the early involvement of the brain isthmus * in paretic dementia. In the first place, the morbid changes are frequently observed to be further advanced here than in the cerebral cortex. In the second place, not only the paralytic phenomena indicating an affection of the nerve nuclei and the emerging roots, but also certain emotional disturbances, point to the oblongata and pons as the seats of at least a functional disturbance. Reference is here particularly made to the fact that paretic dements exhibit emotional manifestations opposed to their true emotional state; thus, while relating a pleasant reminiscence, such a patient may burst into tears, and, while complaining of an affront or of persecutions, smile. Similar disturbances are noted in organic disease of the medulla oblongata.†

This view is also supported by the fact, that at certain periods in the progress of spinal disease, when the morbid process reaches the oblongata, mental symptoms develop, which are strikingly like those of paretic dementia, and imitate both the "melancholic" and "maniacal" types of that disease.

Grouping together all that is known about the interesting affection we are considering, and amalgamating the clinical and anatomical facts with the physiological theory, the conclusion follows, that *paretic dementia is a progressive deterioration of the central nervous system, chiefly affecting the brain, and the result of a chronic inflammatory process of an angio-paralytic nature, whose essential element, the vaso-motor*

---

* The medulla oblongata, pons, and peduncular region.
† In a dispensary patient with complete anarthria, the tongue lying motionless at the bottom of the mouth, this was noted as a salient symptom of the organic disease (tumor of the medulla?) from which she suffered.

weakening, is due to overstrain of the encephalic vaso-motor centre, the exacerbating and remitting course of the malady, being the clinical expression of the struggle of that centre to regain its equilibrium. *

Owing to the multiplicity of lesions found in advanced cases of this disease, it is but rarely possible to establish a relation between localized lesions and special symptoms. As to the most essential symptoms from the alienist's standpoint, the mental disturbances, the fact that in paretic dementia there is a diffuse or multilocular disease of both cerebral hemispheres, is in such striking accord with what we know of the seat of the mind that we may assume that this disease is the basis of the progressive mental failure. A further localization of the strictly mental symptoms is impossible, and we may content ourselves with the assumption that the frequently lacunar character of the amnesia is in harmony with the fact that the cortical provinces are not all equally diseased, some districts being relatively intact, and others destructively involved.

It is a significant fact that those movements which depend on the voluntary motor association of smaller muscular peripheries suffer earlier than those which depend on the combination of the coarser groups. This appears to be in parallelism with the fact that the lesions, at first of slight extent, involve the smaller cortical areas, and subcortical associating fibres more completely than the larger areas and deeper tracts, which are seriously involved only in the later stages of the disease. It is, however, very difficult to obtain much valuable material in support of the localization theory in paretic dementia. If it were only the cortical gray matter that is diseased the subject would be difficult enough, but there are also found changes in the lower segments of the nerve-axis which must have an important influence on the motor and sensory functions, and which must be taken into account.

Thus the sclerotic patches in the course of the cranial nerves, the raphe, and other important tracts, the degenerative changes in the nerve nuclei of the oblongata and cord, as well as the disease processes in the cord, account for many of the ataxic, paretic and paræsthetic phenomena of

---

* The experiments of Mendel, who produced paralytic symptoms in dogs, by centrifugal rotation, aggravated in case of preliminary mercurialization, sustain this view.

the disease. The speech defect in paretic dementia may, as in the patient whose medulla oblongata is figured on page 224, be of the anarthric type, that is, not truly aphasic, and hence not referable to the lesion of the hemispheres. The only really valuable contribution to the study of the areas of specialized function in the cortex is that made by Fürstner, who found that visual disturbances of the kind experimentally produced by Munk, by destruction of the occipital lobes in monkeys, are found in paretic dementia when there is pronounced lesion of the posterior part of the cerebral hemispheres. A case in point coming within the writer's observation has been referred to in the last chapter.

CHAPTER XIV.

Syphilitic Dementia.

Syphilis is an important etiological factor in certain cases of insanity. In most of these its influence is rather of a psychical than a somatic character, as in the syphilitic hypochondriasis and self-accusatory melancholia observed by Erlenmeyer. Here the only direct influence of the disease is manifested through the associated impoverishment of brain nutrition; and although anti-syphilitic treatment is most successful in combating these conditions, this does not prove that the brain state underlying them has any specifically syphilitic characters. The same remarks apply to the insanity sometimes noted with the secondary fever of syphilis, and which has not thus far been demonstrated to be associated with a palpable and constant change in the brain or its vessels.

It is different with syphilitic dementia—a chronic mental disorder which, from its association with anomalies of motility and speech, bears a great resemblance to paretic dementia and to dementia from organic brain disease. Here demonstrable brain lesions, standing in a constant relation to the symptoms, are found in the majority of cases. In some cases syphilitic pachymeningitis or leptomeningitis and gummy infiltration of the membranes preponderate; in others multilocular gummata of the brain substance, or the luetic arterial change, with its results, are the most notable morbid anatomical features. In one case described by the writer some years ago, over a thousand perivascular nodules, single and branched, varying in size from that of a pin's head to a pea, were found scattered through the cerebral tissues and associated with a surface infiltration of the cortex. The pre-lethal symptoms in this instance were in no way distinguishable from those of a maniacal exacerbation in a deteriorated paretic dement.

As a rule the symptoms associated with these morbid conditions are characterized by *variability*, and an almost pathognomonic feature is the accompaniment of the earlier mental symptoms by pronounced paralysis of single muscles, or of definite muscular groups supplied by one or two cranial nerves. It is an ordinary history of the inception of syphilitic dementia, that after such prodromal signs as headache and vertiginous or syncopal attacks, the patient may awake with ptosis, facial paralysis, aphasia, or paralytic weakness of a leg or an arm. These symptoms rapidly disappear, and may as rapidly recur. With this the patient exhibits lacunæ* in his memory, an undue irritability, and intolerance to alcoholic liquors.

Clinically it is not always possible to make a sharp discrimination between syphilitic dementia and paretic dementia proper, for syphilis plays an important rôle in the etiology of the latter affection. Indeed, Snell claimed that seventy-five per cent of the cases of paretic dementia coming under his observation were due to syphilis, a claim which is in remarkable parallelism with that recently advanced by Erb, with regard to locomotor ataxia, in which a similar relation is supposed to exist. Other authors find a smaller proportion of syphilitic paretic dements than Snell. Mendel found that of 201 patients 117 were syphilitic, and Ripping could only detect a syphilitic element in about twelve per cent. It is to be borne in mind that the mere co-existence of a syphilitic taint does not prove a given form of insanity to be syphilitic; but the fact is significant that, of *all* syphilitic lunatics, one half are paretic dements, or suffer from the allied form of disease we are here considering.

In the writer's experience syphilis is an etiological factor in the production of various forms of progressive dementia

---

* Erlenmeyer calls attention to this as a constant feature, and cites instances where a special series of events were blotted from the mind, leaving other recollections intact. This is also found in paretic dementia.

in about one third of the cases among the pauper insane of New York.* Its existence could be determined in fourteen per cent of the paralytic patients in private practice. Of these eighty per cent had typical paretic dementia, and the remainder the true syphilitic form of dementia.

Numerous attempts have been made to establish some criterion, on the strength of which to be able to distinguish syphilitic from typical paretic dementia ; but these do not always hold good. Thus, it has been supposed that the association of true locomotor ataxia with paretic dementia, proves the syphilitic origin of the case. But there are non-syphilitic tabetics who become paretic dements, while there are syphilitic dements who are not markedly tabetic. On the other hand, observers have suggested the empirical test of an anti-syphilitic treatment, which was supposed, if successful, to demonstrate that the disorder was of syphilitic origin.* But we know that *spontaneously* remissions occur in ordinary paretic dementia, while in unquestioned syphilitic cerebral disease the most energetic anti-syphilitic treatment may fail. The writer has found that both mercury and iodide of potassium appeared to be of service in cases where a luetic infection could be positively excluded ; and Ripping goes so far as to claim that he has seen in-

---

* It was impossible to obtain accurate information, and the estimate here given is probably under the truth.

† This procedure has been very faithfully carried out by Dr. Allan McLane Hamilton, who says: " When an apparently strong man comes to us with a history of fugaceous aches and pains, inconstant spasms, and disordered subjective sensations—notably among which are subjective cold —we should not immediately make light of his troubles, and even dismiss him for change of air and scene, but empirically, if our history of cause is not clear, place him upon proper anti-syphilitic remedies." ("Syphilitic Hypochondriasis ; Alienist and Neurologist," vol. i., No. 1, page 79). Ripping (Ueber die Beziehungen den Syphilis zu den Geisteskrankheiten mit und ohne Lähmungen, *Allg. Zeitschrift fuer Psychiatrie*) makes the following comment: "Such views as those I above cited from Allan McLane Hamilton, it may be safely claimed, have no upholders, among German physicians at least." As to this remark it may be said, that it manifests a tone of criticism which is becoming rather too common in Germany. Science is international, and if a faulty or fanciful view be announced by an American physician, that is no justification for raising an international question. In the present case the implied inference that such views as the one criticised are shared by the American medical profession would be in the highest degree unjust; for " it may be safely claimed " that no other American writer, nay, that no other writer anywhere, has ever announced or held such a view as the one which Ripping—properly enough—finds fault with.

sanity in syphilitic subjects recovered from, although no anti-syphilitic treatment of any kind had been applied.

After the prodromal period referred to, the course of syphilitic dementia is progressively, and usually very slowly, toward a fatal termination.* Delusions are not prominent, and rarely expansive, though, in paretic dementia from syphilis, the unsystematized delusions of grandeur may be as well marked as in non-syphilitic cases.† Sometimes the terminal period of an at first clinically well-marked syphilitic brain disorder is in no way distinguishable from typical paretic dementia.‡

It is possible that with the progressing accumulation of clinical and pathological material, syphilitic dementia will share the fate of "syphilitic meningitis," which is now known to differ in no essential respect from ordinary meningitis, except in those rarer instances where the specific gummatous character prevails. It is to cases corresponding to the latter that the term syphilitic dementia should be limited.

---

* In one case of syphilitic dementia, closely approaching true paretic dementia, a remission occurred which lasted four years, the patient, a physician, returning to his practice. The case has since been lost sight of.

† Fournier erroneously claimed the contrary, and termed paretic dementia from syphilis " pseudo-paralysie generale." Foville showed that in syphilitic paretic dementia the delusions may be as expansive and varied as in any case, and some of the most characteristic delusions related in the last chapter were observed in syphilitic patients. Then, too, it must be recollected that in non-syphilitic cases, delusions may be absent; for this reason alone Fournier's claim is to be considered faulty.

‡ Since the above was written a striking confirmation of this occurred in the writer's consultation practice. He was called to examine a patient who had for a year been under the treatment of one of the best clinicians in New York for syphilitic cerebral trouble. For a period of several months he improved considerably on anti-syphilitic medication. When examined by the writer, he exhibited a lachrymose hypochondriasis, with some of the motor symptoms characterizing paretic dementia. The diagnosis of this disease was made. Two days later he had three epileptiform attacks, and subsequently to these manifested gross amnesia, hilarity, extravagant projects, and delusions of physical strength.

CHAPTER XV.

DELIRIUM GRAVE.

There is a comparatively rare form of derangement, approximating in many respects to maniacal delirium, and yet distinct from it in many essential features, which has been variously termed *typhomania, mania gravis, phrenitis,* and *acute delirium*. This disorder differs from the simple psychoses in the fact that it depends on a stormy pathological process; and while the motor excitement and the angry fury of the patient seem to be clinically only a higher degree of maniacal furor or melancholic frenzy, there are somatic signs which justify the pathologist in comparing *delirium grave* to the cerebral disturbances which follow severe and exhausting febrile processes, such as pneumonia and typhus. As Schüele has aptly said, the symptoms of a spurious maniacal furor mark the first period of the disease, while the symptoms of grave cerebral exhaustion characterize the second.*

This disorder is more common in females than in males; this is related to the fact that the puerperal state is often found to stand in a causal relation to grave delirium. It is preceded and undoubtedly caused by profound nervous or physical exhaustion and overstrain. Schüele has known it to result from excruciating physical suffering. In most cases it is noted that the patient has for a long period of time been in a feeble state of health, and that some extra strain on his nervous system, such as a business crisis,† an alcoholic excess, an emotional strain,‡ or the puerperal state, precipitates the breaking out of the delirium.

The mental signs may be briefly characterized as resembling the highest degrees of maniacal furor and melancholic

---

* "Handbuch der Geistes-Krankheiten," von Dr. Heinrich Schüle ("Ziemssen's Handbuch"), vol. xvi., 1878).

† In the case of a lawyer, who for many months had suffered from insomnia, this disorder exploded after the preparation of an argument.

‡ Abandonment of seduced and pregnant girls is a prominent element in the history of a number of cases.

frenzy.* They differ from these states in their mode of development. While maniacal and melancholic frenzy are preceded by the ordinary and readily recognizable symptoms of typical mania and melancholia, the outbreak of grave delirium is either sudden or preceded by a state of impaired consciousness of a kind not found in mania or melancholia proper. Thus, some patients in this state exhibit a panphobia like that of febrile delirium, while others wander about aimlessly, staggering as if drunk.

The ideation of grave delirium is much more incoherent than of frenzy, and is usually the expression of an angry or a frightened state. While in the beginning the patient may still articulate sentences, his speech rapidly deteriorates, and he is finally unable to pronounce syllables. As far as the expressions of the patients permit us to judge, they have hallucinatory visions of the day of judgment, of conflagrations, of bloody scenes, or of those connected with the exciting cause. Sometimes a set phrase is repeated over and over again; usually it has some relation to the emotional calamity provoking the outbreak of the disease. The seduced girl will count as if hearing the bells that toll out the hour of an assignation, and then suddenly break out in a piercing cry or a silly laugh. The business man, who has become delirious after a period of business worry, repeats figures or names of articles of trade, of firms, or of stocks, in an incoherent jumble.* Delusions of grandeur are

---

* Schüle speaks of a melancholic form of acute delirium whose symptoms and anatomical basis are said to be the very reverse of the maniacal form. The writer is unable to recognize in the description anything but a stupid melancholia, developing on a basis of extreme physical exhaustion. Neither the temperature, nor the mental signs, nor the relatively better prognosis of Schüle's melancholic form, support his view that it is due to as grave and active a pathological process as grave delirium.

† The recorded cases of "meningitis from over-study" are in fact cases of grave delirium. They are brought about as much by the emotional strain attendant on competitive examinations, as by the mental effort itself. It is never a strong mind nor a healthy body that suffers in this way. "The mental hygiene" sensationalists who periodically "enlighten" the public through the columns of the press, whenever an opportune moment for a crusade against our schools and colleges seems to have arrived, are evidently unaware of the existence of such a disease as grave delirium, and ignorant of the fact that the disorder which they attribute to excessive study is in truth due to a generally vitiated mental and physical state, perhaps inherited from a feeble ancestry. Our school system is responsible for a good deal of mischief, but not for meningitis.

exceptional. In one case, that of a woman without a previous history, whom the writer saw through the invitation of the city physician, Dr. Hardy, there were expansive though vaguely expressed sexual ideas; she recognized every male and female visitor as one of a large number of husbands.

With these deliria there is great restlessness; the patients make aimless efforts to escape from those around them, or kick and strike in all directions. Sometimes rhythmical motions are observed; one patient under the writer's observation continually rolled his head from side to side day and night (as far as watched) for a period of three days. Grinding of the teeth, strabismus, contraction of the pupils, and convulsive movements mark the transition to the second period of the disease.

There is in many cases absolute insomnia, and while the general nutrition of the patient suffers appreciably, the temperature rises to over 100° F. and may reach 105° or 106°. The pulse becomes frequent (130 in one case), soft, compressible, and the sphygmographic trace indicates extreme cardiac enfeeblement.

The second period of the disease is analogous to the post-maniacal reaction which follows the outbreaks of simple mania. There is now extreme mental and physical depression. The patient lies apathetic, mute, collapsed; has a staring or startled look; does not recognize what is going on around him; or, if he shows any signs of mental life, these are limited to incoherent expressions and purposeless and feeble movements. If the patient does not die in this condition he passes into a state resembling, as Jessen remarks, the convalescence from typhus, without the favorable termination of the latter. It was this feature that induced Luther Bell to designate the disease " typhomania."

The severity of grave delirium is specially manifested in certain somatic sequelæ of the excited period. The hair falls out, the skin desquamates and is cyanotic, the nails exhibit an atrophic zone corresponding to the period when the disease was at its height, the spleen is slightly enlarged, the intestinal tube relaxed, symmetrical atrophy of certain muscular groups occurs, and the reflexes, which at first were exaggerated, become diminished. In one of Jessen's patients anæsthesia became so extreme that he gnawed off the ungual phalanx of one of his fingers. That author[*] claims that

---

[*] " Ueber die klinische Aeusserung der Reactions-Zustände acuter Delirien. *Allg. Zeitschrift f. Psychiatrie.* 1880–1.

pemphigus-like vesicles appear in the otherwise apparently healthy skin, especially of the dorsal faces of the hands and feet. He believes this to be a comparatively constant sign; it was absent in two out of the five cases observed by the writer, while phlegmons and spontaneous gangrene were additionally noticed in one of the cases that had pemphigus. Most of these conditions are due to the vaso-motor paresis which marks this period.

The majority of the patients affected with grave delirium die in the delirious period after an illness of a few weeks; in those who do not die at this period the excitement continues unabated for four or five weeks, the subsequent symptoms of stupor increase, and the history closes with a fatal coma. Complete recovery never occurs; in rare instances the patients emerge from this severe disorder with a slight mental defect, in others paretic and terminal dementia supervene.

The morbid anatomy of this disease consists in an intense hyperæmia of the brain and meninges. This is constantly found in patients dying in the excited period of the disorder; in those who die in the stuporous period, the hyperæmia is sometimes obliterated by a collateral œdema; but in all the brain appears swollen, the cortical ganglionic elements are granular or opaque, stain poorly, and their periganglionic spaces, like the adventitial lymph sheaths, are literally crammed with the formed elements of the blood. In the single case examined *post mortem* by the writer white streaks were found on either side of the larger vessels in the pia. Microscopic examination showed that they were due to an accumulation of leucocytes, whose preponderance suggests an inflammatory nature of the lesion rather than the condition of venous engorgement claimed by Krafft-Ebing. A most positive sign of inflammation was found in the case referred to: the arterioles were surrounded by an area staining in carmine with a beautiful pink flush, probably the expression of a molecular infiltration, while layers of newly-formed fibrin were found in and around the adventitia.

That grave delirium is the result of a vaso-motor overstrain analogous to that supposed to exist in paretic dementia is supported by the etiology, the manner of origin, and the somatic sequelæ of this disorder.

CHAPTER XVI.

Chronic Alcoholic Insanity.

Alcoholic excesses play an important *rôle* in the production of insanity. Ordinarily the insanity which results from such excesses belongs to the groups already described. Thus a typical acute mania or melancholia may follow a prolonged debauch, and the influence of chronic alcoholism as a predisposing factor in the etiology of paretic dementia and delirium grave is well known. There are also certain mental disturbances of a character peculiar to alcoholism, which are not ordinarily ranked with insanity, but which have all the elements of the psychoses—such are the various states of drunkenness itself and *delirium tremens*.* In addition there are various forms of dementia associated with motor disturbances, which depend on the organic changes produced by alcohol in the brain and its membranes, and which appertain to the group of "dementia from organic disease."

While organic changes are common in the brains of subjects who had been addicted to alcoholic excesses, and while the dementia just referred to is with its accompanying motor and sensory symptoms present, in however mild a degree, in all persons suffering from advanced alcoholism, not all forms of mental disorder found in such subjects properly belong to the group of dementia or insanity "from organic disease." Just as epileptic or hysterical insanity may develop as an epiphenomenon on the epileptic or hysterical neurosis, so a special form of alcoholic insanity may become engrafted on the alcoholic neurosis. It has distinct clinical as it has special etiological characters, and alone merits the designation of CHRONIC ALCOHOLIC INSANITY.

Before proceeding to characterize this psychosis it is well to survey the extensive pathological territory of inebriety of which chronic alcoholic insanity is but a province. The inebriate generally exhibits moral turpitude, indifference to his interests and his family, morbid irritability, emotional depression, to overcome which the libations which provoked it are repeated and prove temporarily remedial, and above

---

* A detailed description of these conditions is here omitted, as they are usually treated of *in extenso* in works on general neurology and clinical medicine.

all, a marked enfeeblement of the will. This enfeeblement of the will is at first manifested in the inability of the inebriate to resist the temptation to drink. Numerous cases are on record where prosperous business men and capable men of letters, feeling this abulia, voluntarily went to an asylum for inebriates, and within its walls carried on their labors as well as before they had formed the alcoholic habit. But, with the continuance of the vice, the volition becomes impaired with regard to other matters as well, and the confirmed and deteriorating inebriate becomes the tool of others. He attends fairly well to duties of a routine character, but is devoid of initiative, or, if he has it, is inconsistent and easily diverted from his purposes. With this there is noted a general impairment of all the intellectual faculties, the memory is gravely weakened, and the reasoning powers become clogged.

With these mental symptoms there are positive signs of the disorder of a somatic character. The most important and constant sign is the alcoholic *tremor*. This tremor has the peculiarity that it *decreases* under the influence of alcoholic beverages, and is most marked when the patient is perfectly sober. It is best observable in the hands, tongue, and lips. Crampi and clonic spasms sometimes occur in the extremities, and muscular anenergy is frequently complained of, being most pronounced in the extensors of the leg. In extreme cases an actual paraparesis, or even paraplegia (Wilks), probably independent of a structural lesion, may ensue.

Hyperæsthesias and anæsthesias are usually among the later symptoms of chronic alcoholism, and not unfrequently has the diagnosis of incipient paretic dementia of the ascending or spinal type been made on the strength of an anæsthesia of the legs and feet, coupled with the lax facial innervation, enfeebled memory, hesitating speech, and tremor of chronic alcoholism. Occasionally, lightning-like pains and analgesia are observed, and in severe cases almost any form of sensory disturbance may be found, from the unilateral hæmi-anæthesia observed by Magnan to the amblyopia noted by Galezowski. The latter condition may even be associated with atrophy of the optic nerve.*

---

* In the case of a gentleman whose will was made the subject of litigation, and who undoubtedly during his later years suffered from alcoholic dementia, the distinguished ophthalmologist, Knapp, discovered atrophy of both optic nerves some years before marked mental impairment had set in,

Most chronic alcoholic patients are anæmic and badly nourished, for there is usually more or less hepatic and renal trouble, there is always gastric or gastro-duodenal catarrh and general degeneration of the vascular system. The latter manifests itself in the well-known fatty degeneration of the cardiac muscle, and in ectasis of the capillary and atheroma of the larger blood-vessels.

The facial appearance and attitude of the chronic inebriate are characteristic. There is a general laxity of muscular tone, the body is inclined forward, the knees bent, the eyes dull, and the face generally defective in expression. Most authorities state that the pupils are dilated, but the writer's experience, however exceptional it may be, is to the contrary.

The above may be looked upon as the prominent signs of a constitutional deterioration of a neurotic character; and an analogy may be detected between the progressive mental enfeeblement of the epileptic, which is in relation to the frequency of the fits, and that of the inebriate, which is in relation to the frequency of his libations.

On this chronic alcoholic constitution as a background, the well-characterized psychosis which is the subject of this chapter may develop just as epileptic insanity crops out on the surface of the epileptic constitution. It is noted that positive signs of mental derangement are found in a great many inebriates met with in general practice. Hallucinations, chiefly of vision, are very common with them, and are almost without exception of a frightful character; they may lead the patient to the commission of brutal crimes in subjective self-defence. In addition, many inebriates entertain, if not delusions, at least unfounded suspicions of marital infidelity. When these symptoms become constant and prominent, the psychosis first described by Marcel,* is before us. The patient, after a brief prodromal period marked by congestive attacks and headache, and under the influence of the characteristic hallucinations of alcoholism, becomes the subject of delusions of persecution, and very rarely there may be superadded expansive ones. Krafft-Ebing † happily draws the line between these delusions, which are exclusively determined by hallucinations, and

---

* Marcel, "De la Folie Causée par l'Abus des Boissons Alcooliques." Paris, 1847.
† *Op. cit.*, p. 186.

the persecutory delusions of monomania, which are sometimes associated with hallucinations, and become confirmed by, but are never provoked by them.

The persecutory delusions of alcoholism relate to the sexual organs, to the sexual relations, and to poisoning. This fact is so constant a one that the combination of a delusion of mutilation of the sexual organs with the delusion that the patient's food is poisoned, and that his wife is unfaithful to him, may be considered to as nearly demonstrate the existence of alcoholic insanity as any one group of symptoms in mental pathology can prove anything. With this there are unpleasant hallucinations. The patient, who fears that he is about to be castrated, hears people commenting on the fact that he has a loathsome venereal affection, or that his penis is too small for its purposes, and smells seminal discharges which are drawn from him at night. Delirious exacerbations are likely to occur in consequence of the patient's morbid fear, and in brutal fury he may hack the wife, whom he suspects of infidelity, to pieces.

There is this peculiar feature about the delusions of insane inebriates, that their acts are not consistently regulated by their delusions.* Thus one patient may live in comparative tranquillity with a wife whom he suspects of committing adultery in the boldest manner and before his face night after night. Another, under the influence of the same delusion, may, in mortal fear of being poisoned in the delusive paramour's interest, kill his wife in a fit of blind fury. Lennon, the New York murderer, under the influence of similar insane ideas, cut up his wife in a regular checker-board pattern, and generally the crimes of these dangerous lunatics are as remarkable for their cynical brutality as their delusions are noted for obscenity.

Besides the hallucinations related to the delusions of sexual mutilation, impotence, and marital infidelity, there are others of the same kind, as those found in acute alcoholic delirium: the patient sees mocking faces, snakes, insects, dead bodies, paving stones precipitated on him, and frequently will be found sustaining a dialogue with some absent friend,† and stop in the midst of conversation

---

* This applies to the fully-developed disease only.
† A chronic form of opium insanity which is in every respect analogous to chronic alcoholic insanity, the delusions of persecution being based on visions of supernatural instead of such of a sexual character, has been observed in three instances by the writer. A marked feature of one of

with persons actually present to answer those who are miles away or mouldering in their graves.

Sometimes the hallucinations are of the character of a photopsia; one patient exhibited at Meynert's clinic saw lights streaming in through a closed door one night, and heard a confused noise (tinnitus) in the hall. In mortal dread of robbers and murderers he seized a hatchet in defence, fled from room to room, and finally feeling a head on a sofa, brained its possessor—his own father.

The periods marked by anxious hallucinations are usually but imperfectly retained in the patient's memory; sometimes there is complete amnesia, and one fundamental difference between monomania and alcoholic insanity is the constancy of some degree of enfeeblement of the memory with the latter affection. Occasionally the sufferer from alcoholic insanity may be found in a state of stupor; but this differs from the superficially similar symptom of stuporous insanity in the fact that the patient can be readily aroused from it and made to answer questions.

Chronic alcoholic insanity is more frequent in countries where spirits are consumed than where malt liquors or wines are chiefly used. It is in accordance with this fact that it is greatly on the increase in countries where the lighter liquors are being supplanted by the stronger ones. This has been noticed to be the case in Germany and France; regarding the latter country, Voisin reports the suggestive fact that while in 1856 only 99 patients suffering from the various forms of alcoholic insanity entered the Bicêtre, in 1860 the number had already risen to 207. It must not be believed that persons indulging in malt liquors and wines are exempt from alcoholic insanity. While this disorder is rarer here, quite typical cases have been observed by the writer in persons who never touched a drop of any other liquor.

---

the cases, that of a physician, was the sustaining of dialogues with absent persons. It is remarkable that every form of alcoholic mental derangement is imitated by the opium psychoses. There is an acute opium delirium analogous to *delirium tremens*, a chronic delusional insanity due to opium like the form described in this chapter, and in one case, that of another physician, first treated at a private home for opium *habitués*, the writer witnessed an attack of maniacal furor which could in no respect at the time be differentiated from the exacerbations of paretic dementia. In the quiet period there were noted ataxic and paretic symptoms. The patient recovered at Bloomingdale. This case is analogous to the alcoholic "pseudo-paralyses."

The prognosis of this form of insanity is very unfavorable, as there is a pronounced tendency to dementia. Complete cures are rare, and if the affection has lasted any length of time, impossible. The higher the mental status of the patient the better are his chances, but asylum treatment must be instituted early if they are to avail him. In one case the writer has found that the delusions of marital infidelity disappeared under moral treatment and a reduction of alcoholic beverages at home, and a most interesting case of complete recovery from a delirious and hallucinatory variety of alcoholic insanity has been recently reported by Sander.*

CHAPTER XVII.

Chronic Hysterical Insanity.

Like epileptic and alcoholic insanity, the other main types of this division, hysterical insanity is found to be associated with a fundamental neurotic character. The patients are changeable, emotional, fretful, careless, and superficial in their behavior and thoughts; they are extremely vain and egotistical, and desirous of notoriety or sympathy, or both. To be the sufferer from an equally interesting, rare, and hopeless nervous disease is the ambition of some; to be considered the most abused woman on earth is the ambition of other hysterical patients. If the ordinary means fail to excite attention such patients will resort to extraordinary ones to excite sympathy. The imitators of Louise Lateau, who produced artificial stigmatization, and the hospital patients who drove hundreds of needles into various parts of their bodies, are familiar instances of this fact. A patient of the writer's suborned her servants and nurses to give false testimony to the visiting physician; her vigorous fancy, a quality shared by her sisters in misfortune, enabling her to sustain their assertions with an appearance of truthfulness and conviction which at the time was real. There is no doubt that the "tale told too often" is finally believed in by the patient,

---

* *Psychiatrisches Centralblatt*, Aug. and Sept., 1877.

and that the potent influence of the mind over the body must be looked to, to explain why material and objective symptoms appear subsequent to the pretence being made.

A patient with this hysterical character may develop psychoses quite analogous to those found in epileptic and alcoholic patients. Just as we have transitory epileptic psychoses, so we have a transitory hysterical psychosis manifesting itself in deliria of fear. Just as we have maniacal and melancholic states with epilepsy and alcoholism, we have them in hysteria, and similarly we find a protracted psychosis in hysteria analogous to the alcoholic disorder discussed in the last chapter.

A tendency to simulation and theatrical behavior is characteristic of these various forms of hysterical disorder. They are particularly marked in the chronic form of derangement about to be briefly considered.

In chronic hysterical insanity an intensification of the described hysterical character is the most constant feature, a silly mendacity is frequently added and develops *pari passu* with advancing deterioration. Sexual ideas are common and manifested in two opposite extremes: either there is excessive sexual ardor, which may be so intense that the patients experience the orgasm spontaneously, or,—and this is in the writer's experience far more common—there is an absolute horror of anything that remotely suggests the sexual act, a feeling which is the basis of a hatred of the husband frequently exhibited by these patients.

Hallucinations are frequent, and usually of the kind described by Wundt as fantastic hallucinations. They are analogous to the hallucinations of hypochondriacs, being the outcome of the patient's fancy and fears. In the case of one patient, who was an excellent artist, visions of countless lovers in the costumes of all ages and peoples interspersed with horrible visions of hell, with all the paraphernalia attributed to that region by the older masters, were the most prominent symptoms. In some cases these visions and analogous illusions provoke ecstatic and visionary states. Krafft-Ebing says: " On this basis there develop deliria of a mystic union with God and of celestial visions. The patients see heaven open, indulge in enthusiastic preaching, speak in strange tongues, prophesy, etc." This applies to the episodial deliria of monomania with a hysterical tinge, and not to hysterical insanity proper. Here there may be found ecstatic states, but they resemble rather the deliria of hystero-epilepsy than the visionary deliria, and where such ideas and acts are found in true hysterical insanity as those described by Krafft-Ebing they are like those of the hysterical insane epidemics of the middle ages, imitatory phenomena.

In some patients obstinate mutism is observed. By skilful cross questioning it will be speedily found to be wilful; a comical series of questions will make a patient who has not winced under the wire brush smile; the suggestion of a vaginal examination will make her blush; and a skilfully provoked petulant answer to an invidious remark will demonstrate the patient's simulation.

Illusional transformations of sexual sensations are a fruitful cause for insane ideas in hysterical lunatics. Most of the accusations of rape made against physicians and dentists, and of almost daily occurrence in asylums, are made by insane hysterical patients.

The prognosis of this form of insanity is unfavorable as to the ultimate termination of the case. Temporary recoveries are noted and are as suddenly established as many of the other transformations of the hysterical state. But a recurrence is very probable, and with each recurrence deterioration becomes more marked.

CHAPTER XVIII.

Epileptic Insanity.

Most of our psychiatrical and medico-legal authors, in discussing the medical or legal relations of epilepsy and epileptic insanity, limit their attention to, firstly, the condition called epileptic mania; secondly, to epileptic dementia, and, thirdly, to the peculiar change of character which many epileptics manifest.

This series is, however, far from perfect, and fails to include many important conditions which are allied to and dependent on epilepsy, and which, on the one hand, may require special medical treatment, and, on the other hand, merit the serious attention of every thorough and conscientious medical jurist.

It is an opinion quite prevalent with many, that the epileptic, unless chronically demented, and aside from the period just preceding and following the attack, and the attack itself, is always sane from a medical, and competent and responsible from a medico-legal point of view. This view is held by many general practitioners of medicine, and by most English medico-legal writers. On the other hand, there are those who, as soon as they find the slightest indications of epilepsy in the person under investigation, instantly jump at the conclusion that, *ergo*, that subject cannot be of sound mind or responsible for any transaction performed by him. This view, as the reader will already have anticipated, has had its origin among those who have been or are frequently called by the defence in criminal cases, where insanity is the last resort of the defendant. Both views constitute utterly erroneous extremes, but they are not only erroneous, they are and have been damaging to the cause of justice, inasmuch as interested or possibly unscrupulous medical witnesses have been able to fall back on such views enunciated in published works, in support of testimony which has too often defeated the true purpose of the law.

Aside from EPILEPTIC DEMENTIA, a mental degeneration which is intimately dependent on the frequency of the convulsive attacks, and which, as Esquirol has graphically delineated, may determine stupor, imbecility, or actual idiocy, according as these attacks begin later or earlier in life, aside also from those attacks of furious madness, or purposeless automatism *replacing* the convulsive attack, and which may be regarded as *psychical equivalents* of the convulsion, there are forms of more or less protracted insanity which follow some individual epileptic attack or break out in the interval, or finally extend over the entire interval, which are to be strictly distinguished from these forms.

It was the observance of the new forms, without any differentiation from other varieties, that led Calmeil to say that those epileptics not yet insane are very irascible, very impressionable, inclined to false interpretations, and to exaggerate the importance of petty affairs. This description is probably based on cases of commencing intervallary alienation, and it would be erroneous to extend it to most epileptics living without asylums. The same remarks apply to Baillarger's statement, that the characteristics given by Calmeil often precede the outbreak of complete insanity.

Both these authors seem to have distinguished but imperfectly between actual intervallary insanity and the ordinary change of character discovered in the interval. Delasiauve has also doubtless confounded ordinary epileptic dementia with post-epileptic or intervallary conditions when he speaks of patients afflicted with "stupidité des épileptiques" as performing automatic acts, looking like drunken men, etc.

Falret opened the way for a rational classification with the following *dictum:* "A remarkable phenomenon which frequently complicates the incomplete attacks of epilepsy, or the interval between two perfectly developed attacks, deserves mention. The patient seems to have come to himself; he enters into conversation with the persons who surround him, he performs acts which appear to be regulated by his will, and seems, in one word, to have returned to his normal state. Then the epileptic attack recommences, and as soon as it has ceased and the patient has recovered his reason, it is found, to one's surprise, that he has not preserved any recollection either of his words or acts which were said and done in the interval of the two attacks."

Under the head of "Petit mal intellectuel" (not to be confounded with petit mal ordinarily so called), the same author describes a condition which may continue for several hours or several days after the post-epileptic stupor has subsided, in which the patient becomes sullen, deeply dejected, very irritable, and feels an utter inability to fix his thoughts or to control his will.

Under the head of the "Grand mal intellectuel" he describes an analogous but longer lasting condition coupled with alternate stupor and attacks of furious excitement.

As Samt correctly remarks, the recognition of these forms was an important step in advance; but these do not exhaust our knowledge of the possible forms of post-epileptic insanity. He includes both the *petit mal* and *grand mal intellectuel* of Falret under the head of acute post-epileptic insanity, and defines the latter as insanity immediately following the convulsive paroxysm, and taking an acute course. He subdivides this acute form into:

1st. *Simple post-epileptic stupor*, which may be complicated with dreamy deliria, or with illusional or hallucinatory confusion and verbigeration.

2d. *Post-epileptic morbid conditions* of fear or fright, either simple or complicated with *délire raissonante* or great excitement. The latter form corresponds to Falret's *grand mal intellectuel*. While stupor is usually present in this form it may be so far in the background that some of the cases under this head merit being characterized as cases of partial "frightful" post-epileptic delirium.

3d. *Post-epileptic Maniacal Moria.*—This form is rare, and simulates ordinary acute mania to such an extent that even the expert may be deceived. It is only the irascible character of the mania and the suspicious manner of the patient, and, as the writer believes, the treacherous and malicious character of his violence, which enable one to distinguish this disorder from the ordinary attacks of the acute maniac, who, under appropriate associations, is good-natured and manageable, aside from his episodical furor.

Under the head of *chronic protracted epileptic insanity* he describes many cases which are evidently related to the post-epileptic forms. On the other hand, the writer has observed some cases in which gradually increasing verbigeration, delirium of a religious tinge, or maniacal attacks with or without intervals of stupor, confusion, and automatism, preceded the outbreak of a convulsion or its equivalent. Just as the forms characterized in Samt's classification were designated post-epileptic, these latter, noticed by the writer and which are far from infrequent, deserve being designated as *prodromal* or *pre-epileptic*. If the chronological relation of the mental disturbance be made a principle of classification, much confusion could be avoided by adopting the following:

1. THE EPILEPTIC PSYCHICAL EQUIVALENT, which *replaces* the convulsive attack.

2. The ACUTE POST-EPILEPTIC INSANITY, which almost immediately follows the convulsive attack (including the ordinary post-convulsive stupor as a part of the attack), or similarly succeeds the psychical equivalent of such convulsive attack. Samt states that he has observed a similar condition in connection with epileptiform uræmic convulsions in two cases.*

3. The PRE-EPILEPTIC INSANITY, which precedes the outbreak of a convulsive attack or its equivalent, and increases up to the moment when the paroxysm explodes.

4. The purely INTERVALLARY EPILEPTIC INSANITY, which, neither immediately following nor preceding a paroxysm, occurs in the interval between such. It is possible that all such cases are, after all, equivalents of imperfect convulsions,

---

* *Archiv. f. Psychiatrie*, vi., p. 143.

but as long as the relation cannot be clearly established it is well to provide a category for the reception of such doubtful cases.

It is possible for all these forms to occur together, and in addition there is very apt to be a background of protracted epileptic dementia to complicate the picture. It is only when epilepsy is recent that the above forms are found in an unmixed state; as the disease progresses we are very apt to find that the post-epileptic *grand mal intellectuel* of Falret and Samt is in intimate association with a "replacing" attack of violence. Such cases, lasting with their correlated stupor, delirium, and confusion for entire weeks, figure as "epileptic mania" in our asylum records.

A very marked case of the *grand mal intellectuel*, occurring in a recent case of epilepsy, interesting because of its mal-recognition and subsequent termination, and which came under the writer's notice, may illustrate some prominent features better than any hypothetical description. From the history it is evident that pre-epileptic insanity had been also present.

On the 29th of December the writer was hurriedly called, in the evening, to a police officer who was stated to have "fits" at his residence. On arriving, he found the patient, a powerfully built man, standing up in the middle of the room, his relatives holding him. The patient was muttering unintelligibly, but recognized that a stranger had come in, though supposing, at first, that the latter was the police surgeon of his precinct, with whom he was personally well acquainted. As the number of persons surrounding him, with the intention of restraining him, was evidently a source of excitement, the writer ordered all but a few to leave, and, slapping him on the shoulder, told him everything would be "all right" if he would sit down. He obeyed in a dazed and bewildered manner. During conversation with him he seemed to awaken out of his dreamy state several times and then would attempt to arise, but could be easily prevented by manual restraint and would speedily forget his intention and continue the interrupted conversation. He looked suspiciously and furtively around, and seemed to be suffering from a general oppression and vague fear. His pupils were moderately dilated and the face considerably congested. To an ordinary beholder he would, for considerable periods, give the impression of perfect mental equilibrium, speaking about the details of his duty and the personalities of higher police officers in a quiet, deliberate, and apparently intelligent manner. But he seemed to enter into such conversation more with the idea of getting rid of the questioner, and of the restraint which was imposed on him to prevent his sallying forth to the street. He was dressed in a morning gown, but had thrown his police coat over it and put on the police hat and was trying the different doors, from which the writer had had the keys removed after locking them. He then endeavored to go out of the window, laboring under the idea that, as he would be dismissed from the force if absent continually from duty, he had to get out of the house somehow. It turned out that all this time his diseased condition had been recognized at headquarters, and he had, to his own knowledge, been excused from duty several days previous. He now became violent, but still discriminated between the members of his family, whom he treated both with physical violence and profanity, and the physician, whom he treated with profanity only. When his wife reminded him of his discourtesy, as a *ruse* to divert his attention from his ideas of escape, he said, "Oh, ——, the doctor knows how it is himself." One of the children whispered to the other to close a door which had not been locked. He seemed to hear this and started for it. The writer followed and closed with him to prevent his passage. In attempting to overcome the obstacle which was made to his passing he fell down ; then he said, "Very well, I knew I was going to be murdered," and could not get up till the writer assisted him. The writer turned him right about in raising him and the patient continued his search for the open door, but went in the opposite direction and returned to the room from which he had started without noticing it. He became considerably excited about the absence of his shield and watch. His wife refused to say anything about them for fear that he wanted them to go on his imaginary duty with. He became more and more excited, but would pass to other topics, and rested in the chair from physical weakness, having fallen to the ground on several occasions. When his wife told him that she had his watch and shield he seemed satisfied, and began to talk as if he were in the station house, spoke to the writer as if he had been one of his colleagues, and related incidents and arrests in a wearisome, monotonous way. It was found now that his tongue was tremulous and deviated to the right side, the facial muscles of the left side were more firmly contracted than the right, but there was no noticeable facial deviation. He again wanted to go out with his hat and in a red shirt, and had entirely forgotten the fact that he had been excused from duty, as was shown by conversation. The sedative which had been ordered now arrived, and the writer readily induced him to take sixty grains of bromide of sodium with five grains of chloral. After fifteen minutes he seemed a little more rational, recognized that he was at home, and was induced to go to bed, after it was proven to him that it was night and not noontime, as he was supposing. His previous history was as follows : Half an hour before he had been visited, he had, while standing, "suddenly craned round" on his left side, his head "twisted" to the left, his eyes "rolling" in the same direction, and he was "perfectly stiff," then he had violent spasms, and "worked" with these several minutes; after "a short spell" he got up and acted as if drunk, continuing to manifest similar symptoms to those which are above described, but had one attack of furious violence before the writer came. Two days previous (the 27th) he had been relieved from duty and sent home, under some pretext, as it was not easy to reason with him. That he had had an attack of mental confusion was evident, as one of his brother officers subsequently delivered the watch and badge to his wife. He was "flighty" on arriving home, and on several occasions supposed himself at the police station instead of at his house, and reproached his wife with being in the officers' waiting room, and counselled her to go home. On previous occasions he had suffered from violent neuralgia, which increased in severity until he became unconscious or stupid, after which, the neuralgia disappearing, he became delirious. This had occurred within the last five years probably a dozen times, but the only pronounced epileptic attack which he had had was the one following which the writer saw the patient. He had been a drinking man, and the police surgeon had made the diagnosis "alcoholism." During the night, after receiving the sedative, the patient slept fairly, but awoke twice, and on one occasion went into the street to patrol, and was brought back by another policeman, and the police surgeon again took charge of him. The writer saw him the following morning, but did not treat him, as his regular attendant had seen him shortly before. The patient was found half undressed, eating his breakfast, his face extremely turgid and congested, his mind very much confused. He could enter into ordinary conversation for a few moments consecutively. So far the writer's observation went. Having occasion to attend another member of the family, he learned that the case had terminated in a very abrupt and unexpected manner. The patient went out in full uniform at ten o'clock of the second morning following and patrolled Fiftieth and Fifty-first streets, without exciting any attention, his behavior not appearing strange at all. His actual "beat," however, was Grand and Houston streets, three miles distant, and toward noon he went to Fifty-third Street, stating it to be Grand Street. The police surgeon had, meanwhile, ordered him to be taken to a certain hospital. This was accomplished by deception, a neighbor getting him to arrest her little boy, whom he took to the hospital in triumph. Once there, he was placed in a strait-jacket and breathed his last in this apparatus that same evening.*

This was a case in which pre- and post-epileptic insanity were combined. Other cases differ only in exhibiting either more violence or some predominant delusion or hallucination. The writer may refer, as an example of pre-epileptic mental disturbance, to the case of a little girl described in an earlier journal article, who manifested an hallucination which gradually increased in intensity until the convulsive paroxysm exploded. In this instance the hallucination finally became intervallary, and disappeared entirely with the disappearance of the epilepsy.

The following case illustrates the career of an epileptic, marked by numerous characteristic attacks of epileptic insanity. Dating from his thirty-ninth year, the patient, a prosperous and intelligent business man, for the fifteen years of his remaining life had epileptic convulsions at intervals of from one week to several months. His business associates observed that he forgot important business transactions, claimed to have signed vouchers which he never had signed, did not recollect having signed others which he had signed, and became abstracted and dreamy, on one occasion undressing in the office. Several years prior to his death he voluntarily relinquished the responsible position he had occupied for an humbler capacity in the same business. For some days following each epileptic attack he was unable

---

* It is a question whether the fatal termination was not precipitated by the improper application of restraint.

to attend to his business affairs, and this became more and more noticeable as time advanced. On one occasion he had an outbreak of furious mania, breaking and destroying everything within reach; this was followed by a state of alternate excitement and stupor, he yelled that he was being murdered, that people were setting the house on fire, and it required the force of several men to hold him in bed ; he, on one occasion, got out on the staircase in his night-shirt under delusion of mortal danger. This lasted for several days, when he was transported to the lunatic asylum. The coach driver induced him to enter the carriage under pretext of taking him out for a drive. When he entered the asylum he was indifferent with reference to the trick played upon him. This indifference is a characteristic feature of epileptic insanity, of this variety. Later he was alternately clear and excited, at other times in a drowsy condition. After an asylum sojourn of thirteen days he left the institution physically improved, but in a dazed and dreamy state. In the signature to a will, made a few days later, extraordinary tremulousness and irregularity were manifested. The lines were broken, one "s" looked like an "e," the scrawl was almost illegible, and the name "George" appeared as if it had been written Georger" or "Georgia." His ordinary handwriting was a good, clear, average business hand, and he had been in the habit of signing himself "Geo.," abbreviating the "George." In the signature there was a gap and covered-up break between the first three letters and the last three, as if the decedent had started to write his usual abbreviation, and completed the full name, probably on suggestion. Nineteen days after making this will he was readmitted to the asylum after another attack of violent insanity similar to the one preceding his first admission. The same coachman drove him to the asylum under the same old pretext of driving him to Coney Island. On his reception in the asylum he was found stupid, presenting marked tremors, and for several days he had to be fed with the stomach pump, he refusing food under the delusion that he was being poisoned. His tremor continued, gradually increasing, while his stupor deepened to coma, and he died six days after his admission.

The immediate prognosis of epileptic insanity is favorable as regards the more acute explosions. The protracted forms are sometimes recovered from, but here mental enfeeblement is more likely to ensue than in the former. The safety of society demands that epileptic subjects should be under some surveillance after being discharged from an asylum, for the epileptic psychoses may break out with great suddenness and lead to the most deplorable results at any time.

CHAPTER XIX.

PERIODICAL INSANITY.

Periodical Insanity is characterized by the recurrence of mental disorder at more or less regular intervals; the attacks being separated by periods during which the patient presents a state of apparent mental soundness.

Periodical insanity is in most cases what Krafft-Ebing terms a degenerative insanity, being the manifestation of an hereditary or acquired vice of the constitution, and shares the bad prognosis as to recovery with other degenerative disorders, such as monomania and epileptic insanity. Like these forms it is, in the vast majority of the cases, hereditary, and may in exceptional instances arise after an injury to the skull, or from prolonged alcoholic excesses. Occasionally the outbreaks of the disease are coeval with certain physiological periods; this is notably the case with those periodical derangements of females, which either precede, concur with, or follow the menstrual period, and which are sometimes designated as *menstrual insanity*. Inasmuch as the menstrual condition is not the true cause of this insanity, but merely an exciting factor—the real cause being the hereditary neurotic vice—menstrual insanity cannot be considered a separate clinical form; in the majority of cases it is only a variety of periodical insanity whose periods coincide with and are determined by menstruation. There are other mental disorders in females influenced by menstruation, and the acceptation of the term "menstrual insanity" would hence involve much confusion.

The general feature in which all the periodical insanities agree is, as indicated in the definition, their more or less regular recurrence; a recurrence as marked as, and in many respects analogous to the recurrence of epileptic fits. Just as the epileptic fits are merely the periodical exacerbations of a deeper constitutional condition—the epileptic state; so the attacks of periodical insanity are the manifestations of a chronic morbid state of the brain. And where this illness is not the expression of an hereditary taint, it is provoked by such causes which, like traumatic injuries and alcoholic excesses, may imitate the evil effects of heredity, and artificially produce a disposition to nervous and mental disease.

The regular recurrences of the morbid explosions in periodical insanity have induced not only the ancients to suspect a relation between them and the influence of the lunar changes, but within the year a German alienist (Koster) has published an elaborate treatise to prove that this recurrence is in periods of seven days or in multiple days of seven, determined by the apogee and hypogee of the moon. There may be a dependence of this kind, but the writer is unable to consider it a direct one, but rather as one possibly determined by the general bodily condition at such periods. Barometric and seasonal variations appear to exercise a much more palpable influence on the outbreaks of periodical insanity.

The theory that periodical insanity is the expression of a degenerative taint is supported first by statistics, inasmuch as the majority of the patients have a bad family history; secondly, by the frequency with which somatic signs of degeneration, cranial anomalies, and other evidences of disturbed development are found in them; and thirdly, by the fact that the beginning of the disorder coincides with certain physiological periods, such as puberty and the climacteric, while its exacerbations often follow other physiological periods, such as menstruation; this, it is now generally admitted, is a feature of the degenerative psychoses.

A very important characteristic of periodical insanity is the similarity of the manifestations in the different attacks with the same patient for long periods. Whatever form one given attack takes, that form is destined to characterize the subsequent attacks for many years. The earlier attacks are sometimes abortive and do not resemble the later ones; and, as the disorder progresses, the attacks become, as a rule, more severe; but for any period extending over a number of years the attacks are so similar, that the same morbid propensities, the same imperative conceptions and impulses, the same delusions, hallucinations, nay, the same insane language, occur with a regularity which is not the least striking feature of periodical insanity. This is not unlike what is sometimes observed in epilepsy, where for long periods the same aura, the same form of attack, and the same post-epileptic phenomena are found with each explosion. It is not without some bearing on this similarity that the sufferers from periodical insanity—at least in the experience of the writer—show epilepsy in the direct and collateral family lines more often than other insane patients do.

The intervals between the periodical outbreaks are not always entirely lucid, but rather sub-lucid. The patients are reasonable, capable of attending to their affairs, and a few may exhibit nothing abnormal even to an experienced alienist. But most are what is called "nervous," the female patients particularly are apt to be markedly hysterical, and a morbid irritability is quite a common feature. In advanced periodical insanity the patients exhibit a permanent change of character; they become indifferent, their emotions are blunted, their mental energies decrease, and morbid irascibility becomes more prominent. In this respect there is another close resemblance between the epileptic neuroses and the periodical psychoses, for while in the earlier periods of epilepsy the inter-epileptic states may present nothing noticeably abnormal, as the disease progresses an epileptic change of character usually becomes a more and more marked feature of these intervals.

With the exception of certain cases classed as circular insanity, the inception and termination of the periodical outbreaks are more abrupt than in simple mania and melancholia, which these outbreaks may otherwise resemble. In addition the deliria if present are apt to be of a reasoning character, while moral or affective perversion, and certain propensities and impulses not ordinarily found in the simple insanities, serve to indicate the character of the disorder. Aside from these signs it is only the history of the case, revealing the periodical recurrence of similar attacks, which serves to justify a diagnosis that the disorder is probably a periodical one.

PERIODICAL MANIA generally begins abruptly, though sometimes it is inaugurated by a brief period of depression. More frequently signs are observed which Krafft-Ebing happily compares to a pre-epileptic *aura :* the heart palpitates, while vertigo and fluxionary head symptoms and neuralgic signs are precursors of the maniacal explosion.

The latter is marked by angry rather than pleasant excitement, by moral perversion rather than by sanguine exaltation, and by what the French term *delire des actes;* namely, a tendency to continuously perform acts impulsively, such as sexual excesses, indecent assaults and exposures, alternating with thefts, incendiarism, and errabund * tendencies, rather than by the ambitious, teasing, and jocose acts of simple mania. In females the tendency to cast aspersions on other females is pronounced, and almost characteristic. With these symptoms, which are termed deliria of acts, occasionally hallucinations, more commonly illusions, and rarely delusions, may be added, there are frequent outbreaks of angry excitement, and these are of a violent and dangerous character. They are sometimes provoked by the alcoholic excesses to which periodical maniacs are so likely to resort; and it is observed, even in the free intervals of periodical mania, that alcoholic beverages are not borne well, a moderate indulgence leading to disproportionate disturbances of consciousness and of the will power.

Instances are recorded where a single morbid propensity has been the most prominent and constant feature of periodical mania. Certain sexual aberrations (p. 40) are particularly apt to be manifested in this way; as in the case of a lady observed by the writer, who exhibited violent fits of jealousy, in one attack leading to a sanguinary suicidal attempt growing out of a sexual perversion of a platonic character. Most of those patients described as kleptomaniacs are periodical maniacs, in whom the propensity to steal predominates over the ordinary symptoms of mania.

From the cases of periodical mania in which kleptomaniac and other morbid impulses predominate over exaltation, the transition to those forms in which the morbid impulse is the sole manifest symptom is natural. Almost any one of the known forms of morbid impulse may appear in periodical phases, but this is particularly the case with the morbid craving for drink, which seizes on its subjects at certain intervals with such intensity that the ordinarily quiet, orderly, refined, and sensitive patient, losing all sense of propriety and shame, gives himself up to unrestrained and ruinous debauchery. This distressing condition is known

---

Mania errabunda is a term which has been indiscriminately applied to periodical, pubescent, and paretic insanity whenever the tendency to roam about aimlessly has been a marked, however temporary, feature of the insanity.

as DIPSOMANIA. It is to be distinguished from inebriety and alcoholism; for the inebriate is not driven to his excesses so suddenly and irresistibly, nor does he cease them as abruptly, as the dipsomaniac. In the inebriate the motive grows out of appetite and habit; in the dipsomaniac it is a blind craving which, if it is not stilled by alcoholic beverages, will seek some other outlet. Often these patients develop a morbid craving for certain narcotics, and we may thus have a periodical craving for opium analogous to the periodical craving for drink, and as distinct from the ordinary opium habit as dipsomania is from inebriety.

As a consequence of his blind indulgence in drink during his diseased periods the dipsomaniac may become the subject of acute alcoholic delirium, or of chronic alcoholism, though the latter is rare; these conditions are to be looked upon as results, and not as essential features of dipsomania, which is to be defined as *a form of periodical insanity manifesting itself in a blind craving for stimulant and narcotic beverages.* The relationship of dipsomania to the other periodical neuroses is well illustrated by the instance—not to go beyond cases already cited—of the lady suffering from periodical exacerbations of sexual perversion, who had a father and two brothers dipsomaniacs, and one sister suffering from periodical neuralgias, another from periodical gloomy spells.

PERIODICAL MELANCHOLIA presents no distinguishing marks from ordinary melancholia in its individual attacks. Its periodicity is its sole criterion. It is worthy of note that periodical melancholiacs are the most persistent, cunning, and successful of all suicidal lunatics.

Periodical insanity does not always manifest itself under the guise of a single form of derangement. There is a subdivision known as CIRCULAR INSANITY (*cyclothymia*) *which is characterized by the alternation of mania and melancholia in regular recurring cycles.* In the marked cases, for example, a profound melancholia is followed by a violent mania, this by a lucid interval, and then the melancholia, mania, and lucid interval return again and again, in the same order, comparably to the succession of the cold, hot, and latent stages of an attack of intermittent fever. In some cases there is no free interval, the mania begins when the melancholia ends, and the latter is immediately followed by mania. This is the variety to which Falret first applied the term *folie circulaire;* while Baillarger subsequently distinguished those cases in which a more or less perfect and prolonged lucid interval is interpolated under the designation *folie à double forme.*

The order of each cycle varies in different patients : the mania may precede the melancholia or *vice versa.* Both may be of a mild type, and both may be very severe; or one may be slight and the other intense. A furibund maniacal attack may open the scene and be followed by a mild depression; and a simple exaltation may be succeeded by a profound melancholia, with anxious delusions, hallucinations, and suicidal inclinations. On the other hand, a mild melancholia, not exceeding the limits of a moderate degree of inertia, may be followed by violent agitation and destructiveness; while a melancholia, so intense as to approach the degree of cataleptic stupor, may give way to a sanguine exaltation of spirits, scarcely meriting the name of a mania. It is such cases as the latter which constitute connecting links with the ordinary forms of periodical insanity above considered.

As a rule the mania and melancholia correspond to each other in intensity. Where the cycle is of brief duration, lasting a few days or weeks, both are apt to be very well marked; where it is of a duration of months both are apt to be of a mild type. In some cases the patients seem to be oscillating between extreme moods, which show an alternation like that of circular insanity, throughout their lifetime; for weeks and months such subjects are sanguine, loquacious, energetic, indulging in expensive and ambitious schemes, and then during the next few weeks or months they are just the reverse; they seem deprived of all hope, are taciturn, inactive, regret their extravagances, and undo what they have undertaken. Such individuals are a constant source of anxiety to their relatives, and of danger to themselves. It requires but a slight circumstance to lead them to the wasting of their fortune, to other extravagant acts, or to develop an attack of furious frenzy, during the exalted period; while during the period of depression they may allow a flourishing business, undertaken in the exalted mood, to go to ruin from inertia, or even commit suicide. It is in cases of this kind that we are least likely to have a lucid interval. Where the maniacal and melancholic stages are most clearly marked, on the other hand, we may find an equally well marked period of unquestionable mental health separating the morbid periods. Instances are related where a patient has been maniacal one day, melancholic the next, lucid the day thereafter, then maniacal again, and so on. The writer believes such cases to be exceedingly rare, the shortest cycle he has seen in over fifty cases lasted from ten to twelve days, and ordinarily each stage covers from a week to a few months. It has been noted that in some cases the alternation corresponded to the seasons: the patients being melancholy in winter, maniacal in spring, and lucid during the summer, developing melancholia again in the fall.

As a general rule—not, however, without numerous exceptions—the shorter the cycle the more intense are the symptoms, and the better also are the prospects of the case. Some German observers have found that the patients gain in weight during the maniacal, to lose in weight in the melancholic period; but this is not a constant phenomenon.

The differential diagnosis of circular insanity can usually be made only by learning the history of the case. The characteristic feature which serves to distinguish it from other forms of insanity is the alternation of the opposed conditions of mania and melancholia, and this alternation can be gleaned only from the history, or detected by keeping the patient under prolonged observation. During the maniacal stage, as a rule, it is impossible to discover any difference from an ordinary case of simple mania, while during the melancholic stage it is equally impossible to recognize any feature not to be found in simple melancholia.

That it is of the highest importance to discriminate between an ordinary mania or melancholia and the maniacal and melancholic phases of circular insanity, must become evident when it is borne in mind that the prognosis in the former affections is in the highest degree favorable, while in circular insanity it is most unfavorable. In some cases, the mania and melancholia may present a "reasoning" character, and thus lead to the suspicion that the insanity is circular aside from the confirmation furnished by the history. But even the observation of an entire cycle does not establish the existence of this form of insanity; for it is well known that a simple mania may be preceded by depression, or pass to recovery through a stage of stupor, and thus resemble such a cycle. It requires the demonstration of several cycles to make the diagnosis of circular insanity complete.

We have further strong reasons to suspect the existence of this disorder, if during the free interval the patient is noticed to be morally perverse. As in periodical insanities, generally, the lucid, or rather sub-lucid, intervals of circular insanity are often marked by anomalies of character; we find these patients in these periods of this disorder to intrigue against their surroundings merely for the love of intrigue and the delight which they experience at annoying others. Neither the true maniac nor the melancholic patient ever manifests this. The former delights to tease, but usually in a good-humored way, not from malicious inclinations; the latter prefers to be let alone.

Occasionally we are aided in our diagnosis by narrowly watching the transition between the mania and melancholia. Usually this transition is very abrupt and complete; the patient goes to bed melancholic and rises maniacal; it is uncommon for the maniacal and melancholic symptoms to balance each other, so as to constitute a para-lucid transition.

Circular insanity generally begins at or about the age of puberty, is, like other periodical insanities, more frequent with females than with males, is intractable to treatment, and while it does not ordinarily lead to dementia, some mental deterioration is manifested in its subjects sooner or later. The reported cures are few, and, as far as can be gathered, the diagnosis was not well established in the majority of these. It is to be borne in mind that the hysterical psychoses as well as malarial neuroses may exhibit an exquisite circular type of insanity. The writer has seen this latter phenomenon, and succeeded in controlling the disorder with quinine and calomel; but he regards such a case as a cured malaria which manifested itself under the mask of a cyclical insanity, and not as a true cyclothymia, which is the expression of an essentially cerebral and deeply-rooted vaso-motor neurosis.

As previously stated the periodical insanities are more frequent in females than in males. Among other causes which account for this difference in the sexes is the fact that uterine disorders frequently act as exciting causes of the malady in predisposed subjects; this may account for the few reported cures effected by gynæcological treatment. Of 2,297 male pauper patients, the author found five per cent suffering from periodical and its sub-group circular insanity. Unfortunately some cases of so-called "recurrent mania" were included in the computation, so that the correct figure would probably be nearer four per cent.

CHAPTER XX.

The States of Arrested Development.

By many the conditions known as idiocy, imbecility, and cretinism, have been considered to occupy a position separately from insanity proper. To-day we know that the typical psychoses of the neuro-degenerative series may arise on the basis of the same or similar developmental defects as those which are so characteristic of the states of arrested and perverted development. We also know that this fact is in harmony with the observed "transformation" of the ordinary forms of hereditary insanity into idiocy and imbecility in the course of hereditary transmission; and that the clinical manifestations of the latter are sometimes in the same direction as those of insanity proper. For all these reasons it appears inexpedient to make a sharp separation.

It is customary to distinguish three grades in this group. To the subject deprived of all higher mental power, and who is unable to acquire the simplest accomplishment, the term IDIOT is applied. He who is capable of acquiring simple accomplishments, but unable to exercise the reasoning power beyond the extent of which a child is capable, is designated an IMBECILE. Finally, there is a large class of subjects who are defective as to judgment and in whom this defect is of similar origin to, though not as intense as, that of the imbecile and idiot, who are termed FEEBLE-MINDED.

There is a complete series of transitions, beginning at the lower end with the non-viable anencephalous monster and passing up through the brain-monstrosity, the microcephalus, the idiot, the imbecile, and the feeble-minded, to the normal person. This transition is at once structural and physiological. (Appendix IV.)

In idiocy there is usually, in addition to the mental defect, some deficiency in the peripheral organs or their functions. Many idiots are deaf or mute or both, some are blind, and anæsthesia as well as anosmia have been observed. They learn to walk late or not at all, and those who learn to walk have a shambling, shuffling gait, which, in the case of the microcephali, is said by Vogt to resemble the mode of progression of the anthropoid apes when erect. The skeleton is usually poorly developed, rachitis is common, and the somatic functions generally are imperfectly performed; the sexual organs particularly are found to be rudimentary or deformed.

On comparing a large number of idiots the reflection forces itself on the observer that three different sets of causes of arrested development are active in producing this condition.* In some cases we find that one of the parents of the idiot has an abnormal cranial shape or premature ossification of the sutures, and is himself or herself insane, epileptic, hysterical, or feeble-minded. Here a transmission and intensification of the ancestral defect is to be assumed to have taken place. In another group of cases we find that the parents were originally mentally healthy, but that the fœtus has been injured or has acquired some constitutional vice, such as syphilis *in utero*, or suffered from some brain disorder such as epilepsy, eclampsia, or meningitis in infancy. Ireland has found idiocy resulting from brain disease as late as the tenth year. In the third group, the smallest, an *atavism*, that is, a reversion to the hypothetical ancestry of man, has been suspected. This claim of scientists must not be confounded with the paradoxes involving formally similar views which best flourish in a soil untilled by either anatomical, physiological, pathological, or clinical observation. It is a fact which may retain some degree of that same historical interest which, as the writer has stated, he believes will cling to the views emanating from the laboratory of the Utica Asylum (page 96), that a recent course of demonstrations, which were illustrated by patients from the Ward's Island pauper asylum,† called into being the following gem of combined psychiatry and zoology, which should not be lost:

---

* Dr. Ireland classifies idiocy as follows: 1. Congenital. 2. Microcephalic. 3. Eclamptic. 4. Epileptic. 5. Hydrocephalic. 6. Paralytic. 7. Cretinic. 8. Traumatic. 9. Inflammatory. 10. Due to deprivation of the senses. The microcephalic idiots are always congenital idiots, while paralytic idiocy is really mental impairment from organic disease, and, as the subject may regain mental power, should not be classified with idiocy. The last group is not a real idiocy, any more than Casper Hauser was an idiot, because he had had no opportunity to learn.

† Clinical Lecture on Dementia, Idiocy, and Imbecility: being the third of a course of four lectures upon the diagnosis of insanity. Delivered at the New York City Lunatic Asylum, Ward's Island, by A. E. Macdonald, M.D., Medical Superintendent.—*N. Y. Medical Record*, Dec 20, 1879.

"Here is a negro whose feet look as if they were formed to clutch the limb of a tree, and it does not require a great stretch of the imagination to picture his ancestors, in no very remote generation, jumping from limb to limb of some African forest.

"And with this return, if we may so call it, toward the appearance and form of other animals, there is an equally perceptible return in habit and action. The place of intellect seems to be supplied by instinct, and by it the behavior is apparently often governed. Thus in a recorded case, an idiot girl, having, while alone and unattended, given birth to a child, turned, with the instinct of an animal, and gnawed the umbilical cord. *Commonly there is a consistent imitation of the habits of some one animal*, and its posture and movements will be assumed, and its habits copied even to the extent of showing a preference for whatever forms its natural food. I have read of a case where a woman lived and acted like a sheep, and ate grass; and I know of a case where a young man has all the habits, and a good deal of the appearance, of a well-conducted horse. He harnesses himself to a wagon every morning and trots about all day, switching a tail which he has fabricated out of old rope, and so great is his consistency that he never fails to shy at a wheelbarrow."*

Bucknill mentions cases of lunatics (mostly imbecile) who believe themselves changed to toads, to oil-flasks, jump and flutter like frogs and bats, making all the while a sound like these animals. Esquirol reports that in a certain convent the monks believed themselves to be cats, and at a

---

* The writer takes it for granted that there are no embellishments in this account, although he has frequently seen the patient and never observed the last symptom, which, whether it existed or not, had, it is needless to say, no bearing on the question of a reversion to the equine instinct. Huxley, Darwin, and Haeckel ought doubtless to appreciate the friendly assistance thus afforded them by Dr. Macdonald; but they will find the subject of the descent of man somewhat complicated by his theory, for it is, to say the least, difficult to believe the ancestry of the human being to have started pentadactylous, become artiodactyle (sheep, according to him), monodactyle (horse, same authority), and then pentadactylous again! As to the negro, "Cuffy," whose feet are the strongest support of the Doctor's theory, they happen to be the seat of a symmetrical deformity which is in a direction altogether the opposite of a simian reversion. The great toes are long and stand out at wide angles from the line of the next toes, and are *less* apposable than in the normal human foot, while they should be more so than in the latter to justify the flowery language of the quotation. The italics are the writer's.

certain hour of the day went through the performances of skipping about and caterwauling. The writer has seen patients who acted like and believed themselves to be steam engines, windmills, *et cætera*, and, if we were to apply the same argumentation running through the above extract, to these cases, we might say, it requires no great stretch of the imagination to picture his (the patient's) ancestor, in no very remote generation, as an oil-flask, as a cat in some New York back yard, as a frog in some swamp of the carboniferous epoch, as a toad in some muddy flat of New Jersey, as a steam engine in Birmingham, as a windmill on some Netherland dyke, or, finally, as a bat flitting through the darkness of some ruined castle on the Rhine, and so on *ad infinitum*.

The presence in idiots of gyri found in the anthropoid apes and negro (Zwickelwindung) and of muscles which in normal man are usually rudimentary or absent, are facts that lend some color to the view that in some cases idiocy may be due to an atavism. It is to be insisted on, however, that no atavism can ever imitate or reproduce the links of progressive development. Just as in normal development the branchial slits, the coloboma oculi, the caudal appendage, the cloaca, the supernumerary digital rays, the thirteenth and fourteenth ribs, the Wolffian body, and the carnivora-like claws and foot-pads of the human embryo imitate certain structural peculiarities of the lower creation without being exactly like them; so the cerebral and skeletal peculiarities of atavistic idiots resemble without accurately reproducing those of the ape.

It is sometimes observed that the appearance of idiots strikingly suggests a reversion to or imitation of certain ethnic types. In some cases Caucasian idiots reproduce to a perfectly wonderful degree the Mongolian features. It is to cases of this kind that Dr. Mitchell and Dr. Fraser give the name of Kalmuck idiocy. The writer has observed the same resemblance in an imbecile murderer, and in the three dwarfed idiotic brothers who are now on exhibition in a "Museum" of this city, and whose photographs accompany a paper on the subject, published by Hammond in his "Neurological Contributions." The thick lips, large fleshy tongue, bullet-shaped occiput, curly hair, and dark skin of the negro are found in another group, and Dr. Down, who first called attention to the ethnic types in idiocy, claims that not only the Mongolian and Ethiopian but also the Malay type may be found in Caucasian idiots.

It is in idiots of the "atavistic" group that we sometimes find the so-called instinctive faculties tolerably well developed, in contrast with the majority of those suffering from arrested brain development. Usually, however, and contrary to the current belief, the lower faculties of the mind suffer as well as the reasoning powers, and this is in harmony with the fact, that it is not alone the higher centres that suffer with defective cerebral development, but that the thalami, the cerebellum, and the cerebral isthmus of idiots are often found to be defective, or asymmetrical, or both.

It is probably due to a deficient trophic innervation that idiots so frequently suffer from cataract, and it must be recollected in this connection that the morbid anatomy of idiocy and imbecility is not necessarily limited to anatomical defects, but that progressive structural lesions may develop in the idiot's or imbecile's brain.*

While the sexual function is usually in abeyance in idiots, there are cases where it has been fairly well carried on. Thus, John Rouse, the celebrated microcephalous idiot on Randall's Island, is known to have had sexual relations with low women, and to have manifested at times a strong sexual appetite; and several instances are reported where idiotic girls have been impregnated and delivered of children.† The history of the confinements of idiotic mothers illustrates very finely the erroneous nature of the popular view that the idiot is necessarily a creature of strong or perfect instincts. While idiots have been confined and have like animals lacerated the umbilical cord with their teeth—an admirable provision against hæmorrhage—and evidently have done so without reflection and judgment, there are many more cases on record where both reason and instinct seemed to be altogether in abeyance. Chambeyron, in his translation of Hoffbauer's treatise on the medical jurisprudence of insanity, relates the case of an

---

* Thus Luys (*L'Encéphale*, Mai, 1881) has found that the cortical nerve cells in idiots undergo necrobiotic changes, and Brückner (*Archiv f. Psychiatrie*, xiii. 1) and Bonneville (*Archives de Neurologie*, 1, p. 81) have found a peculiar "tuberous sclerosis" which involved the cortex in numerous patches.

† Unfortunately the history of the children has not been satisfactorily traced.

idiot whom he confined, whose vocabulary was limited to the sounds "ta-ta," and who, although her pelvis was well formed and the presentation a good one, did not know enough to assist the expelling power of the uterus with her abdominal muscles; and could not be made to imitate the movements necessary, although made before her by other women. She simply fingered around her genitals in a purposeless way, and after the child was born she took no notice of it whatever.

The imitative tendencies are often very strong in idiots; in imbeciles they may be utilized to make good artisans of the subjects; in idiocy they lead to destructive and tragical results owing to the utter absence of any higher intelligence. Thus some twelve years ago an idiotic boy in Maine killed the child of the people who cared for him, hung it up, and dressed it exactly as he had seen a sheep dressed; and another, referred to by Gall, butchered a man precisely as he had seen a hog butchered.

In the lowest form of idiocy speech may be altogether absent, or limited to a few inarticulate sounds, in others a few words and short sentences may be acquired. While some idiots show no spontaneity whatever, and have to be fed like infants, others are ravenous eaters. A few exhibit explosions of furious and blind violence and morbid impulses.

Idiots rarely reach maturity.

The study of imbecility and its lesser degree, feeble-mindedness, is of much greater practical importance to the alienist than that of idiocy. It is a popular and erroneous belief that an imbecile is one entirely void of ideation. It thus happens that imbecility is often overlooked, when the subject is in the lower walks of life, and that, as Georget very happily observes, "it is above all in the inferior walks of society, where the individuals need but little intelligence to carry out simple labors, and to fulfil limited social obligations, that only those are considered imbecile who are not even able to lead a horse or to watch a herd." A degree of imbecility which would scarcely be observable to the laity in a hod-carrier, would be very manifest in a school-boy and academical scholar. The writer has known imbeciles, who had to be removed from school because unable to keep up in their studies with children many years their juniors, to become successful mechanics and good copyists. Even in asylums there are many imbecile inmates who are employed in the garden, the kitchen, as aids in the hospital wards, and who are occasionally more methodical and reliable in their limited sphere of action than the attendants placed over them are in theirs. The imitative tendencies which are more common here than in idiocy are often utilizable, as has been previously observed, in making of the imbecile not only a good, but sometimes a very skillful mechanic. His mechanical skill is mainly shown in the direction of imitation and reproduction, and but rarely in a new or untrodden field. It is similar with other mental processes. Imbeciles sometimes have an excellent memory for simple facts. There are instances on record where imbeciles have known the dates of the birth, marriage, and death of every person dying in a certain community for thirty years, or the time of departure of every railroad train that had left a certain station in the same time. But the imbecile is unable to form or to unravel those complex combinations of which simple impressions are the component units. The mental state of the imbecile has been very well expressed by the statement that those mental co-ordinations acquired in the course of a higher civilization have not been formed in him. (See page 101, foot-note.)

While the imbecile is defective as to his reasoning capacity, his emotional state may present every analogy to that of healthy persons, or approximate that of other forms of insanity. Thus, there are imbeciles who are mild, affectionate, good-natured, and even philanthropical; on the other hand, there are imbeciles who are treacherous, suspicious, and cruel. Moral defect is a prominent feature of some cases, and this condition may be the chief manifestation of mental deficiency. There are subjects whose reasoning powers are fair, whose memory is excellent, who are perhaps, accomplished in the arts, but in whom the moral sense is either deficient or entirely absent. The term *moral insanity* of authors should be limited to this class of subjects, and a much better term to use would, in the writer's opinion, be MORAL IMBECILITY.

Morbid projects, imperative impulses, and morbid egotism are found in some imbeciles, and in such cases it may be difficult to decide whether they appertain to the group of imbecility or of original monomania. There are numerous other points in which imbecility proper, and monomania, which may in some respects be considered a "partial imbecility," approach each other. Several of these have been referred to in the chapter on the somatic signs of the predisposition to insanity (page 88); a most remarkable one is the fact that one-sided talent, other than that resulting from the imitative tendency referred to, is sometimes found in imbecility, just as a similar condition is found in monomania. Thus imbeciles have been known to manifest a marked aptitude for the arts, such as music. This latter has been especially noted by Meyer in imbeciles presenting the *crania progenia*.

The one-sided development of special faculties, and the positive signs of alienation, such as moral perversion and anomalies of character, are characteristic rather of the hereditary than the acquired cases of imbecility.*

Both imbecility and idiocy are sometimes marked by other disturbances of the nervous functions than those comprised in the mind. Epilepsy is a frequent accompaniment, and may be very bizarre in those cases where it is the result of a cerebral defect, involving a few muscles or one extremity, or being associated with a choreiform aura.

Both idiocy and imbecility may be dependent on early epilepsy, but more frequently the mental defect and the convulsions are collateral phenomena, both depending on defective development.

Spastic symptoms, contractures, strabismus, peculiar speech defects—manifested in the inability to pronounce certain consonants—and stuttering are also noted, and it is the occasional presence of all these signs in hereditary monomania (Originäre Verrücktheit) that gives additional force to the view that there is no absolute line of demarcation to be drawn between the various forms of the degenerative nervous states.

The course of idiocy and imbecility is usually unmarked by any changes, and these conditions are therefore, as a rule, stable ; occasionally progressive deterioration is caused by epileptic fits, and where hydrocephalus and

---

* Krafft-Ebing asserts that the one-sided development of special faculties is never found in acquired imbecility. It seems more correct, while regarding them as characteristic of congenital, to admit their possible development in acquired forms. Exceptions undoubtedly exist ; there are well-established cases where acquired imbecility has chiefly manifested i self in the moral sphere, although moral imbecility is as a rule found in the transmitted forms. In the case of Louisa W——e, exhibited before the N. Y. Neurological Society, whose imbecility developed with a scarlatina, moral perversion was the most prominent of the constant symptoms. An analogous case has been reported by Hughes.

THE STATES OF ARRESTED DEVELOPMENT. 283

meningitis are the causes, exacerbation of the morbid process may lead to further impairment of the mind. In the latter cases the cerebral disorder may directly lead to a fatal termination. Unless syphilis is the cause, and rarely then, therapeutic measures are incapable of doing any good,

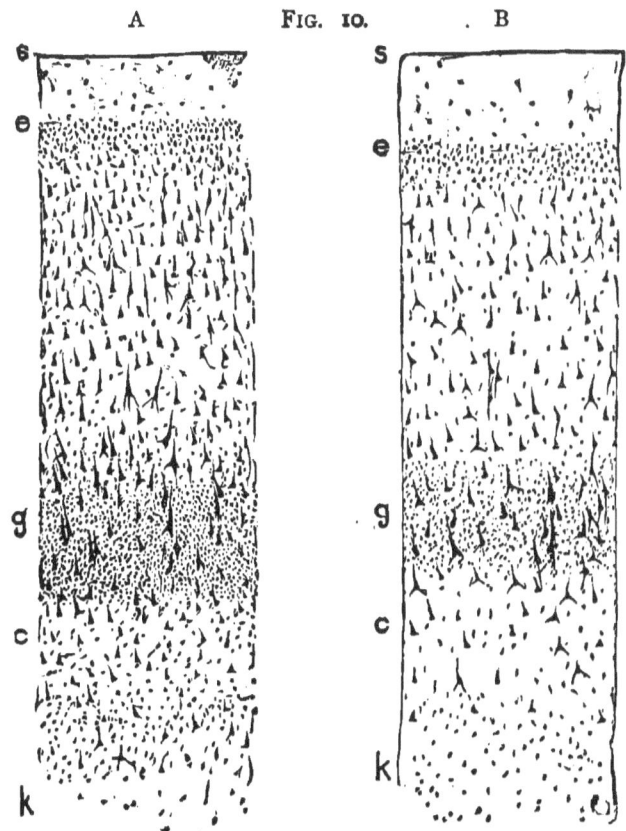

and the prognosis, as far as the development of the mind is concerned, is as bad as it can well be. As to the life of the patient it is usually shortened by intercurrent diseases, to which the feeble body of the imbecile, like that of the idiot, readily succumbs; in a few cases the bodily health may be good, and the subject reach old age.

A peculiar form of idiocy is found associated with a disorder endemic to certain mountainous districts, particularly the Alps, Pyrenees, and Cordilleras. This is a constitutional deterioration manifesting itself in pronounced anomalies of the entire physique. There is usually great physical deformity, the head appears swollen, the features are coarse, the nose depressed at the root, the belly is distended, and the cheeks puffy owing to a hypertrophy of the skin and subcutaneous cellular tissue. With this the thyroid gland is commonly enlarged. A more disgusting object than such

FIG. 11.

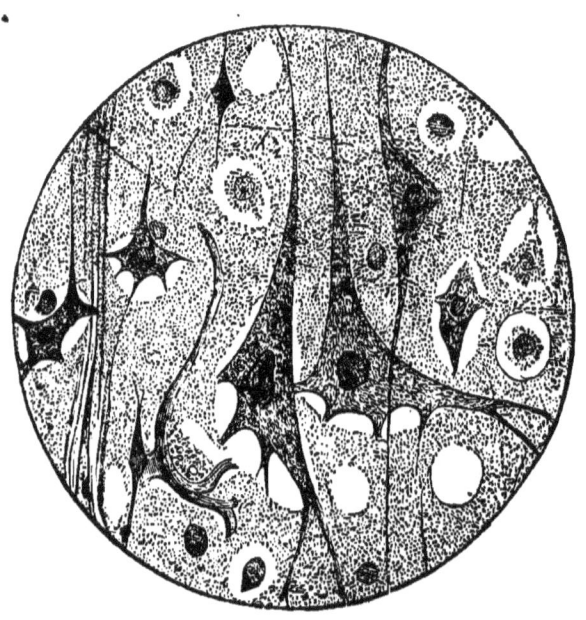

a *cretin*, with his childish expression, yet old-looking features, the deformed body and the enormous lobulated goitrous appendage, cannot well be imagined. True cretinism has not yet been observed in North America, but a similar condition has been noticed by the writer in three children of parents living in our swamps, and is probably the expression of a paludal cachexia.

The mental phenomena of cretinic idiocy are like those of ordinary idiocy. Similar physical defects are found, in addition to those which are characteristic of the cretinic state itself. Thus, cretins, like idiots, are liable to epileptic convulsions, their dentition is imperfect or retarded, and

THE STATES OF ARRESTED DEVELOPMENT. 285

the teeth decay early, and in extreme cases walking may be impossible.

To enumerate all the interesting and significant anatomical conditions, particularly of the skull and brain, found in the various forms of idiocy and imbecility, would require a special volume. In addition to those anomalies mentioned in general terms in the first part of this work, the following, found by the writer, may be briefly referred to: In two imbeciles, one of whom had a hypertrophy of the brain, that organ presenting one of the heaviest weights on record (68 ounces), there was found disproportionate thickness of the outer or barren layer (ependyma formation of Mey-

FIG. 12.

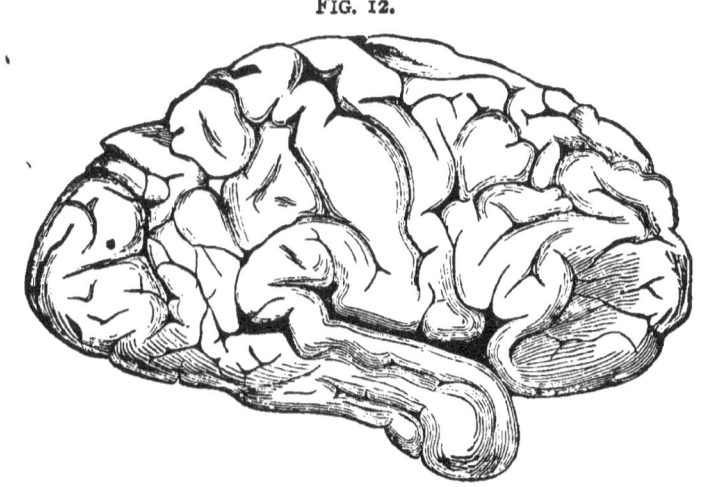

nert) of the cortex, over the entire expanse of the two cerebral hemispheres. With this there was a relative sparseness of ganglionic elements in the other parts of the cortex, particularly noticeable in the granular layers. In figure 10, A and B respectively illustrate the difference between the normal cortex and that of the imbecile; it is a suggestive fact that one of the chief respects in which the cortical structure of man differs from that of the lower animals is in the relative reduction of the barren layer at the expense of those which are rich in ganglionic elements. In the case where the relative overgrowth of the barren layer was most marked of the two referred to, much more so than in the one from which the illustration was taken, the walls of the blood-vessels were found to be sclerotic, and there was a general preponderance of connective tissue elements over the nervous structures throughout.

Another condition was found by the writer which at present he is unable to offer a satisfactory explanation of. This consists in the presence, in large numbers, of nuclear bodies, surrounded by a little granular protoplasm, and contained in clear round spaces of the neuroglia (Fig. 11). They differ from the similar bodies found in paretic dementia and in other organic diseases of the brain in the fact that they incorporate the finely granular protoplasm referred to, which resembles that of nerve cells. The same kind of bodies are found in large numbers and in special layers in the brains of the lower mammalia, and their presence in the cortex of the imbecile may indicate an arrested state of histological development.

Among the anomalies in the type of the convolutions which dave been studied in so large a number of cases, the most interesting are the deformities of the occipital lobe. The latter is often shortened, or the seat of *microgyria*, that is, abnormal smallness of the gyri, a condition which is sometimes associated with deficiency of the splenium of the corpus callosum, as was the case in one of the imbeciles who were examined by the writer.* In several Caucasian imbeciles pathologists have found the gyrus of the cuneus running superficially, as in the chimpanzee, and the so-called "Affenspalte" (ape-fissure) of the convex surface of the brain (Fig. 12) has been so termed because it imitates in disposition the opercular fissure of the apes. It is, however, not a perfect homologue of that fissure, though its presence, when it is due to the fusion of the external occipital with the internal perpendicular occipital sulcus, is a significant sign of disturbed cerebral growth.

CHAPTER XXI.

PARANOIA (MONOMANIA)—PRELIMINARY CONSIDERATIONS.

Probably no word in the nomenclature of mental science has been so confusedly used and has led to so much misunderstanding, and consequent protest against it, as mono-

---

* "The Etiology of Insanity," *loc. cit.*

mania. By this term the great Esquirol, its originator, designated that form of insanity in which, while the memory, the conceptions, and judgments generally are not destroyed, and no pronounced emotional disturbance exists, yet the patient is controlled by some expansive delusion or ambitious project. Esquirol's failure to appreciate the fundamental feature of monomania led to his separation of the same class of lunatics, whose delusions are of a formally sad character, under the term lypemania (melancholia), thus placing them in the same group with the emotionally depressed patients. The untenability of this distinction must be evident when it is borne in mind that the expansive delusion of him who to-day believes that he is a king, may to-morrow become masked by the depressive delusion, that he is persecuted by the usurper of his throne; or, to cite a case which may be better illustrated in American asylums, the projector of some insane invention, who, before being committed to an institution, revels in bright anticipations of the prospective income to be derived from it, after his interdiction will develop the depressive delusion, that the invention or its secret has been stolen from him, and that, to prevent the pressing of his just claims, he has been immured in an asylum on the certificates of conspiring physicians. (see page 28–31.)

Esquirol's pupils and contemporaries recognized that so-called "partial insanity"—which term, when divested of the erroneous conceptions still clinging to it, is a fair vernacular rendering of " monomania"—not only manifests itself in disturbances of the conceptional sphere, but sometimes also in morbid impulses and affective perversion. Here those alienists who delighted in burdening the infant science of psychiatry with new systems of classification found a fruitful field for innovation. Whatever the direction in which a lunatic manifested his most prominent symptoms, that direction determined the coining of a new term! Persons who exhibited a tendency to homicide were termed homicidal monomaniacs; those who enjoyed thieving were classed as kleptomaniacs; those who delighted in conflagrations were denominated pyromaniacs,* and so on, till no unusual act committed by the insane had been left uncanvassed. The designations "Gamomania," or "the insane

---

* It must be stated here that some of these terms, having become associated with fixed clinical conceptions, are still in use, and have acquired a definite meaning and position in science, secondarily.

desire to marry;" and "Frauenschuhstehlmonomanie" or the "mania for stealing women's shoes," are imperishable monuments of this folly. From the dignity of a clinical conception, "monomania" sank to the level of a convenient label for trivial and often incidental symptoms; and it is not to be wondered at that the earlier German and English alienists, finding this term loaded with such ballast as that alluded to, abandoned it altogether. Thenceforth, until the later revival in classification took place, monomania was crowded into the mixed group of the secondary mental states under the non-committal designation, "chronic mania" or "chronic delusional insanity."

The French writers, uninfluenced by the assaults made by Falret and Delasiauve,* continued to use the term, making the proper restrictions and differentiations from time to time. Baillarger pointed out the essential distinction between the melancholia (lypemania) and monomania of Esquirol, while Marcé showed that the gay and expansive delusion is not a universal characteristic of monomania, inasmuch as many patients with depressive delusions present the same perverse logic, and particularly the same antecedents and terminal history so distinctive of monomania as thus far recognized. He spoke of patients of that class as affected with sad monomania,† in contradistinction to the older recognized form which he classed as gay monomania.‡ With slighter modifications this sense in which Marcé employed the term became generally adopted, and monomania to this day, as a name for a form of insanity, holds its place in France—the land where it originated, and where it passed through the ordeal of a now historical discussion and criticism.

While to Morel belongs the credit of having first called attention to the intrinsic feature of the monomaniac's delusion, its systematized character,§ the Germans, who adopted the distinction of monomania in its limited and proper sense comparatively late, subsequently did much to

---
* Whose proposed "*Pseudomonomanie*" shows that he recognized the necessity for some such classification as that attempted in the group of monomania.
† *Monomanie triste.*
‡ *Monomanie gaie;* it is noteworthy that the eminent American clinician Rush, probably the most advanced psychiatrical thinker of his day, in some degree anticipated this distinction in his *Tristimania* and *Amenomania*.
§ Hence his synonym, "Manie systematisée."

establish it on a firm etiological, pathological, and clinical foundation. To Snell* and several of his contemporaries we owe a clear demarcation of the systematized insanities as a whole, under the designation *primäre Verrücktheit.*" The main fact which they determined was, that these systematized insanities are of *primary* origin, and not secondary to some other psychosis, as Griesinger and the followers of that eminent alienist believed. The latter taught that all the pure forms of chronic insanities are secondary transformations from an acute primary mental disorder of the nature of a simple mania or melancholia. This is indeed true of secondary confusional insanity (Chapter VIII.), which is sometimes the intermediate stage in the transformation of acute melancholias and manias into terminal dementia; but it was precisely the failure of the older German alienists to discriminate between this secondary form and true monomania that led them to predicate the same clinical development characteristic of the former for the latter as well, on *a priori* and speculative grounds. Griesinger shortly before his death recognized this error, and formally accepted the doctrines of Snell. To-day the German alienists, by a resolution of their association, stand unanimously committed to the recognition of a primary form of chronic insanity known by them as *primäre Verrücktheit*, and equivalent to the *monomanie* of the French, or *Paranoia.*

If we cast a glance at the earlier literature with reference to the category of patients who are classed as monomaniacs, we will find that even the popular mind appreciated in a crude way the distinctness of the morbid ideas of such subjects from the ideas of those suffering from other forms of insanity. The English word "cracked" aptly expresses that there is but a flaw and a relative shifting of the elements of the understanding, not a general confusion and annihilation of them. Where language has been used accurately such patients have neither been termed foolish nor crazy. From distant times the Germans have employed the expressive term of "*fixe Idee*," to designate the delusions and projects which are so prominent features of monomania, and which are popularly rendered by the word-

---

* It is a significant support of the views announced in this work that, in the absence of any other English term, "monomania" is a proper equivalent for the "primäre Verrücktheit" or "Wahnsinn" of the Germans. Snell himself stated that *Monomanie* is an acceptable German synonym for the latter.

pictures "a bee in the bonnet" and "a screw loose." The noun "*Verrücktheit*" is derived from the vernacular adjective *verrückt*,* which is a good metaphorical equivalent of the English "cracked," and perhaps a better designation in so far as it directs attention to the prominent feature of monomania, the mal-association of special mental components.

The authors of the most comprehensive English treatise on insanity † exemplify the unsatisfactory views prevailing on this subject in the following lines: "We heartily wish 'Monomania' had never been introduced into psychological nosologies; for if understood in a literal sense its very existence is disputed, and if not, the various morbid mental conditions it is made to include by different writers lead to hopeless confusion. With one author it means only a fixed morbid idea; with another only partial exaltation; while a third restricts it to a single morbid impulse.

If words were to be eliminated from the vocabulary merely because they do not literally correspond to their acquired and accepted meaning, more than half of those in the medical dictionary, and about nine tenths of those employed in the special branch of mental medicine, would have to be replaced by new ones. With the abuse of a term investigators have little to do; and while the distinguished authors cited are fully justified in their criticism of the onesided and narrow interpretation referred to by them, the reflection cannot be suppressed that they have themselves not accepted that view of monomania which has so much facilitated classification in France and Germany, and which view does not harmonize with their later statements: that monomania may be either "delusional" or "emotional," and may under certain constructions include both "melancholia without delusion," and "exaltation without incoherence." That monomania is a just conception of psychiatry is nowhere better shown than in the fact that the authors of this, the largest recent work on Psychological Medicine in the English language, are compelled to use it repeatedly, and find no better designation for a typical case illustrated in the frontispiece of their volume than "Monomania of Pride." Strictly, monomania may be construed to mean a fragmentary mania, or insanity on a single topic.

---

* Literally, shifted from its place.
† Bucknill and Tuke: "A Manual of Psychological Medicine," 4th edition, 1879, p. 50. In the present treatise all these objections are endorsed, but only as applying to the *abuse* of the term.

Based on this construction, authors have stated that there is no such mania, and that there is no insanity on a single topic, an objection that must be sustained at least for the majority of cases. But in medicine we cannot afford to be strict constructionists of terms to the extent of quibbling on the etymological signification of a prefix. On the same grounds we might show the terms insanity, mania, melancholia, and hallucination to be improper and inapplicable. Yet they are sanctioned by usage, and, like phthisis, tabes, hysteria, etc., against which terms similar objections may be urged, will remain in our nomenclature to the end of time. With the limitation that the prefix shall be understood to denote that the insanity extends in a special direction across the mental horizon, monomania may well retain its place in our vocabulary. The term chronic partial insanity, which might be used in its place, is objectionable on the single ground that its use would necessitate the incorporation with monomania of other forms. The term "delusional insanity" is objectionable, for the reason that delusions are not essentially features of all varieties of monomania, for they may be entirely absent in that variety which is characterized by imperative conceptions, and they are found prominently in other chronic as well as in acute forms of derangement.

Sankey, in his excellent "Lectures on Mental Diseases," writes: "The popular opinion about the existence of monomania, I need scarcely add, is a very erroneous one. The French writers use the term "monomanie" in a much more restricted sense; but, to avoid confusion, it is better to avoid the term altogether." It is to be regretted that this author, one of the best and clearest of English writers on insanity, had not investigated more closely to what patients the French apply the term monomania, and adopted the sense in which they use it, under some equivalent discriminating term. If he had done so he would not, on the very same page containing the quoted words, have confounded cases of the most widely differing forms of insanity. In fact his group of the "Chronic Insanities" is about as practically useful and as clinically sound as would be the uniting in ophthalmology of glaucoma, microphthalmus, retinitis pigmentosa, and cataract in one group of "Chronic Ophthalmia."

A third, rather popular author, Blandford, is responsible for the following opinion, which, it is not saying too much, could never have been written if he had fairly represented the views of the continental writers, even to the extent accomplished by the previous writer referred to. "Probably what is most commonly called monomania is chronic insanity, where the patient is removed from deep depression on the one hand and gay or angry excitement on the other, and when the bodily health has assumed its ordinary level, and all pathological marks have by time been effaced. The distinction between mania and monomania is for the most part verbal. Formerly all insanity was called melancholia, nowadays it is spoken of as mania, and if chronic as monomania. There is nothing pathological in such a nomenclature, and it only serves to draw us away from the due consideration of that pathology of the disease we have to consider and treat. We may retain such terms as acute delirious mania, acute melancholia, acute dementia, general paralysis, because they denote a certain set of pathological symptoms occurring in individuals of various ages, requiring special treatment and capable of receiving a similar prognosis. We may, if we like, retain besides, the general terms mania and melancholia, but beyond this we need not go; any further distinction should be made, not according to mental peculiarities, but according to the pathological causes or conditions of the case."

It is not true that "what is meant by monomania is chronic insanity, where the patient is removed from deep depression on the one hand and gay or angry excitement on the other." Some of the most violent scenes in the asylum corridor are enacted by these patients under the influence of episodical states known to the Germans as *Primordialdelirien*, to the French as "*delire vesanique.*" Nothing could be more unfortunate than the statement that monomania is "chronic insanity . . . when the bodily health has assumed its ordinary level and all pathological marks have by time been effaced." If we are to understand Blandford as meaning by "pathological marks" all the indications of the pathological mental state, then we must conclude that monomania and sanity are synonyms, for we have sanity when all the pathological marks of insanity (the most characteristic and essential ones being the mental signs) are effaced ! But if we are to understand him to mean by "pathological marks" the somatic signs of insanity, we can only conclude that he has involved himself in a profound contradiction with the best established facts of mental pathology. It was the great Morel who taught that monomania is frequently associated with somatic signs *of degeneration*. Now such somatic signs as the monomaniac sometimes shows are never, under any circumstances, effaced. This fact is accepted by v. Krafft, Sander, Meynert, Westphal, Lombroso, Schuele, and generally throughout Germany, Italy and France. In view of the fundamental misappreciation of certain well-established principles of modern mental medicine demonstrated in the quoted extract, it would be simply fruitless and superfluous to discuss such paradoxes as that the "distinction between mania and monomania is for the most part verbal," and that it draws us away from "the due consideration of that pathology of the disease we have to consider and treat." The facts of the case happen to warrant diametrically opposite conclusions.

It is more difficult to give a full definition of monomania in a few words than to give the characteristics of any other form of insanity in a single sentence. It is altogether too complicated a symptom group to admit of a brief definition. It is a constitutional insanity, almost without exception hereditary, or based on an inherited or acquired degenerative taint; it involves the highest logical processes primarily, but does not warp them all equally, or some of them, in fact, at all. On first sight it would appear to merit the designations of ideational and intellectual insanity,* proposed by Maudsley and Hammond. When we inquire more deeply, however, we find that here, as in every other group of mental disorders, attempts to classify the clinical forms according to metaphysical distinctions are predestined to failure. While the prominent feature of these insanities consists in a series of ideational aberrations, imperative conceptions, delusive interpretations, systematized projects and actions, or, finally, a tendency to morbid speculation, yet we find that they are not always free from anomalies of the perceptional sphere, nor free from disorders of the will.

The general intellectual status of these patients, though rarely of a very high order, is moderately fair, and often the mental powers are sufficient to keep the delusion under

---

* The term intellectual insanity, in the absence of any positive justification, has not the advantage of age, it is based on a metaphysical distinction, it does not cover the entire ground of monomania, and it covers a great deal of ground outside of monomania. Except as an abstract symptom designation it cannot hold a place in clinical psychiatry.

control for the practical purposes of every-day life. While many are what is termed crotchety, irritable, and depressed, yet the prominent mental symptoms of the typical cases of this disease consist of the fixed delusions. Since the subject matter of the delusion is of such a character that these patients consider themselves either the victims of a plot, or as unjustly deprived of certain rights and positions, and narrowly observed by agents of their foes, delusions of persecution are added to and incorporated in the fixed ideas, and they become sad, thoughtful, or depressed in consequence. In such cases the erroneous diagnosis of "melancholia" is sometimes made, because with many, depression, whatever its motive, is equivalent to melancholia. Now such a case is no more melancholia than is dyspepsia. The patient is depressed as the logical result of reflections growing out of his morbid train of ideas, and his sadness and thoughtfulness have causes which he can explain and which are all intimately allied with that peculiar faulty grouping of ideas which constitute the *rendezvous*, as it were, of the conceptions of the patient. Nay, the process may be reversed, the patient, developing a hypochondriacal state, imagines himself watched with no favorable eye. Because he is watched and made the subject of audible comments (hallucinatory or illusional) he concludes that he must be a person of some importance. Some great political movement now takes place, he throws himself into it, either in a fixed character that he has already constructed for himself, or with the vague egotistical idea that he is an influential personage. He seeks interviews, holds actual conversations with the big men of the day, accepts the common courtesy shown him by those in office as a tribute to his value, is rejected, however, and then judges himself to be the victim of jealousy or of rival cabals, makes intemperate and querulous complaints to higher officials, perhaps makes violent attacks upon them, and, being incarcerated in a jail or asylum, looks upon this as the end of a long series of persecutions which have broken the power of a skilled diplomatist, a capable military commander, a prince of the blood, an agent of a camarilla, a paramour of some exalted personage, or, finally, the Messiah himself.*

All through this train of ideas there is seen running a

---

* It may be well to state that these lines are reproduced without alteration from a journal article published a year prior to the assassination of President Garfield.

chain of, however faulty, reflections and inferences; there is no absolute gap anywhere. Indeed, if the inferences of the patient were based upon correctly observed facts and properly correlated with his actual surroundings, his conclusions would sometimes be perfectly correct. With regard to matters unconnected with their morbid ideas, monomaniacs present the ordinary powers of memory and judgment, and exceptionally may even be mentally productive.* One can only say that there is a flaw, a break in the logical apparatus, and a weakening of the logical inhibitions; not an utter confusion, as in the terminal incoherence of other forms; not an absolute loss of power, as in the demented, imbecile, and paretic; not a fundamental emotional disorder, as in the maniac and melancholiac.

There would be, strictly speaking, as many varieties of monomania as there are patients suffering from the disease if we admitted the principle of classification adopted by the earlier French writers; for every monomaniac develops an aberration of judgment or conceptional error differing from those of other monomaniacs in some respects. Unless such differences are due to differences in the mechanism of the mental malady they should not serve as grounds of subclassification. As, however, many of the terms which have grown out of the unfortunate tendency previously criticised cannot be violently eradicated from our nomenclature—particularly, too, because they are in part acceptable designations for the special *direction* in which the insanity is manifested in well-defined groups of monomania—it may be well to briefly characterize and to illustrate one of them as a sample at this point, holding to the reservation previously expressed as to the value of such groups.

MEGALOMANIA† is a term applied by the French to monomania manifesting itself in delusions of a socially ambitious

---

* This is exemplified in the case of that Professor Titel referred to by Griesinger ("Mental Pathology and Therapeutics"), who, laboring under the delusion that he was the pope, yet continued to deliver his lectures at the University; as well as by the case observed by the writer, of a patient whose characteristically insane writings are cited on page 77, and who made a useful invention adopted in our navy. Another instance is that of the religious reformer, Vanini, burned at Avignon, of whom an interesting account has been written by the Marchioness Clara Lanza.

† *Monomanie vaniteuse* (French), *Grössenwahnsinn* (German). These terms were employed before the great distinctness of paretic dementia was recognized, and hence have been erroneously applied to the unsystematized ambitious delusions of that disease.

character. The patients suffering from this variety of monomania are the nobility, princes, political reformers, inventors, and poets of the asylum ward. In one marked example of this kind, which the writer observed at the New York City Insane Asylum, the patient's history presented an exaggerated reflection of the social ambitions which characterized the different communities in which he lived; thus strikingly exemplifying the common observation that insane delusions are caricatures of the opinions and aspirations of the time and place where they are developed. In an absolute monarchy the megalomaniac believes himself a prince or duke, in a semi-civilized community a prophet or a Messiah, and in a republic a political reformer. In England the patient referred to laid claim to the estates of a great family, on the ground of a remote and doubtful relationship to an earl; in France he threw himself into the midst of the revolution of 1848, and was actually created a general by Cavaignac; in New York he competed for the office of comptroller, one of the fattest in the land, and before his insanity was complicated by another form of mental disease he devised panaceas for quelling the troubles in Ireland, and corresponded extensively with Disraeli, the British ambassador, and Prince Albert on this and other topics.*

Patients of this group exhibit a characteristic tendency to associate all prominent occurrences and personages with themselves. This is well exemplified in the following record from private practice, which also illustrates the main features of the *episodical delirium* sometimes interrupting the ordinarily calm mental state of such patients:

The writer was called to examine a lady, the wife of a pawnbroker, aged about forty-four, and visited her in a general practitioner's capacity at the request of a relative who is a physician; she received him without reserve and exhibited considerable volubility; the great difficulty experienced in her examination was to keep her to the line of inquiry adopted, as she manifested the common tendency to wander off to other topics. She also showed marked insanity of manner, and by the time the writer was compelled to break off the two hours' interview she had worked herself to a high pitch of excitement and become vituperative in regard to her husband, accusing him of the worst crimes. She was theatrical in

---

* The full history of this patient is detailed on page 345.

her language, in her attitude and gestures, used long words, extracts from celebrated works of fiction, and gave a highly colored account of her intimate relations held with the minister of war, high military officers, and the higher aristocracy while in Vienna, her native city. Her own statements revealed that these intimate relations consisted in her leasing her husband's villa to the former, in bringing one of her sons to a military school under the control of the higher military officers mentioned, and in subscribing with several ladies of the aristocracy to some charitable undertaking. From her husband's account it was evident that the patient had always manifested a longing to enter the higher walks of society, and attached great importance to the most trivial transactions between members of the aristocracy and herself. The noticeable outbreak of her mental disorder occurred in Vienna seven years ago, after her husband lost something like a million florins in real estate speculation. At that time an electrotherapeutist treated her with castoreum, and told the husband that the disorder was nervousness. She was also examined by Leidesdorf, but his opinion had not been satisfactorily recollected by the husband. At that time she had nocturnal hallucinations and marked delusions. On coming to this country she became quieted, though numerous documents dating from this period attest that she still entertained ideas that her husband had wronged her financially and in his marital relations, considering her high social and financial merits. A year ago she went to Hamburg with about two thousand dollars' worth of jewelry, which, as she remarked with an affectation of cunning, she had entrusted to the purser of the vessel without a receipt or other voucher, in order that her relatives might not get hold of it. In Hamburg she discovered that her son-in-law robbed her and his wife of money and other objects, and it appears that she was confined for six weeks in the insane asylum there. She had frequently, within the last year, entered her husband's place of business, and, in presence of customers, accused him of the foulest crimes, of having hired a prostitute the first night of their marriage, of having stolen her patrimony to purchase an order of the government, and finally proceeded to indict her whole family, with the exception of one son and a nephew, as being engaged in a conspiracy against her. The son in question presented insanity of manner far more strikingly developed, even, than his mother, was conceited, arrogant, and excitable, cut his father short when he gave me the hereditary history of the family, and behaved altogether in such a manner that the writer thought it best to inform him that it would require the greatest care on his part to avoid following in his mother's footsteps. This evidently produced a good effect, he ceased to agitate against his mother's examination by a second physician, and was duly asserted by her to have joined the conspirators. The nephew was a weak-minded young man. and the evident tool of his intellectually vastly superior aunt. She had instructed him to put the brakes on when she became excited, for he took away a document which she was about to present to the writer, and which a glance showed to be a characteristic production, being full of underlining marks, exclamation points, and other symbols. From receiving the writer in the capacity of a physician to treat her "nervousness," which she evidently did merely to introduce the subject of her alleged wrongs, she begged him to take charge of her matters in the capacity of a legal adviser. Her memory was excellent, her judgment on all matters not connected with her delusions sound, and even brilliant, in this respect corresponding to the other patient's here spoken of. Aside from her son and nephew, on both of whom heredity had set its stamp, she had a brother dying with paralytic insanity at an Austrian asylum, a sister affected with similar symptoms to her own in another institution, and an uncle also deceased in an asylum. After the visit she became more and more excited, kept the neighbors awake at night with her declamations, and when the second physician arrived to examine her she spat in his face.

This history serves to illustrate the facts to which attention has previously been called, that delusions of persecution are common sequelæ of delusions of aggrandizement, in monomania, and that the two may exist side by side. A demarcation between monomanias in which persecutory delusions predominate and those in which the ambitious ideas prevail, it is impossible to defend ; they are merely the same disease, manifesting their symptoms under different external guises, whose formal character is often fostered if not determined by external circumstances, such as, for example, an arrest, or an asylum incarceration following some extravagant act. The French appear to have been determined in their classification of certain delusional lunatics as megalomaniacs by the vain attitude of the patient. This "*attitude orgueilleuse*" is, indeed, a most characteristic sign of the malady in certain cases; everything, from the erect and stiffened position of the body, the upturned head, and the supercilious sneer, to the dramatic gestures, and the air of condescension displayed in conversation with others, typifies the dominant egotism of the patient. Occasionally, however, a gross delusion of the most ambitious kind may exist without displaying itself in the external appearance or conduct. A historical instance is that of an inmate of a Parisian asylum, whom a government commissioner, after a prolonged and careful examination, was about to discharge as a person of indubitable mental soundness, and who displayed nothing abnormal until, in accordance with the *modus* of discharge in use there, he was requested to sign his name to the discharge papers, when he signed himself "The Christ."

In another instance which came under the writer's own observation, while the patient's self-esteem was great and his ideas were expansive, he claiming to possess a large and valuable piece of land, and that the United States Government owed him several millions, yet, as a colleague expressed it, his manner and address were simply such as would be expected of a courtly, prosperous man of business. In great contrast, side by side with him in the same ward, was a patient who, notwithstanding the fact that he had marked delusions of persecution—such as that Dickens had written "Bleak House" to injure him, and that he was annoyed by human and superhuman agencies—manifested an egotism and conceit in his manner and in his writings which could not well be excelled. It is often a difficult question whether to range such cases under the head of megalomania or *persecutory monomania*; and it is best to designate them as "monomania," adding the qualifying clauses "with predominating delusions of persecution," or such "of grandeur," as the case may require.

Not only the expansive and depressive tinge of delusive conceptions, but also their intrinsic contents, have been and are still made the basis of sub classification. Where the delusions grow out of perverted visceral sensations, or relate to the condition of the body, we have the so-called *hypochondriacal insanity*, which in its most elaborate development constitutes the *nosomania* of Guislain. When the ideas of the patient exhibit a hysterical tinge, *hysterical insanity* has been asserted to exist. The predominance of religious delusions has given rise to the term *religious monomania* or *theomania;* while those cases where the morbid ideation centers around some real or imaginary object of platonic love pass under the designation *erotomania*. Now the delusions of the *erotomaniac* and of the *theomaniac* may be entirely expansive; and in reality erotomania and theomania are in their inception merely varieties of megalomania; later in the course of these disorders persecutory ideas are apt to arise, and may predominate to such a degree that the patient then presents the picture of persecutory monomania to perfection. Add to this the facts that the religious delusion is merely an accidental feature, inasmuch as it could not exist if the world had always been and were to-day agnostic, and that the same patient may unite in his delusive conceptions the features of hypochondriacal, religious, and erotic monomania, and the reflection will naturally follow, that however well-established the great group of monomania may be, nothing is gained by subdividing it into varieties based on features which are so changeable, and, in part, so accidental as the contents and tinge of the delusions.

For the rehabilitation of "monomania" in its wider sense as a form of insanity, including at least a quarter of the chronic insane population, no better reason can be assigned than that there is no other single English term which is sanctioned by usage and by our foremost authors on the subject which covers the conception of an insanity characterized by a flaw in the understanding manifesting itself in a special direction. Yet, while adopting the term, and devoting further chapters to the consideration of the derangements falling under it, grounds have been advanced in the present chapter which justify the ignoring of the more minute classification of the monomanias attempted in the first half of this century.

CHAPTER XXII.

PARANOIA (MONOMANIA), ITS COURSE AND VARIETIES.

Paranoia is a chronic form of insanity based on an acquired or transmitted neuro-degenerative taint, and manifesting itself in anomalies of the conceptional sphere, which, while they do not destructively involve the entire mental mechanism, dominate it.

While "monomania" or paranoia is in most instances based on an inherited taint of insanity or a transmitted neurotic vice, it may also develop after any deep or sudden injury to the nervous system. Thus there are rare cases of monomania which have developed after typhus fever, after head injuries, in conjunction with alcoholism, and even after functional perversions of the cerebral faculties occurring in the course of dreams. There can be very little doubt that great emotional strain and the continual harping of the mind on one subject is also an important factor in the etiology of many cases of monomania, and while usually physical disease or a faint neurotic taint may be found even here, there are exceptional cases where the most careful examination fails to determine their previous existence.

In that larger number of cases in which an original neurotic taint is found, there are usually noted before the actual outbreak of the disorder, anomalies of character, of the general nervous functions, and of the somatic constitution. As these indicate the essential basis on which typical monomania bursts out into full bloom, warn us of the danger ahead, and sometimes constitute unfailing criteria of insanity in disputed cases, they are invaluable to the alienist. The somatic signs in question have been discussed in previous chapters. It may suffice at this point to refer to those which have been found in a form of monomania which, from its early development and its patent dependence on gross structural defect, has by Sander been termed "originäre Verrücktheit." In this form the subjects are noted to be peculiar from infancy, they entertain vague aspirations, are excessively egotistical, and the non-recognition of their supposed importance or abilities leads them to consider themselves the subject of persecution. In others the egotism is so great that the most ridiculous failures are not capable of dispiriting them, but, on the contrary, are accepted as proofs of a divine mission which is to be carried on in a state of perpetual martyrdom. Hallucinations frequently develop as the disorder progresses. With these there are symptoms of neural disorder, similar to those found in imbecility and idiocy: some have epileptiform, others choreiform, and still others quite peculiar and indefinable movements of an "imperative" kind. Peculiarities in pronunciation and inability to pronounce certain consonants in childhood have been noticed in others. There are in addition defects in the bodily conformation of a similar kind, though usually of lesser degree, than those characterizing idiocy and imbecility; the head is often asymmetrical or deformed, the teeth are sometimes badly developed, and there may be club-foot, strabismus, and atrophy of one side of the body. The writer has, in several cases, found anomalies of the cerebral architecture associated with those of the skull in such cases; but the most instructive case on record, which may serve at once as a description of this variety of monomania and as an illustration of the anatomical conditions found with it, may be here cited from Muhr:*

Ludwig Schw—— exhibited a defect in the development of one half of his body and clumsiness in movements from infancy. He did not learn to walk until very late, and then it was with great difficulty that he could be induced to assume the erect posture. He preferably walked so as to support himself by a wall or other objects on the side which was defective. His figure was tall and slim, and a stoop was noticeable throughout life. He was known to be left-handed from childhood, the right side being affected with atrophy. The right foot was shorter and sparer, the right hand was smaller and more unserviceable than its fellow. Strabismus was observed from the period of adolescence; both ocular axes converged inward and downward, more so in the case of the left than of the right eye. Vision was defective on the former from an early period, and he was in the habit of twisting his head round so as to utilize the other eye in reading; ultimately he became completely amblyopic on both sides. He had no sense of music whatever. Notwithstanding these corporeal defects and the cerebral defect to be detailed, this person attained a considerable

---

*Anatomische Befunde bei einem Falle von Verrücktheit, von Dr. Muhr.—*Archiv f. Psych. u. Nervenkrankheiten*, vi. p. 733.

degree of education. He passed college and the university where he studied law, and, aside from this his legitimate study, he devoted attention to many collateral branches, receiving the best testimonials in the respective departments. Besides, he obtained a fair knowledge of the Italian, French, English, and Sclavonic languages. He was a perfect master of Latin, had studied the whole ecclesiastical literature in that language, gave instruction himself in the different languages mentioned, and was constantly engaged in some one or other literary production. He shifted round a great deal among the different vocations that were open to him; he was at one time studying with a notary, at another engaged in the Department of Finances, at still another, in the Trade Chamber. All this time he manifested a morbid desire to enter a cloister, and at his thirty-sixth year he succeeded in being admitted to the novitiate in a Benedictine institution, where he stayed, however, only one year. It was his habit to endeavor to attain any desired position by the circuitous channel of patronage, and from his boyhood up he had exhibited an anxious desire to attract to himself the attention of influential patrons. His moral sense was entirely perverted. He bored the dignitaries of the church with reformatory projects, was in consequence recognized as a lunatic, and sent to the asylum. Previous to this period his insanity had been so little suspected that at one time he had actually been appointed curator for the property of his insane sister, Marianna Schw——, by court, a position he held till the latter's death, that is, from 1860 to 1861. He had a decided antipathy to the normal gratification of the sexual desire, and was a persistent onanist to his dying day. He had remarkable views concerning marriage, and wrote a treatise in defence of polygamy, sustaining his position by means of voluminous quotations from ecclesiastical authors. Hallucinations could be demonstrated to have existed, dating from his twenty-fifth year. At that time he received the revelation that he was the Saviour of the world by means of hallucinations; he concealed this revelation for fifteen years, absolutely. Just previous to this he had delusions of persecution with a strong sexual tinge and of a repulsive and disgusting character. He then began to consider all remarks, even of the most indifferent character, made by strangers as applying to himself, and interpreted them as of great importance. He claimed that the pope, in declaring (1864) "that the Star which was destined to relieve the Church of its persecutors was about to appear," referred to him. In consequence of his delusive interpretations of remarks made by strangers he became involved in conflicts in restaurants, having gone the length of pronouncing the strangers intoxicated. In short, delusions of an expansive character were continually alternating with those of a depressive kind. Between this time and the asylum sojourn he had at different periods hallucinations of a ceremonial nature, such as that, in 1863, the Archangel Michael appeared with a flaming sword amid the sound of trumpets, to notify him that he should prepare himself for extraordinary events. In 1864, after a prolonged period of fasting and praying, he saw God himself in a vision, telling him that he was called to occupy the pontifical chair; at the same time, he added, the vision informed the faithful in the church at the time that "this is my beloved Son, him shall ye hear." The following year he saw the Virgin Mary, and, observing that he was becoming bald, the next day declared this to be the tonsure which she had conferred on him. Previous to his admission to the asylum he had an occasional perception of the fact that his actions were erratic, and that he was considered a fool, a fanatic, and even a lunatic by the priests whom he was in the habit of persecuting with his attentions. In 1870, having been persuaded by the clergy whom he communicated his projects to, to consult an alienist, he was pensioned off by the government, under which he held a position at the time, and received at the asylum of Feldhof. Here he became gradually more perverse in his logic, and ceased to correct his delusions. While at the asylum he entered into unnatural sexual relations with a weak-minded epileptic, and the two were finally discovered together in a water-closet, and transferred to separate parts of the institution. Whenever he had his periods of religious exaltation he indulged in onanism, and he was on one occasion found reading the legend of a female saint and masturbating at the time. He also had hallucinations of sexual congress with the female saints, accompanied by seminal emissions. He reached the age of 47 years.

There was a bad hereditary history in this case on both sides. On the paternal side nothing is known definitely as to the great-grand and grandparents of the patient. The latter had six children, of whom Charles (the father of the

patient) was abnormal from infancy and notorious for his
eccentricity. He died suddenly of cerebral hæmorrhage.
The next three children do not present a clear history;
there were grave indications of heredity, however. The
last two (females) founded families said to be mentally
healthy. On the maternal side an uncle was a priest, with
a notable absence of all musical sense, and dying suddenly
with cerebral hæmorrhage. His sister, the mother of the
patient, is said to have been of a normal mind. Aside from
the patient, Ludwig Schw——, there were six children in
the family; the first, of a romantic disposition, went to
America and was never heard of again. The second was
the aforesaid Ludwig. The third, a girl, had a beard on
her chin, exhibited expansive delusions, and died in the
asylum. The fourth was born blind and died as a child.
The fifth died insane. The sixth is living; he is an en-
gineer, right-handed, cannot use his right leg as well as his
left, is without any mental defect, and has several sons, not
yet grown up to manhood, and, so far as ascertained, well
developed, with the exception of Emmerich, aged 14. The
latter, a bright boy, is left-handed, the right extremity being
generally less voluminous than the left, and the right hand
measuring half an inch (1.5 cm.) less in the metacarpo-
phalangeal line than its fellow. He does not use this hand.
The right half of his cranium is noticeably smaller than
the left, particularly as regards the parietal boss.*

The examination of the brain showed it to be far below the average of the adult male brain in size and weight;
the most remarkable feature was the notable and unusual
asymmetry of its two halves. Dr. Muhr attributes this
asymmetry to the reduction in volume of the left half of
the cranial cavity and of the left internal carotid artery.

The general impression created by looking at the skull
was, that the actually longer right half had clasped round
the shorter left half so as to accommodate the general
shape of the skull to an ellipsoid. It was also observed
that the region of the sagittal suture was carinated, the
suture itself being but little serrated. The temporal ridges
rose unusually high so as to come within 48 mm. on the
right and 36 mm. on the left side of the sagittal suture. The
frontal region was retreating, and the supraorbital margins

---

* A most remarkable deviation from the rule that the bodily and cerebral defects are on opposite sides.

were very prominent. The left half of the cranial cavity was reduced in every dimension, the ideal horizontal axis of the skull would have to have been represented by a curved line. As if to compensate for the defect on the left side of the cranium the face deviated to the other side, so that the vertical ideal plane of the skull was slightly S-shaped. The left cerebellar fossa was scarcely one third as capacious as the right. The right transverse sinus was shallow and the right jugular foramen narrowed, while the left sinus was well formed and its foramen of fair dimensions. The openings of the carotid canal as well as the canal itself were abnormally small on the left side, the left optic foramen being also smaller than its fellow. The clivus was very steep owing to the foreshortening of the entire skull. The clinoid processes were unsymmetrical, the right ones being the larger.

This unusual degree of cranial asymmetry was the direct cause of the atrophic state of the left side of the encephalon, and in its turn due to the asymmetry of the great vessels supplying the skull and brain with blood, according to the interpretation of the matter made by Muhr.

On passing to a study of the brain itself, it was found that its deformity was partly due to a relative atrophy of all the parts of the encephalon lying to the left of the median line, partly to an actual absence of certain parts.

The two halves of the cerebellum, for example, were so unlike each other that one half might have been supposed to have been derived from a different brain than the one from which the other half was obtained, if they had been seen separately. The difference was most marked on the inferior aspect; the left cerebellar hemisphere was a round lump, with gyri not running parallel, as in the normal cerebellum, but diverging in every direction, and the lobules appeared twisted on themselves. The flocculus was rudimentary, the amygdala altogether absent on the left side. On that side the gyri (folia) of the inferior and posterior face numbered eighty, on the right or better developed side only sixty. This discrepancy was due to the fact that in the former enumeration are included not only those folia which reach and are counted in the surface of the normal cerebellum, but also the collateral branches which here extended to the surface, owing to the atrophic condition of the terminal lamellæ.

The left cerebral hemisphere was in every respect unlike

its fellow. Shorter and less massive, it were particularly its parietal occipital, and temporal lobes, which appeared much reduced, and this reduction extended to the half of the corpora quadrigemina and the basal ganglia of that side. The left optic tract and right optic nerve were smaller than the right tract and left nerve; the left corpus mammillare appeared to be absent, the right well developed. The right olivary eminence was smaller and the left auditory nerve narrower than their fellows.

The fissure of Rolando was situated further backward on the left hemisphere than on the right. The left parallel fissure opened into the transverse occipital; there was a small concealed gyrus to indicate their separation. The right parallel fissure ran into the post-central. On the left side the calcarine and parieto-occipital fissures failed to meet. Altogether there was observable a difference in type of the fissures of the two sides, these being more of the longitudinal pattern on the left—more of the transverse pattern on the right side. The length of the corpus callosum was 64 mm.; that of the right cerebral hemisphere, 140 mm.; of the left, 120 mm.; the width of the former, 60 mm.; of the latter, 50 mm.; of the right cerebellar hemisphere 50 mm.; of the left, 33 mm. The right olfactory tract was narrower than the left.

In another case, equally carefully studied by Kirchhoff,* where the mother and grandmother as well as an uncle and aunt on the maternal side had been insane, the patient had had epileptic attacks up to her seventh year. She appeared to be mentally and physically healthy from that time up to her twenty-fourth year, when an emotional disturbance provoked monomania of a persecutory charactery. The sole of the right foot was found narrower and longer than that of the left, the right kidney showed the embryonic lobulation, while the left was normal, the skull was microcephalic, and the left hemisphere did not cover the cerebellum ; in addition the cerebral commissures were defective, and there was a porencephalic defect in the left gyrus lingualis.

The bearing of the somatic anomalies found in Muhr's, Kirchhoff's, Sander's, and the writer's cases * of original

---

* *Archiv fuer Psychiatrie*, xiii. p. 168.
* The Somatic Etiology of Insanity: being the W. &. S. Tuke Prize Essay of the British Medico-Psychological Association, published as the supplement of the *American Journal of Neurology and Psychiatry*, 1882–83.

monomania must be analyzed from an embryological point of view (page 86).

Even in cases where the somatic signs are not as patent as in the group to which Sander has applied the designation of *congenital monomania* (originäre Verrücktheit), anomalies of the character are frequently noted in early life. Thus one lady now under the writer's treatment, suffering from that abortive form known by the French as *folie du doute avec délire du toucher*, had been remarkable as a girl for her timidity before visitors, hiding away in closets at some times, and at others insisting on having a certain dress before appearing among her own cousins. Others, particularly females, exhibit hysterical peculiarities and emotional disturbances. Many of these patients are morally imbecile; they are devoid of an inherent moral sense, and this defect is usually found to be associated with, if not dependant on, a morbid egotism which recognizes no rights on the part of others, including the patient's relatives, that the patient feels bound to respect, except from motives of expediency.

The general nervous and mental state of patients predisposed to monomania has been aptly characterized by the Germans as "reizbare Schwäche"—irritable weakness It manifests itself in early life by a tendency to convulsions and to delirium in the course of slight febrile affections.

While attention has been called, by more than one of the classical writers on insanity, to the fact that the intellectual faculties in this condition may be in part intact, that the memory may be excellent, and the judgment on ordinary matters unimpaired, yet there is seldom any higher ability. These subjects may succeed in a routine calling, but they are rarely capable of a sustained mental effort in an original direction; the remarkable exceptions recorded prove the rule. Usually their conceits are bizarre rather than productive, their reasoning paradoxical rather than logical, and their argumentation tricky and shrewd rather than substantial. To the laity such subjects often appear to be brighter than the common run of mankind, because commonly oddity is mistaken for brilliancy, and unblushing pretence for merit. When the whole career of such subjects is followed up it is found to be exceedingly checkered, and vagabondage, theft, and fraud are often prominent incidents of their lives. Sexual perversion has been noted, and it may or may not be related to this subject that female patients belonging to this group sometimes have a beard on the chin.

In a limited number of patients the continued existence of the monomanical character is the chief or sole manifestation of mental abnormality. Pinel the nephew described such subjects as "turbulent, indocile, quick to anger, committing outrageous acts; which they are always ready to justify by plausible reasons, and who are to their families, their kindred, and their friends continued subjects of anxiety and grief. They are continually doing wrong, either by neglect, by malice, or by wickedness. Incapable of mental or physical application, they destroy, subvert, and unsettle everything with which they are brought into contact and which they can injure." Hallucinations and fixed delusions may be entirely absent, but the overbearing egotism of these patients, which leads them to the most fanciful and extravagant undertakings, can be regarded only as the expression of delusive opinion. From the fact that they are capable of reasoning readily and with a show of ability, and manufacture plausible excuses for their behavior and acts, they have been termed *reasoning monomaniacs*. This designation is hardly accurate; the patients reason with facility, it is true, but they are as unreasonable in their beliefs, and more so in their acts sometimes, than the delusional monomaniac. The writer suggests terming this disorder *monomania sine delirio*, for it bears the same relation to monomania with delusion that *melancholia sine delirio* bears to melancholia with delusions, or which hypomania bears to typical mania.

An interesting case of this kind recently came under the writer's observation. The son of the member of a manufacturing firm in one of our larger cities, laboring under certain suspicions regarding a business associate of his father's, took a bomb into the place of business, intending to "clean out" the firm, and particularly to kill the obnoxious member of the latter. He exploded the bomb, demolishing the entire store front, killing his uncle, wounding the object of his wrath, and injuring himself mortally. He then drew a revolver, and discharging it against himself, made sure of his own death. It was a cousin of this unfortunate person who came under the writer's care. He had acted oddly for some time, not committing a single act, however, which he could not explain away, or render justifiable in the eyes of the laity. The writer examined him after an escape from an asylum in which he had been placed, and from which he had managed to free himself with considerable cunning and deliberation. Beyond a bad conformation of the skull, an abnormal dentition, spasmodic action of the facial muscles, the *insane manner*, and a delusive opinion which accurately imitated that of the bomb-throwing relative, the patient manifested no sign of insanity whatever. In view of the threats he had made, and the projects he had expressed, but which he dissimulated at the time of the examination, the writer, with a well-known fellow-physician of New York, decided to place him in an asylum. About this time there was a "liberation epidemic" of alleged sane persons. A lawyer became "interested" in the case, and a general practitioner, as well as an alienist of repute, were requested to examine the patient by the "liberators." Both reported favorably, grounding their belief not only on the calm and rational behavior of the patient, but in great part on the statement of an assistant asylum physician, to the effect that he had observed nothing "wrong" about him. The patient was accordingly discharged and sent "abroad;" that refuge for those of the insane whom their liberators wisely hesitate to assume the risk of attempting to manage. Exactly six weeks later the same patient was sent to the same asylum (on the certificates of the two physicians who had reported him sane) in a greatly excited condition and with symptoms of early deterioration. The commitment to the asylum had undoubtedly anticipated a repetition of the tragedy which was enacted by the cousin afflicted with the same form of insanity, suffering from the same delusive belief, and harboring the same morbid project five years before. The opinion given that the patient was insane, vindicated as it was by the subsequent history of the case, rested on what to the laity and to some of our colleagues would have appeared trivial if not fanciful grounds; and the case is detailed chiefly because it illustrates the importance of the somatic signs in determining the existence of the subtle form of insanity which is the subject of this chapter, and which, in some of its phases at least, may be aptly designated a hypertrophy of character in an abnormal direction.

In many cases of *monomania sine delirio* a periodical outburst of excitement, coupled with impulsive acts or marked by hallucinations, is noted. These constitute a transition to impulsive and delusional monomania, and the periodical recurrence of the insane explosions is an illustration of the relationship existing between all the forms of insanity which are the expression of a continuous neurotic vice, of several of which the periodical recurrence is characteristic.

The symptoms of classical monomania may be numerous and varied, or they may be few and limited in range. In some patients, usually encountered without asylums, a single imperative conception or impulse, or a delusive suspicion which may never become organized into an insane belief, may be the sole mental symptom. Some of these cases may be looked upon as abortive cases of monomania, and as within the border land separating sanity from insanity, while others are undoubtedly insane in every sense of the word. Vague fears are experienced by some patients; they dare not, as in one of the writer's cases, go to a certain part of New York city without experiencing a nameless terror; others imagine that they can do harm by glancing at objects in a certain way. One patient presenting the latter symptom, who could never glance at a crucifix without imagining that some "flash-like action" was excited by her on it, could not be convinced by the clergy, notwithstanding her devoutness, that it was an absurd belief on her part. The *folie du doute avec délire du toucher* of Le Grand du Saulle, is a variety of abortive monomania.* In a patient presenting this symptom, who came under the writer's care since the present volume has gone to press, the fear of defilement ordinarily found in this disorder was replaced by a fear of doing herself an injury. Her sister-in-law was compelled to sift her food to see that no glass had gotten into it; she would not eat with a fork, and finally dreaded a spoon. She did not dare to go out on the street without a veil on her face, because, frail as the fabric of the veil was, it gave her a certain sense of security against the countless dangers she feared. On one occasion the writer ordered her to take the veil off, to give it to him, and to go home on foot; the relatives were ordered to use stern measures; the examination of her food, which had been done to humor her, was forbidden, separate cooking

---

* The "Grübelsucht" of Griesinger is a similar affection: the patient is tortured by myriads of reflections and queries about commonplace matters. The prognosis is excellent, two patients of the writer's recovering completely within three months, and without the change of scene which proved so beneficial in Berger's cases.

was suspended, and the result was that the morbid fear grew fainter, the patient began to resume her use of the various articles she had previously dreaded to handle, and even went out on the street alone and unveiled. But how much deeper the disorder may lie than the surface symptoms serve to indicate, was shown by this case, in which, with the improvement of the symptoms for which the writer was consulted, the more serious ones of insane mannerism and logical perversity which to the laity had been less noticeable on account of the preponderance of the morbid fears, came to the front.

While the phases of monomania just described may be regarded as rudimentary or abortive forms of that malady, the fundamental disorder may be as great as in delusional monomania, if we may infer the nature of the cerebral disorder, at present undiscoverable with our methods of examination, from the somatic anomalies found in monomania sine delirio, and in monomania with imperative conceptions. The same mal-configuration of the skull, as well as the same facial expression as in delusional monomania, are sometimes found in cases falling under the head of the abortive forms.

Delusional monomania is the most frequent form of this disorder. It is the "primäre Verrücktheit mit Wahnideen"* of the Germans, the Wahnsinn of Snell, and the chronic delusional mania of some English and American authors. The delusions of this form of monomania are alone sufficient to characterize this disorder, and when found serve to establish the diagnosis; they are of the systematized variety (page 26).

It is only exceptionally that the delusions appear abruptly, and those appearing in this way constitute connecting links between those rudimentary delusions which the writer considers the imperative conception to be, and the true systematized delusion. Usually unpleasant visceral sensations, hyperæsthesia in the ano-genital region, strange sensations flashing up from the latter "through the spinal cord to the brain" (serving as the basis of delusions of sexual congress with God, devils, or men), or a feeling of dryness in the throat or a bad taste in the mouth (serving as the basis for delusions of poisoning), are experienced,

---

* The Germans speak of a secondary "Verrücktheit," applying this term to some sequalae of mania and melancholia, with which there is dementia marked by stabile delusions.

and the patient, in endeavoring to account for them, builds up his insane belief. Sometimes hallucinations or dreams contribute additional material.

Usually the outbreak of the disorder coincides with some one of the physiological periods, such as puberty, the second climacteric, pregnancy, and the puerperal state. It is sometimes precipitated by sexual excesses, more frequently by masturbation, and occasionally by visceral diseases and fevers. Its development is usually gradual, and is comprised in the progressively firmer and more extensive organization of the delusions. Sometimes this advance is by fits and starts.

Delusions of persecution are the most common ones in delusional monomania. There is a marked difference between these delusions and the delusions of persecution found in melancholia. While the melancholiac believes that he is pursued or punished because he is a weak, cowardly, bad, or criminal person, the monomaniac believes that he is persecuted from motives of envy, and, as a rule, he develops exalted delusions of his personal importance or worth side by side with those of persecution. The dread of his persecutors may, however, throw the monomaniac into as violent and dangerous a frenzy, howbeit of a different kind, as the melancholiac. Hallucinations, particularly of hearing, which are very common, which may be limited to mere "voices," or consist of distinct sentences, precipitate the explosion of these episodical deliria. Thus Dubourque, the "Fourteenth-Street assassin," who had inherited from his father the malformed cranium and insane expression of a monomaniac, as well as the delusion that an uncle had died leaving him several millions of dollars, which were withheld by the Government of the United States, had heard as a child that the neighbors proposed poisoning him, the heir to this great estate. While at work painting a transatlantic steamer, after his father's death, he heard remarks by his fellow-workmen to the effect that, now that the older claimant was dead, they would waylay him that night and thus exterminate the family of heirs. On three occasions he stabbed persons who, he alleged, were pursuing him and crying out to kill him on the street. On the second occasion he had stabbed a police officer. On the first he told such a plausible story of an assault and his earnestness of manner was so impressive that the police magistrate before whom he was brought discharged him.

On the last occasion, in broad daylight, he was passing through a crowd of ladies on Fourteenth Street who were out "shopping" in that thoroughfare. The expression, attitude, and walk of the man were such that many of the passers-by avoided him. Suddenly he drew a compass, such as is used by artists, and began stabbing right and left. It is not known how many persons he stabbed, but three women were dangerously and one was fatally wounded. To the writer, who examined him at the request of the district attorney, he said that from all sides he heard the cry, "There goes the man who is going to take all that money out of the land; he is going next week. Kill him!" and that then he had drawn his weapon to defend himself.*

FIG. 13.

No better example of the insane expression could be selected than that of this subject of *monomania of persecution.* (Fig. 13.)

The beliefs of monomaniacs are almost as numerous as the patients. They are all characterized by this feature, that the occurrences in the outer world are anxiously examined by the patient with a view of tracing their connection with himself. Accidental remarks by others, initials in the personal columns of the daily papers, bill-posters,

---

* He shammed amnesia of the occurrence at first.

biblical passages, and certain phrases in sermons are interpreted as having special reference to him. Sometimes a mere exclamation, coughing, sneezing, or the turning around of a person on the street, are supposed to be signals by which the patient's enemies recognize each other. He believes that he is ridiculed ; that the clergy point at him for the purpose of degrading him in the eyes of his fellow-men ; that he is accused of unnatural or foul crimes, and that persons on the street hoot at him and are employed to do this by detectives, free-masons, Jesuits, or their business and professional rivals. If an official, he suspects his subordinates or colleagues of putting compromising letters among his papers, with the intention of intriguing him out of his position. The result is that these patients are involved in conflicts with the authorities, bring libel suits or claim rewards for unearthing conspiracies which they believe were directed against their superiors and themselves conjointly. For the tendency to seek the protection of courts, or to punish their real or supposed foes through them, the Germans have devised the term *querulous monomania*.* As Krafft-Ebing says, these patients are defective in their ethical qualities; their alleged possession of a sense of right, which they emphasize on all occasions, is in reality reducible to a tremendous egotism which allows them to recognize in law only a means to their personal ends. The subjects of querulous monomania are early remarkable for their obstinacy, their tendency to interfere in the affairs of others, and their impatience of contradiction. Involved in a lawsuit through some extravagant claim or supposed sense of injury, they become seized with a perfect furor for litigation, buy books of law, and may become even well versed in the details of that science. Defeated in one court, they will try another, and appeal again and again. Such persons are well known in several of the courts of this city, and usually treated as harmless lunatics. But they do not all sink into the quiet state so finely exhibited by an old lady who some years ago was a sojourner at the surrogate's court of this city, and regularly handed a document to the clerk, to have it as regularly returned. Some, infuriated at their failures, become convinced that the judges are bribed or under the advice of

---

* " Querulanten-Wahnsinn;" " Irrsein der Querulanten und Prozesskrämer." or Insanity with tendency to litigation.

secret societies, and they resort to invectives and libels, send lengthy documents to the press, or even make assaults on their supposed opponents and their agents.

Sometimes hallucinations and illusions preponderate in the mental sphere of the depressive monomaniac. He hears a thousand foes, feels scorpions and parasites crawling around him, which have been let loose by his destroyers, he tastes poison, blood, the flesh of corpses in his food, and there are concealed foes or animals in his intestines. Women generally complain of being raped, others of accusations made against them affecting their chastity; and men, as in one case of the writer's, complain that their foes are drawing out their semen through the nose by an invisible influence in the shape of an "ascending vapor." Where the delusions assume this character, and relate to bodily states, the designation *hypochondriacal monomania* is applicable. A most interesting instance of this variety is that on which the mediæval belief that there was a "wandering Jew" was based. At various periods (as it happened, about once a century) persons appeared, wandering anxiously from place to place, asserting and believing that they were this mythical personage. In a French asylum two hypochondriacal monomaniacs were observed who entertained the same belief, which arose in this way : One organ after another was suspected to be the seat of a fatal disease; the patients, believing that these diseases should naturally have terminated their existence, came to the conclusion that they were not mortal ; but, as the only human being not mortal and still on this earth was supposed by them to be the wandering Jew, they believed themselves to be necessarily identical with him.

A remarkable phenomenon, and one bearing out the statements of the last chapter as to the essential sameness of all the monomanias, is what Krafft-Ebing designates as the *transformation* of the disorder ; namely, a rapid and sometimes sudden change of the delusions of persecution into those of aggrandizement. This transformation, unlike that occasionally observed in paretic dementia (p. 200), which is without logical motive, is the result of inward reflection and reasoning. The patient, who has been all along depressed because of the machinations of his foes, now concludes that he is a person of some importance, and may even take a delight in his martyrdom within the walls of an asylum. He believes himself to be a king, a prophet, or a religious and social reformer. Occasionally an intermediate trance-like or ecstatic state is observed ; or the patient, who interpreted the glances of people on the street as inimical, now notices an approbation or admiration in their looks and expressions, and, instead of the attacks, sneers, and libels previously detected in the daily papers and on bill-posters, he discovers obscure hints as to his great services to the state, his birthright to the throne, or to some great estate in them.

That variety of monomania which is characterized by the erotic delusions already described in the chapter on delusions (page 27) is termed *erotic monomania*. While the ideas of the patient are in the main expansive and quixotic, yet ideas of persecution may be developed, in consequence of the failure to accomplish the platonic union with the adored object, the ridicule by his relations and friends, or the incarceration in an asylum rendered necessary by his extravagant behavior.

One variety of expansive monomania, which is undoubtedly becoming rarer, is that denominated *religious monomania*. The patients exhibiting this form usually manifest a certain degree of weak-mindedness or imbecility in childhood. The misconception of religious instruction, or the misinterpretation of sermons, particularly of such as are delivered by popular pulpit orators, leads them to the development of a pseudo-religious and sometimes of a fervid religious enthusiasm. The occurrence of any of the disappointments or hard blows incident to life leads to their complete self-abandonment to religious speculation and the perusal of religious works. Ecstatic and visionary states then occur, and the delusion may develop of being the mother of God, of Christ, or of All Saints. Hallucinations are frequent: the gates of heaven are seen ajar, processions of angels singing anthems, the Virgin Mary beckoning, God pointing out "his faithful son," or " daughter," as which the patient considers himself or herself, are seen within them. Often depressive ideas develop, due to hallucinatory visions of devils and tempting or taunting voices. The devil is then supposed to fill the apartment of the devotee with noisome smells, to repeat the temptations of St. Anthony, or to select some one of the viscera of the patient as the site of his operations against his health and happiness. Usually the patient, who about this time develops ideas of self-exaltation, claims to have triumphed over the arch-enemy, and then proposes to proceed to aggressive measures against him and his servants on earth. More than one insane fanatic of the middle ages has been responsible for the fierce campaigns waged against dissenters and alleged infidels, and not one of the least remarkable incidents of this period of history is the fact that for two centuries Europe poured out its best strength in the crusades, under the influence of the prayers, sermons, and visions of a Peter the Hermit, who undoubtedly suffered from this form of monomania.

Sexual ideas are common in *religious monomania*. The male patients believe that female seducers are sent to them at night by Satan ; the female ones, that they are pregnant by God or some other sacred personage.

The chief danger from these patients lies in the fact that they often suffer from the hallucination of hearing the commands of God to do certain things. It is in obedience to such commands that religious monomaniacs have committed homicide, suicide, or self-mutilation. A remarkable instance is that of Matthieu Louvat, who crucified himself in obedience to such a command.

The prognosis of monomania is very unfavorable. The chief feature to be consulted in reference thereto is the mental power of the patient. The more considerable this is, the more likely is a correction of the delusive beliefs, the delusive suspicions, or morbid fears to take place. Consequently the prognosis is best with those patients who suffer from simple delusions of persecution or of social ambition, worse with erotomania, and worst of all with religious monomania, for here, as has been already stated, a background of original weak-mindedness is generally present. Bad as the prognosis is in this form, cases are reported where the hallucinations and delusions disappeared, and the patient, if not altogether recovering, showed nothing abnormal beyond an extravagant religious zeal and a desire to convert mankind to what he happened to consider, in the excessive egotism of the fanatic, the right faith. An instance of well-nigh complete recovery is that of the author of Pilgrim's Progress. In his case as in the few others in which relative recovery ensued, hallucinatory phenomena were in the foreground and logical perversion in the background.

Monomania, when not cured, remains stationary for years. The logic of unrecovered patients becomes more perverse, they are more frequently found in abstracted reveries than in the earlier periods of the disorder; but mental deterioration does not proceed rapidly, and never reaches the degree of chronic confusional insanity or of terminal dementia, unless there is some inter-current disease. A condition similar to chronic confusional insanity developed in consequence of an attack of cerebral hæmorrhage in a patient shown the writer by Dr. T. A. M'Bride. In four cases in private practice, and in two in the pauper asylum, the complication of monomania by paretic dementia (see " Diagnosis"), first noticed by Höstermann, was observed by the writer. In such cases the paretic dementia is a genuine complication, and not a true terminal state of monomania. Any of the ordinary forms of insanity, mania, or melancholia, may occasionally complicate the case and the diagnosis for the moment, just as any acute affection may occur in subjects suffering from chronic constitutional complaints, and mask them for the time being.

PART III.

INSANITY IN ITS PRACTICAL RELATIONS.

CHAPTER I.

How to Examine the Insane.

We will suppose that the physician is called to see a person whom, from the previous history or the expressed suspicion of the relatives, he considers it necessary to investigate the mental state of. In such a case he should bear the following prominently in mind: The majority of the insane are either communicative, or, if not communicative, readily betray their insanity by their physical appearance, and it is best in the interests of such patients for the physician to visit them in his actual capacity as a medical adviser. On the other hand, there are certain of the insane who are skilful dissimulators, whom the most expert alienist might fail to unravel the real mental state of at any single examination, and who would be put on their guard or led to the commission of dangerous acts by the announcement that a physician was approaching. It may be necessary in such exceptional cases for the physician to visit the patient as if casually, or even to conceal his real character.*

---

* He who has been in that emergency to which asylum physicians, with attendants and other conveniences at their disposal, are rarely liable, of single-handed encounters with homicidal, treacherous, and cunning lunatics, fully aware, as some of them are, of their legal irresponsibility, can have but a smile for the injunction never to resort to "deception"—as it is called—with the insane. Nor is it easy to draw the distinction between the ordering an attendant to watching a suspected simulator through a window, or a crevice, a procedure resorted to time and again by the best French and German alienists, and the visiting of such persons by the more competent alienist himself in a character calculated to throw the simulator as well as the dissimulator off his guard, and to reveal that truth which it may be desirable to establish in the interests of justice, or

The first step in the examination of an alleged lunatic is the study of his features, manner, and attitude. In some of the insane these will not betray the mental state; in the majority, however, they afford such significant indications of the insanity, that the expert alienist may arrive at a provisional and approximate opinion of the form of insanity with which he has to deal, and thus be able to adopt a special line of examination by the inspection of the patient alone.

It is very unwise, however, for the physician, on entering a room filled with people, to walk directly up to the person whom, from his appearance, he suspects to be the patient, and to proceed brusquely with the examination. He may be right in his selection, and accomplish his possible object of impressing the laity with his diagnostic skill. But if the patient should happen to be suffering from a form of insanity of a hypochondriacal or depressive character, the procedure would have a bad effect on him. The patient might, if hypochondriacal, argue that there must be some truth in his hypochondriacal belief, inasmuch as a stranger, on first sight, picked him out as the patient; while he who is suffering from delusions of persecution might discover new building material for the delusion that there was a conspiracy against him in his recognition by a person who had never seen him before. On the other hand, there is—even with an extensive experience—a chance of committing an error. An experienced alienist, who, in almost every case, had been able to pick out the insane member of the family he was called upon to visit—wherever he saw fit to make the attempt—picked out the imbecile brother of the patient as the lunatic. It is true that there was a far more serious congenital mental defect in that brother than in the patient whom it was proposed to have him examine; but, as the latter suffered from an acute psychosis, which had led to a suicidal attempt, this was not appreciated, and the alienist might have been supposed to have blundered. A source of many possible mistakes is the fact that, in case of insan-

---

at least of the individual. It so happens that the recent demonstrative sneering at the procedure here advocated has been by members of a circle composed of men who could readily afford to disregard the most legitimate methods of investigation, because their testimony is rarely regulated by the demands of medical truth, but who with remarkable inconsistency resorted to such subterfuges as taking part in carousals with a paretic dement in order to accomplish his commitment to an asylum.

ity dependent upon a transmitted taint, other members of the family than the one concerning whom the physician is consulted may present peculiarities in behavior and appearance suggesting the existence of insanity, or of the insane disposition. This has frequently been the writer's experience.

In proceeding to examine a patient the physician will be guided in great part by the expression of his countenance, his manner, and the first words spoken if he talks spontaneously. It is obvious that his own demeanor must be very different with various forms of insanity. Indeed it would be absurd to attempt to follow any fixed rule of conduct; though, as a general thing, it is well not to appear searching or anxious in the examination of any alleged lunatic, nor to give the impression that the examiner is particularly interested in the mental features of a suspicious one.

If the patient's countenance expresses distrust or suspicion, it is well to delay the examination until he becomes somewhat accustomed to the physician's presence. Sometimes on arriving at the patient's residence the physician will find him held by others or tied down. In the majority of cases, the physician can risk sending the restraining apparatus and the holding persons (whom the patient often confounds with his supposed enemies) out of the room ; a step which if feasible, will facilitate the further examination by gaining the patient's confidence. In case the insanity is of a violent and dangerous type this procedure will not be necessary, for the actions and words of the patient will then establish the diagnosis sufficiently well for all immediate purposes, and as well as any single examination is calculated to do.

A large number of patients whom the alienist is called upon to examine are not apt to be communicative to a stranger at first. And nothing would defeat the purposes of the examination so certainly as an immediate cross-questioning with regard to mental symptoms. Frequently the patient apprehends that he is considered insane, occasionally even is himself convinced of his insanity; but he is as little desirous of being pronounced insane as an ordinary patient in private practice would be to have the existence of a gonorrhœa or a chancre revealed in the presence of his family. In all such cases one fact comes to the physician's aid, namely, that the insane as a rule are deficient in concentration power and in self-control, and that however firmly they may have resolved not to reveal their thoughts, yet a prolonged examination will evoke involuntary admissions, which, once secured, enable him to reach the very centre of the mental citadel.* He must consequently approach him by a circuitous line, and there is one which, in his character as a physician, he may follow without arousing the suspicions of the patient, or resorting to a subterfuge, namely, that of an examination of his physical state. Indeed this is itself sometimes calculated to reveal important facts; few patients will suspect that an examination of the tongue can refer to their mentality, although a fibrillary tremor or deviation of that muscle may prove of great signification to the physician. The existence of visceral disturbance, of disordered sensations and pains, and of imaginary complaints in some of the insane renders them very willing to be examined on these points. The transition from questions relating to visceral trouble to inquiries about the patient's sleep is an easy and natural one, and appears legitimate even to the most suspicious lunatic. If the sleep is admitted to be disturbed the patient may make avowals which suggest the existence of hallucinations, and the character of these symptoms will often alone suffice to reveal the nature of the insanity. In other cases a few judicious inquiries as to business or family troubles, made on the assumption (on the patient's part) that these may bear a causal relation to his physical disorder, will sometimes lead to "confidential" communications as to alleged conspiracies, antipathies, attempts by others to poison his food, marital infidelity, the ruin of his fortune, the commission of some crime, or of the fact that

---

* This the following conclusion of a dialogue illustrates :
Q. What is it that kept you awake last night?
A. I heard voices telling me that I was a bad woman for suspecting my husband.
Q. What did you suspect your husband of?
A. (Obstinate mutism.)
Q. What made you say that your husband was a bad man, and went with other women, as well as the other things you said about God?
A. I am compelled to say those things against my will. I do not believe that he is a bad man.
Q. Oh I see! You do not think these things are true—
A. Do I [getting excited]. Why they are revelations! God speaks through me. [Here the patient burst out in delusional vituperation against her husband, and although quiet and reserved up to that time, developed a delirious flight of ideas of a combined expansive and persecutory kind.]

he is unable to feel for his family as of yore. As soon as a patient has reached this point the ice is broken, and the mental symptoms may be elicited in abundance, and as soon as he begins to reveal his mental state it is well to let the patient speak without interruption, and particularly to avoid asking leading questions.

There are patients whose affections for their relatives are changed, and others in whom the affections for some one or other or all the members of their family are unchanged. In the former case the patient will be more communicative if examined by himself; in the latter case it is best to have some relative, in whom the patient has confidence, present. Frequently the presence and aid of the family physician is of great service in case the examination is made by a stranger. But even where it is found advisable to conduct the examination of a patient alone, it is well, at some time in its course, to introduce the family, and study his demeanor, and mark his sayings when confronted with those whom he may regard as his foes, his assassins, or his victims, as the case may be.

There are some patients who are really anxious to be examined—not for the mental trouble, which they ignore,—but for imaginary visceral disease. This is particularly the case with hypochondriacal monomaniacs and paretic dements with hypochondriacal delusions. With such patients the examination is child's play; for in every sentence they reveal their mental state, and spread out their delusions unasked before the physician.

The use of physical appliances, the ophthalmoscope, sphygmograph, æsthesiometer, thermometer, etc., must be considered from two points of view: first, their actual diagnostic value; second, the possible effect of their employment on the patient's mind. In paretic dementia and hypochondriacal monomania, for example, the use of these instruments paves the way for the subsequent mental examination. The paretic dement shows that exaggerated appreciation of these appliances already referred to (p. 194), while the hypochondriacal patient becomes reassured by the thoroughness of the examination he so morbidly craves. The melancholiac and sufferer from persecutory delusions may have his fears redoubled by the mere sight of these to him unfamiliar and mysterious objects, and it is therefore best, if the instruments of precision are employed at all here, to use them at the close of the examination.

It is a rule, which goes without saying, that no deception, direct or indirect, is ever justified, unless it is necessary for the good of the patient, the interest of his property, and the safety of his family and of society at large. But only a pretender or one unfamiliar with insanity will demand that no deception should ever be practiced. If a patient asks point blank whether the physician proposes to take him to an asylum, and who it is that has requested him to do so, while it is possible that, in the event of a direct answer, the lunatic may take steps to revenge himself on a member of his family, it would be tantamount to criminal negligence to give a so-called "truthful" answer. Let him who gives it bear the consequences! It is best, in case the patient presses the question of what the physician proposes to do, to claim time for reflection, and, when all necessary steps are arranged, to tell him the entire truth. In some cases even this would be grossly inhumane; as, for example, in the case of a paretic dement, whose property, being already, it is to be presumed, under proper guardianship, and he being about to be placed where he can harm neither himself nor his family, may be permitted to linger out his days in dreamy and sometimes comparatively felicitous unconsciousness of his dread malady and impending death by it.

With patients who are hilarious, such as exalted paretic dements and maniacs, it is well for the physician, although he may not go so far as to assume the character of a "hail fellow well met," to pocket for the time being the stiffer variety of professional dignity if among his "accomplishments." These patients are as quick to form dislikes and antipathies as friendships and exaggerated admiration. They are very apt to entertain as exaggerated a contempt for anything that smacks of what they may regard to be conceit, overbearing pride, and æstheticism; and from contempt to a demonstration with the fists the transition is sometimes very rapid with them.

There is an idea current that patients can be stirred up to reveal their suspicions and beliefs by threats and promises. There are very few lunatics whom the physician is likely to be called upon to examine outside of the asylum who, if not in a stuporous or apathetic state, would not resent the former and despise the latter. It is a mistake to believe that a lunatic can be treated altogether like a child; his perceptions may be as acute, his feelings as sensitive, and his pride as great as those of the examiner. It is with hysterical, pubescent, and masturbatory lunatics only that harsh measures are sometimes indicated and efficacious.

Although, under exceptional circumstances, the physician may, of his own choice, consider it desirable to examine an alleged lunatic without previous communication with other parties, he will in ordinary practice find it of the highest importance to obtain a history of the patient before examining him. It is well to collate all, even the most trivial, observations made by the laity, before seeing the patient; for among them may be discovered facts which in the subsequent examination can be utilized in a more accurate analysis of the case than the best examination without them could furnish. But it stands to reason that the statements of others should never constitute the basis for an opinion unless the physician becomes convinced that they are consistent with the results of his own observations of the patient.

In the examination of the patient's facial expression and attitude the physician should include that of his dress and surroundings. Peculiarities of costume when found may often serve as a basis of comment and inquiry, revealing the existence of morbid projects or of absurd reasoning. On one occasion the writer, on entering the residence of a patient, saw little square patches of wall-paper pasted over different parts of the plastering on the side of the staircase. The patient was very taciturn, but the inquiry as to the unusual appearance of his house led to the revelation of the fact that the patient believed himself ruined, unable to meet the expenses of plastering, and had himself taken scraps of paper at random to cover up cracks and defects in the ceiling and walls.

Some patients, as soon as the ice is broken, exhibit documents relating to their morbid ideas, which often serve to portray the nature of their illness better than any verbal inquiry. To study these is hence of the highest importance. With the chronic insane it is well to induce the patient to reveal the contents of his pockets. In some cases the physician will find that scraps of string, tin-foil, and rubbish are accumulated without any special idea; this usually indicates deterioration. In a few, alleged preservatives against the assaults of demons and imaginary foes are found, and questioning reveals the delusion which has caused the patient to provide himself with them. A large number of patients carry their insane documents about with them, and these are hence obtainable by a personal search, which, as a rule, the patient assists in, or submits to willingly.

In tracing out a morbid idea the physician must not content himself with "drawing out delusions," as a superficial writer advises, under the erroneous idea that the existence of a delusion is satisfactory and sufficient proof of insanity. Any asylum attendant of experience would be an expert on insanity if this were so. The true alienist will always remember that he has an intricate mental mechanism to analyze, and that however much that mechanism may be disturbed, no examination from any one-sided point of view will suffice to reveal the character of the disturbance. It is not the patient's ideas so much that he is concerned with as the manner in which they have arisen and are nursed by the patient. Let him therefore carefully watch his method of reasoning, and bear in mind that those patients who consent to communicate any of their thoughts are usually so preoccupied with the morbid ones that they are only too glad to get a listener, and when they have one, prefer a patient listener to one who gossips. There is no surer means of making patients conceal their delusions than the ridicule to which some examiners resort with the object of "drawing them out." No delusion was ever cured or discovered by ridicule; but, on the contrary, delusions are sometimes thereafter fortified and obstinately concealed. It may, however, be very well to express surprise at, or to affect not to understand certain minor features of the patient's statements, and thus to induce a fuller explanation, and to test his reasoning and recollecting power. It is particularly desirable to have him go over the ground twice, to note inconsistencies in the two stories, to bring these to his attention, with the purpose of testing his memory as well as the systematization of his ideas. In case any one present at the examination ridicules the patient it will materially facilitate the inquiry and gain the patient's confidence to reprove such a person, or to send him out of the room.

In the case of patients who are reluctant to be examined it will often be found of service to turn the conversation on recent events of importance to the patient, his family, or to such of a sensational and political character. It will be not unfrequently found that the morbid ideation or morbid emotional condition of the mind is connected with some important event in the patient's career, such as marriage, divorce, financial gains and losses, and new business undertakings. In other cases, prominent political events, religious revivals, and temperance movements will be found to furnish the keynote to the patient's mental state.

Much has been said about the necessity of verifying delusions: a popular writer, as well as his plagiarists, have laid great stress on the necessity of finding out whether there may not be, after all, a substantial basis for the patient's ideas. While it is well to always do this for other reasons, particularly in cases where an examination is necessarily hurried,* or where the physician anticipates the possibility of having to defend his opinion before a non-expert jury, it may be stated right here that he who, after a *careful* examination of the patient, requires such an examination of his circumstances to find out whether he is insane or not is simply not an alienist. Repeatedly does it occur in the alienist's experience that the facts of a case and the delusion happen to correspond. Thus a salacious woman may be actually unfaithful to an impotent and inebriated husband, who entertains the suspicion of marital infidelity. But that suspicion is nevertheless a delusion, because the patient cannot give the reasons for his belief as a sound person would, nor reason logically on, and react normally to it. He also exhibits a tendency, common to the insane, of attributing to everything, whether trivial or of magnitude, some relation to himself. This selfish tendency, using the adjective in its widest sense, is one of the distinguishing features of insane ideas. An acute maniac claimed that people had put a rope under her bed ; this was true. She added that it was for the purpose of hanging her that night ; which was insane. A person of sound mind, if annoyed by the idea of

---

* An instance of the risks assumed in making a "snap diagnosis" is the following. The writer, being belated at his clinic, and having about half a minute to look over the patients to be introduced to the class and examined by the students, had a slightly intoxicated man brought to him, who complained of poisoned wounds, and spoke of lions and tigers. His speech was thick as a result of continued libations. The writer suspected that it was a case of alcoholism with hallucinations and delusions, and anticipated having a good opportunity for illustrating some important points. Before the class, when a more careful examination was made, it was found that the patient, while addicted to spirituous liquors, was the trainer of the lions and tigers of one of the large circus shows, and had actually been seized and mangled by a tiger, showing the severe wounds made by the animal, which, as is frequently the case with the tiger's bite, had been of a poisonous character.

a rope being under his bed, accounting to himself satisfactorily as to its presence, as this patient might have done if she had not been insane, would have removed it or have had it removed, and neither thought nor said anything momentous about it afterwards. A paretic dement came to the writer's clinic, whose occupation was that of an artist's and anatomist's model. He asserted that he was the best built man in the United States. Having to undress him before the class, as he offered his services in his professional capacity, the fact was revealed that he had a magnificent figure and a wonderful muscular development. But his announcement was, notwithstanding, that of a paretic dement, for further inquiry revealed the fact that the "girls looked at him because he had a peculiar expression in his eyes which they fancied." The sanity or insanity of an idea can be gleaned from its inherent construction, and psychiatry would be no science if the physician were compelled to rely on his ability as a detective of family secrets to exclude fraud and to make a diagnosis.

It may be laid down as a general rule that, in examining a suspected insane patient, the physician should proceed as if he were examining the mental calibre of a sane person, except where the injunctions laid down above require a deviation from this rule. Though disordered, the insane mental mechanism is not always grossly different from the mechanism of the sane mind; and it is particularly the tyro who should hold prominently in view the fact that, in venturing to examine an alleged lunatic, he may encounter as much and sometimes more wit, cunning, and knowledge of mankind in such a lunatic than he is himself possessed of. And while, as a rule, the mind of the insane is diffusely pervaded and weakened by morbid ideas or by impending deterioration, yet here and there the physician may have the tables completely turned on him by a ready patient, if he ventures outside of his province as a physician.

In his demeanor toward all patients the examiner should be gentle, yet firm. He will find the *skill* of a cross-examining lawyer or of a detective very useful, particularly in his inquiries of members of the family in whose statements the truth is sometimes difficult to winnow from the fancies of the laity; but his *behavior* should never approach that of the members of either of these professions. There is a popular delusion that the human eye has an influence over the insane similar to that claimed for the same organ over wild animals; a delusion that the writer has known the insane themselves to ridicule, and which he who attempts to utilize will learn to recognize the absurdity of at the first attempt. An overbearing, haughty demeanor, a patronizing, condescending air and fidgetiness, are all equally to be deprecated, because they will all equally tend to defeat the purposes of an inquiry. He who has the characteristics necessary to constitute a member of a learned profession will require no stage effects to aid in the accomplishment of a serious inquiry; he needs but to act perfectly naturally, that is, with earnestness and scientific purpose.

CHAPTER II.

The Differential Diagnosis of the Various Forms of Insanity.

As briefly hinted in the last chapter, the physician diagnosticating insanity is frequently in a different position from him who is called upon to investigate the existence of a bodily disease. He is met by obstacles of a kind rarely presenting themselves in general practice, and he must adopt the first symptom noticed as a cue, and be guided in the further analysis by incidental circumstances varying with each case. Such a clean-cut routine as that adopted in physical diagnosis, with its neat exclusions and combinations, is not always at the disposal of the alienist, though the methods of psychological diagnosis are analogous to those of general medicine. In the sequel it will be attempted only to indicate the salient points of diagnosis, and not to cover rare and exceptional cases.

Let us suppose that the physician is introduced to a patient who is quiet, passive, and whose attitude is relaxed. He will remember that this condition may be found in the following forms of insanity :

1. Simple melancholia.
2. The melancholic and cataleptic phases of katatonia.
3. Stuporous insanity.
4. Primary deterioration.
5. Apathetic dementia.

DIFFERENTIAL DIAGNOSIS. 331

6. Senile dementia.
7. Insanity of pubescence.
8. Prodromal and depressive phases of paretic dementia.
9. Prodomal period of mania.
10. Dementia from coarse brain disease.
11. Alcoholic insanity and other toxic forms.
12. Periodical melancholia.
13. Melancholic phase of circular insanity.
14. Paranoia with ideas of persecution.
15. Imbecility.
16. Post-epileptic stupor.

Perhaps some salient symptom may be manifested by the patient which alone suffices to establish the narrower diagnosis between these forms, but let us suppose that the physician is compelled to resort to the process of exclusion. He will bear in mind that *absolute mutism* is characteristic only of :

1. The atonic variety of simple melancholia.
2. The cataleptic phases of katatonia.
3. Stuporous insanity.
4. Atonic melancholic phases of circular insanity.
5. Periodical melancholia of the atonic kind.
6. Paranoia, and other forms of insanity with overpowering delusions.

If the patient's expression betrays intelligence and mental activity, or is ecstatic, he is likely to be a paranoiac acting under delusional commands; if he resists manipulation the same is true, or he may be suffering from the toxic forms, or be in a visionary trance. His earlier history will reveal the absence of a deeper emotional condition, and either the statements of the friends or the patient's documents will betray the existence of systematized or visional delusions. If this form can be excluded, which is usually easy, the diagnosis lies between the atonic varieties of simple and of periodical melancholia, or of cyclical insanity on the one hand, and stuporous insanity and katatonia on the other. In the three former spasmodic contraction (tetany) of the facial muscles is sometimes found, in the two latter never; and when this spasmodic contraction of the facial muscles is absent in true melancholic conditions, there is either an anxious or terrified expression in the patient's face. *Per contra*, an absence of expression is characteristic of stuporous insanity and the cataleptic periods of katatonia. The latter is readily recognizable by the characteristic symptom of waxy flexibility; while the patient whose countenance is equally blank, who is similarly mute and inactive, but who does not exhibit waxy flexibility, is most probably a stuporous lunatic.

The absolute diagnosis in all these cases can only be made with the aid of the previous or the subsequent history of the case. Thus, if the patient has had attacks of a similar kind before and at approximately regular intervals, and they have been of the true melancholic character, and of this character only, the case is one of periodical melancholia. If there have been such attacks previously, but in alternation with maniacal paroxysms, it is one of circular insanity in its melancholic phase. It is to be recollected that both periodical melancholia and cyclical insanity are rare affections, while simple melancholia is a very common form of insanity; and that both the former are apt to develop with great rapidity, while it is exceedingly rare for them to reach the degree of atony. The etiological history is also a collateral aid in determining the probabilities of the case. Thus, a melancholia which can be referred to mal-nutrition, to the puerperal state, to pregnancy, to excessive lactation, or to mental depressing causes, is more likely to be a simple melancholia than a periodical or cyclical form; these are developed independently of the exciting causes of simple insanity.

Other points, which aid in the differentiation of these forms are the following : When the patient suffering from stuporous insanity shows any signs of mental life (p. 159) these are not in the direction of raptus-like explosions or deliria of fear, like the remissions of atonic melancholia (p. 145). The katatonic patient, while he presents an emotional disturbance which is altogether absent in the stuporous lunatic, differs from the melancholiac in that he rarely exhibits deep emotional depression, but ideas of importance or a morbidly pathetic state or verbigeration intermingled with his depressive motions (p. 151). There are, however, cases of katatonia which, while in the melancholic phases, cannot be distinguished from true melancholia.

We have seen that if the patient does not speak at all, that the diagnosis—difficult as it seems on first sight—can be made with approximate accuracy even under this discouraging circumstance. Let us now return to the fifteen forms (for stuporous insanity may be excluded) of which motor inactivity and apparent depression are the first indications noted on examination, and with which it is supposable that mutism does not exist.

If there are fixed systematized delusions of persecution (p. 26), the examiner may assume that the case can be only one of paranoia. If there exist unsystematized delusions of a hypochondriacal kind, the diagnosis will lie between the prodromal period of paretic dementia and melancholia. The existence of motor disturbances (p. 207), and the specific character of the delusions (p. 195), characterize the former affection; a subjective feeling of worthlessness and deep seriousness of manner are more likely to be found with the latter disorder. The hypochondriacal paretic is more communicative and more likely to make petulant complaints than the true melancholiac, who is more retiring, secretive, and suspicious.

Among the forms enumerated at the head of this chapter as possibly associated with seeming depression a distinction can be made on the strength of the mental rank of the patient. This is decidedly low in apathetic and senile dementia, in insanity of pubescence, in dementia from coarse brain disease, and in imbecility. Of these forms insanity of pubescence is characterized by the shallow silliness and sham character of the emotions of its subjects as well as by its occurrence at and shortly after the period of puberty. Senile dementia is similarly characterized by its association with the senile state, the amnesia of the patients for recent events, their suspicions, and miserliness (p. 172). Dementia from coarse brain disease is recognizable by its association with the physical signs of ordinary brain lesions. Apathetic dementia has characters which are negative, but the history of a preceding psychosis will demonstrate that the present disorder is only the terminal phase of another and long-past form of insanity; while imbecility is ordinarily marked by somatic signs of imperfect development, or by the history of *early* and *primary* mental defect. In the dement there are almost always flashes breaking through the cloud of mental darkness, proving that there has at one time been more mental power than the patient now has; in the imbecile the mind, as far as it goes, indicates a permanent organization.

Delusions of persecution of the unsystematized kind characterize the different forms of melancholia, the melancholic phases of katatonia, the depressive phases of paretic dementia, alcoholic and other toxic insanities, and the post-epileptic forms.

Melancholia, whether simple, periodical, or in cyclical development, differs from the other forms here mentioned by the depth of the emotions. In periodical and cyclical forms the delirium growing out of the delusions is apt to be subdued and to have a more "reasoning character" (approximating in this respect the systematized delusion) than that of simple melancholia. The patients are also apt to show a moral perverseness and irritability, rare at best in simple melancholia. In addition, the history of previous attacks and of their character, herein already referred to, will serve to establish the nature of the melancholia.

Post-epileptic stupor and post-epileptic insanity with depression are recognizable by the peculiar drunken appearance of the patients, their combined religious and sexual delusions, the previous occurrence of fits, the sudden development of the symptoms, the prevalence of angry excitement *unassociated* with expansive flights of ideas, and the deep disturbance of consciousness. The alcoholic forms are characterized by their association with ghastly hallucinations, the delusions of marital infidelity, sexual mutilation, and poisoning, and the physical signs of alcoholism.

It is impossible in the present imperfect state of our knowledge to indicate the signs by which the chronic form of opium insanity can be recognized. As far as the writer's experience goes absolute mutism may here occur under delusional influences, exactly as in paranoia; while, on the other hand, visions of a religious character are likely, and hallucinations of a kind resembling those of acute mania and confusional insanity are common.

If none of the positive signs which are pathognomonic of the special forms are discoverable in the way of disturbances of the memory, of consciousness, hallucinations, and delusions, the diagnosis rests between:

1. Melancholia sine delirio.
2. The prodromal period of mania.
3. The prodromal period of some cases of paretic dementia.
4. Primary deterioration.

Melancholia sine delirio is recognizable by the continuously gloomy disposition of the patient unassociated with formal disturbance of the intelligence. The depression is entirely subjective. In primary deterioration and the prodromal period of depression of simple mania, on the other hand, the depression is the result of the recognition by the patient that his mental powers are more sluggish than of yore, and that he is unable to exert his will-power or carry on his duties. Between these two conditions the diagnosis can in turn be made by the fact that in primary deterioration there are usually serious lacunae in the memory, which are absent in mania. If this differentiation is impossible, the subsequent course of the trouble will soon establish the diagnosis beyond doubt. Primary deterioration if properly treated may be permanently arrested, or, if we are unable to check it, will progress in the same direction ; the prodromal period of typical mania, however, is followed by the maniacal explosion.

In the vast majority of cases the prodromal depressive stage of paretic dementia can be recognized by means of the criteria given on page 187. The patients exhibit an undue irritability in small matters, and apathy as well as abulia with regard to important ones, a contrast not found in primary deterioration. The peculiar abstraction of the paretic is a characteristic symptom of paretic dementia not found either in melancholia sine delirio nor in the prodromal period of mania ; while the oddities of behavior observed in the former condition are never found in the latter two disorders.

Let us now suppose that the patient exhibits *motor excitement*. This is found in :

1. Simple mania.
2. The maniacal phases of paretic dementia.
3. Transitory frenzy.
4. Agitated dementia.
5. Agitated melancholia.
6. Periodical mania.
7. Maniacal phase of cyclical insanity.
8. Epileptic insanity.
9. Episodical frenzy of chronic insanity.

The differentiation here can be made according to the character of the associated mental state. Agitated melancholia is readily recognizable by the complaints and the panphobia of the patient; agitated terminal, as well as agitated senile, dementia, by the motiveless character of the patient's activity. In epileptic insanity with motor excitement, in transitory frenzy, and in the frenzy of mania and paretic dementia, there is disturbance of consciousness, and for this reason the closer diagnosis between them cannot always be made on the spur of the moment. The history of epileptic fits, the association of religious and sexual ideas, the dreamy character of the excitement, and the frequently present motor weakness, serve to distinguish epileptic frenzy. Angry excitement combined with expansive and aggressive delirium is equally found in maniacal and paretic frenzy, their resemblance being increased by the temporary erasure of the motor signs in the latter state. This is, however, never complete in the later periods of paretic dementia. The diagnosis of transitory frenzy is usually made after the attack is over, and when the existence of epilepsy or of other neurotic conditions has been excluded.

The same rules which were given above for the distinguishing of simple melancholia from the periodical and cyclical forms, apply to the differentiation of simple mania from periodical mania and the maniacal phases of cyclical insanity.

From the fact that paretic dementia imitates various phases of other psychoses, it is of the very highest importance to properly appreciate the exact value of the signs whose presence or absence justifies us in making a positive diagnosis of this affection, or to exclude it. To do so it is necessary to refer to many errors current on this head. It is generally believed that the unsystematized delusion of grandeur—as, for example, the belief of great personal beauty, of Herculean strength, and of vast possessions—is pathognomonic of this disorder. But the same delusions may be found in simple mania as well—though, it is true, rarely in the same extravagant form. With many it suffices to detect a pupillary difference and a tremor of the outspread fingers, in addition to such delusions, to render the diagnosis of paretic dementia a certainty in their minds. It must be recollected in this connection that not only are differences in the pupillary diameters found in simple mania, in melancholia, and in monomania, but also in perfectly sane persons, either congenitally or in the consequence of unilateral eye-strain, as in microscopists and watchmakers, or as the result of the obscure influence of visceral disease; thus the right pupil is sometimes found dilated in hepatic disease, the left in cardiac, splenic, and gastric affections, and a similar sympathy has been found to exist between unilateral pulmonary disease and the state of the pupil on the same side.*

---

* The possible existence of an aneurism should not be forgotten in this connection.

"As to tremor of the outspread fingers, it is a common sign in many other forms of insanity, and is characteristic of numerous nervous disorders which have nothing to do with paretic dementia. In persons suffering from nervous exhaustion or alcoholism it is common ; in monomania an emotional tremor is almost always found with episodical excitement, and the same is true of insanity with excitement generally. In all these cases if the patient's attention is directed to the desirability of keeping the hand steady, the tremor diminishes up to a certain point, while the hand is held out. In paretic dementia the tremor increases the longer the hand is held out, and is more excursive and irregular, though the rhythm is sometimes finer than that observed in the conditions mentioned.

The writer has also met with a case where the diagnosis of paretic dementia had been erroneously made in a patient suffering from posterior spinal sclerosis, and who had had in addition an attack of suicidal melancholia. Spinal-cord affections may be accompanied or complicated by insanity, which may or may not have an essential connection with them. It is particularly in the course of locomotor ataxia (posterior spinal sclerosis) that delirious frenzy or melancholia acutissima may occur and disappear with equal suddenness.*

When paretic dementia is so far advanced that its episodes may be confounded with those of mania, symptoms will be observable, after the maniacal attack has subsided, which leave no room for a doubt as to the diagnosis. The suddenness with which the patient emerges from the attack, his residual delusions, irritability, and amnesia, involving events preceding the maniacal outbreak, are in striking contrast with the usually gradual disappearance of true mania, the lucidity of its convalescent periods, and the complete recovery of the recollections accumulated prior to the illness, and often even of those accumulated in its course. But, in addition, there are observed in the paretic dement under these circumstances speech disturbances, paresis of certain muscular groups, and a *fibrillary* tremor of the tongue which are not found in other forms of insanity (p. 191), except in the allied one of syphilitic dementia and in alcoholic mania.

In the earlier periods of paretic dementia, when the men-

---

* Probably due to an anatomically impalpable involvement of the intracranial vasomotor centre.

tal symptoms are either those of depression and hypochondriasis, which in their differential relations have been discussed above, or of a subdued maniacal kind, the diagnosis may be more difficult, for the physical signs are then not always well marked, though the characteristic facies of this disease (p. 209) is sometimes already present at this time. In its absence the following signs, or the combination of several of them, point to the existence of paretic dementia with almost unerring certainty: 1st. Morning headache of a dull kind, either described as comparable to an encircling band or to a grinding sensation, which may be so severe in some cases as to cause the patient to cry out and to beat his head against the wall; 2d. The *sudden* disappearance of this headache, accompanied by a subjective feeling of lightness and exaltation; 3d. Sudden stoppage of words, or momentary cessation of ideation; 4th. Abstraction in the midst of conversation, with corresponding mental lacunæ; 5th. Congestive attacks, associated with the word-stoppages and fits of abstraction; 6th. Drowsiness after meals; 7th. Dreams of a vivid and usually disagreeable character; 8th. Amnesia for special events or series of events; 9th. Vertigo independent of gastric disturbance. All these signs, with the exception of the second, may be premonitory of focal organic disease of the brain, and it would be impossible to characterize the symptoms of paretic dementia in the early stage better than by saying that, in addition to the prodromal signs of the ordinary *psychoses*, there must be the prodromal signs of ordinary *brain disease* to justify a positive diagnosis. Neither group by itself is sufficient to prove, though it may be strongly suggestive of the existence of, this disease. The occurrence of an epileptiform fit, or of an apoplectiform attack, particularly if there be a rapid recovery from their grosser effects, gives strong support to the diagnosis,* if either of these symptoms be present.

It may be assumed to be a general rule that even in the early stage of paretic dementia there is more mnemonic and logical enfeeblement than in mania or melancholia in the corresponding period.

The countenance in paretic dementia is supposed to be

---

* The occurrence of an epileptiform attack after a maniacal paroxysm, is not in itself proof of the existence of paretic dementia. Worthington) *Journal of Mental Science*, October, 1881) reports a case of *mania in puerpero*, ending in, and apparently "cured" by, an epileptic fit.

one of its most characteristic features, and in the overwhelming majority of cases in which such a countenance is found that disease is present. But there are three conditions in which a similar appearance may be discovered: the saturnine cerebral disorder, monomania with deterioration, and bromine saturation. In the saturnine encephalopathy the expression of the paretic, his facial pareses, and the labial tremor are closely imitated. In monomania with delusions of persecution of long duration there is the same corrugation of the brow (Fig. 13), not the effect of a compensatory effort, as in paretic dementia, but the, as it were, petrified expression of care and dread. The congenital facial asymmetry and defective innervation of one side of the face sometimes found in monomania increase the resemblance. The following points will serve to distinguish the two: When there is marked corrugation of the brow in paretic dementia there is always in addition more or less ptosis in the writer's experience; this is usually absent in monomania, though in subjects of a neuropathic constitution a slight degree of ptosis is sometimes found. In case this leads to a doubt in diagnosis the examination of a photograph taken some years before will reveal its true nature. If present then, it can not now be an indication in itself of paretic dementia. When there is asymmetry or unequal innervation in monomania there is never the exaggerated (snarl-like) action of the muscles of the nasolabial fold so frequently noticeable in the paretic dement when excited or after speaking any length of time. A fine emotional tremor may be present in monomania, but the spasmodic twitches of isolated facial muscles or their fasciculi, so multitudinous in paretic dementia, do not occur much more frequently in monomania than in that portion of the sane population which suffers from insomnia and other accompaniments of nervous exhaustion and over-strain. There are sometimes observed excursive spasmodic contractions in monomania, but they resemble in character the *tic convulsif* and are not fibrillary. They usually involve one entire side of the face, a distribution which, if it ever occurs in paretic dementia, is exceptional. In patients suffering from brominism a combined condition of mental impairment and defective facial innervation is found, which, on first sight, strongly suggests the existence of paretic dementia. The writer has seen double ptosis, a stony stare, lax facial folds, zygmotic, labial, and lingual tremor, in one male (a physician) and in one female patient, which exactly imitated the appearance of advanced paretic dementia as noted after epileptiform attacks. The apparent stupor of the brominized patient, his ataxia and paraparesis, cannot, however, be confounded with the superficially similar symptoms of paretic dementia. The patients have full *ego* consciousness, though their reception, registration, and reproduction of impressions is temporarily impaired; but while it is altogether impossible to remind the paretic dement of certain events which happened some time before, the brominized subject can usually be brought to recollect them on suggestion. The patient struggles against his amnesia, just as he makes frequent efforts, like a drunken person, to straighten out his features and raise his lids. His stagger is very much like the stagger of a person intoxicated with alcohol, a character which the paretic's stagger does not imitate as closely, because, while the paretic's stagger has the combined ataxic and paretic character of the alcoholic stagger, there is not that same struggle to retain the equilibrium which is common to the drunken and the brominized subjects.

There is one feature observed, not in all but in a large proportion of paretics, toward the terminal period of their illness, which is never found in any other condition, except in *delirium grave*. This is an appearance of obliteration of the features and puffiness, which can be characterized by no single term so well as by "sogginess."

Of single symptoms which merit special consideration in their differential diagnostic relations, the delusions of the insane have been already detailed in the first part of this volume (p. 23). The discrimination there made of systematized delusions and unsystematized delusions may be supplemented by the following:

I. Systematized, fixed delusions are found only in paranoia and hysterical insanity.

II. A feeble systematization of delusions, that is, an approach to a compliance with the demands of logic and consistency, is sometimes observed in periodical insanity.

III. The delusions of alcoholic insanity are fixed as to their general contents, but they are not logically systematized.

IV. The delusions of mania are constructive, and frequently display some logical and creative power, but they are never stable enough to become systematized.

V. Fixed, that is permanent, delusions are observed in

secondary confusional insanity without logical correlation, and therefore are unsystematized (p. 170).

VI. The delusions of some imbeciles, without showing logical strength, creative power, and systematization, bear the same relation to the mental mechanism of the imbecile that the systematized delusions of the monomaniac do to his. They are as permanent, and control the *ego* proportionately.

The rapidity with which delusions develop is also of some significance. The systematized delusions of paranoia are the result of a process of reflection, and are consequently of slower growth than the unsystematized delusions of some other forms of insanity; and an apparently rapid growth of them is always associated with preceding visions, which are in turn the subjective confirmations of previous morbid trains of thought. Sometimes the most extravagant delusions of paretic dementia develop immediately after a sleepless night, or after a series of epileptiform attacks.

The *delirium*, that is, the flight of ideas of the insane, presents characteristic features in various forms of insanity. The older German writers distinguished true maniacal delirium from so-called *moria*. Under this latter term they comprised those deliria which are marked by unreasonableness, or are of a shallow and foolish character. Such deliria are found in primary and secondary confusional insanity, in the epileptic psychoses, in katatonia, and in insanity of pubescence. The higher deliria of mania and of paretic dementia in its earlier stages are characterized by an exalted or angry emotional state, those of melancholia by the reverse, and those of delusional monomania by their consistency with the systematized and fixed morbid beliefs. In hysterical delirium there is often noticeable an exuberant fancy with an approach to systematization.

The character of hallucinations and illusions in the various forms of insanity has been detailed in the first two parts of this work.* In a general way it may be stated that visions of an elaborate kind are rarely found in any other form of insanity than in paranoia. More confused visons are sometimes observed in paretic dementia (p. 202). In acute insanity the hallucinations are more changeable, and what were previously characterized as "multitudinous"

---

* Pp. 49—54, 134, 142, 152, 163, 174, 200, 248, 254, 257, 303, 304, and 317.

hallucinations (p. 48) are indicative of alcoholic, epileptic, or opium insanity, and occasionally of paretic dementia.

Let us now apply the foregoing to a few cases, which may serve as illustrations of the methods of diagnosis to be followed in mental disease: I. The physician is called to a female patient about eighteen years old; she is seated on a bed, of which the bedding and blankets are in great disorder, and vociferating loudly. These few facts will indicate to him that he has to deal either with maniacal excitement of some form, melancholic frenzy, or hysterical simulation. On approaching closer, the patient applies some remarks to him, either calling on him to protect her against assassins, or accusing him of being in league with them. All this might occur in either of the three conditions named, but diminishes the probability of hysterical simulation. Her expression is animated, and she continually spits out saliva into her handkerchief, asserting that this is the poison administered by her foes which is driven out by some counter-poison given her by her friends. There is a marked effluvium in the room, due to the exaggerated cutaneous secretions, and the history is given that the patient has been in nearly the same condition for over a week. The physical signs noted, the history, and the character of the ideas positively exclude hysteria. The first delusions expressed by the patient are those of persecution, and, as far as this is concerned, might appertain to melancholia as well as to mania. But the following are inconsistent with melancholia: 1. The animated look; 2. The recognition of a friendly power combating her foes; 3. The consciousness of the patient; 4. Her responsiveness. On listening to the word-delirium it is found to be multifarious, relating to a great many objects; she can be checked for moments and made to answer reasonably on minor topics; finally, on asking her *why* she is pursued and why her parents are "hired murderers," she asserts that in reality she is the Princess Louise, and that they are not her parents. Melancholic frenzy, and indeed melancholia in general, may be now excluded, for they are inconsistent with ideas of exaltation and creative fancy; but the new suspicion arises and must be taken into account, that, after all, the case may be not one of pure maniacal excitement, but of monomania with episodical delirium; for with monomania, particularly in young subjects of that disorder, the delusion of being exchanged for other children, and deprived of a birthright and persecuted, is very common, and almost characteristic of a certain group. On being questioned further, the patient herself tells the physician that she has not entertained this idea more than a few days. She does not claim to be princess of any particular land, and while she asserts that her parents are not her real parents, she is not able to say who the latter really are; speaking of a schoolmate who is also maltreated, as she claims, by her parents, she expresses the suspicion that she too may be in reality a princess brought up by foster-parents. The lack of systematization, and the suddenness with which the delusion arose, disprove the suspicion of monomania, and establish the diagnosis of maniacal excitement. But maniacal excitement may be found in a number of psychoses. The sex and youth of the patient, as well as the absence of motor disturbances, show that it cannot be a maniacal exacerbation of paretic dementia. The patient has never had epilepsy, which fact alone would show the impossibility of its being an epileptic mania, even leaving out of consideration the comparative clearness and consciousness of the patient, the absence of the epileptic expression, and of sexual, terrible and religious delusions. On making inquiry of the relatives the physician finds that the patient has never been insane before; this fact renders it improbable that the case is one of beginning periodical or cyclical insanity. In addition it is found that the disorder exploded with hallucinations and illusions of the identity of others, after a suppression of the menses. The diagnosis can then positively be made of simple mania.

II. The physician visits a patient who is himself desirous of having the state of his physical and mental health examined, but cannot summon up resolution enough to call on the doctor himself. It is only in melancholia, in incipient paretic dementia, or mania in primary mental deterioration, and in hypochondriacal paranoia that the patient is likely to request a physician's services himself. The patient rises in a listless way to greet the physician, there is an expression of the *ennui* of invalidism on his face, and his movements are slow and denote a certain degree of abstraction. His tongue is coated, the appetite poor, and the bowels are constipated. He gives no history of excessive mental strain, but for some months past has failed to attend to his business properly, neglected it, and takes no interest in it. The period of several months is an unusual one for the prodromal stage of simple mania, which may be provisionally excluded therefore. The patient, on being asked why he is unable to call up his former interest in his business, says that he is disgusted because he is not making enough money at it. At the same time he does not show any deep emotion, and the reminiscence of sad occurrences does not affect him. This renders the existence of simple melancholia improbable. On re-directing his attention to his physical state, he expresses the idea that his bowels are grown together, and that no fæces can pass. This characteristic hypochondriacal delusion excludes primary deterioriation, and limits the diagnosis to hypochondriacal paranoia and the hypochondriacal phase of paretic dementia. His wife now calls attention to the fact that he has had a passage from his bowels, which he has evidently forgotten, and he is apparently nonplussed by the reminder. This speaks strongly against hypochondriacal paranoia, for here the fact of a passage would be accounted for by some explanation however absurd, as, for example, that there was an accumulation below the stricture, or a new channel formed. On the other hand, the amnesia displayed with regard to an occurrence of the day before, which, because related to the patient's morbid ideation, he ought to recollect, if a systematic delusionist, much better than even a sane person, strongly suggests the existence of paretic dementia. In addition it is found that the memory has become enfeebled, that the patient complains of a sensation like a tight band around his head, of his sleep being disturbed by dreams of a disagreeable character; his color-field is found limited, while there is pinhole contraction of the pupils, and occasionally a fibrillary twitch in the facial muscles. The case is, therefore, one of incipient paretic dementia.

An important fact to be remembered in connection with this branch of the subject is the possibility of certain forms of chronic insanity being complicated by the more acute forms. Thus typical melancholia and mania may occur in a paranoiac or an imbecile. Here the diagnosis will not be very difficult. It is different with regard to a complicat'on, which the writer has reason to consider a much more frequent one than is ordinarily supposed, namely, that of paranoia by paretic dementia. The writer has seen two cases of this combination in the pauper asylum on Ward's Island and two in private practice. In one of the latter, morbid projects and delusions of the systematized variety could be determined to have been entertained by the patient over forty years before the outbreak of the paretic disorder. The history of one of the asylum cases, which was completed by a *post-mortem* record showing the characteristic lesions of paretic dementia, is so instructive, and in more than one respect illustrates the differential indications referred to in this chapter, that it is appended in full.

At the time when the writer's attention was most closely directed to the patient he presented the physical signs of progressive paresis, and at the same time morbid projects and delusions. While some of these, like his idea that the superintendent was responsible for the introduction of vermin into the institution, and that he was going to build an asylum with a school attached to it, in which the medical officers were to be properly instructed in mental science, notwithstanding the kernel of justification for the latter intention, were unquestionably true paretic notions, others were so elaborate, and involved such an excellent memory and fertility of invention, that, without knowing his earlier history, the writer suspected that the paralytic insanity was engrafted as a foreign element on a pre-existing monomania. It may be said here that the patient had a fair recollection of some of the experiments of Magendie, a good knowledge of anatomy, was a fair logician and ready writer, even to the time when he could hardly enunciate words properly, and was unable to use his lower extremities. He all the while exhibited the characteristic symptoms of querulous monomania. If the peculiar character of the patient's delusions alone justified the suspicion entertained, that the paralytic insanity which terminated in his death was engrafted on a pre-existing mental disorder, it became established more strongly after obtaining an account of his past history, through an article which appeared in one of the daily papers the day after his death. It read as follows: "David Wemyss Jobson died of paralysis, on Friday last, in the Ward's Island Hospital for the Insane. He was about seventy years of age. His body was taken to the morgue yesterday, and, if not claimed to-day, the commissioners will inter it to-morrow in the Potter's Field on Hart's Island. Letters found among his effects show that he was born in Dundee, Scotland, of a family that had been landholders for centuries. He graduated at the Edinburgh University in 1827, and studied both surgery and dentistry. About 1836 he settled in London, and obtained permission from King William IV. to call himself Surgeon Dentist to the King, and this permission was subsequently [also] accorded by Queen Victoria. A year or two after the Queen's accession he gave offence by espousing intemperately the cause of the Lady Flora Hastings, one of her maids of honor, who was afflicted with a form of dropsy, and was wrongfully said to be about to become a mother. The lady soon died, and an autopsy disproved the suspicion. In consequence of Jobson's partisanship the permission to style himself Surgeon Dentist to the Queen was withdrawn, and he lost much practice among the aristocracy. His pecuniary fortunes declined, and he obtained a precarious support by writing for newspapers and magazines. At times he was aided by remittances from his relatives in Scotland.

"At the outbreak of the French Revolution of 1848 he went to France and ingratiated himself with Lamartine and other members of the provisional government, and, without any military education, he managed, according to his own account, to obtain the honorary title of general.* He returned to London, and about 1854 came to the United States, where he found an indifferent support by his pen. He became naturalized, but afterward made repeated voyages to England, and there he continued to write for the press. There are numerous letters among his papers from the secretaries of Prince Albert, Lord Palmerston, the Earl of Derby, Mr. Disraeli, and others, thanking him politely for documents sent, but expressing inability to render them available for any purpose. Jobson was imprisoned for a night in London and fined £50 for insulting Alderman Gibbs, while that official was acting as a magistrate. On another occasion Recorder Russell Gurney sentenced him to hard labor in the Bridewell for libelling Sir James Ingram. Sir Edward Thornton, the British Minister at Washington, was repeatedly written to by Jobson, with reference to panaceas for quelling the troubles in Ireland, and Jobson received several polite letters from him declining to act upon his suggestions. In 1858 a Mr. Ira D. Jobson, of 2 Paton's Lane, Perth Road, Dundee, Scotland, wrote that his uncle had died, leaving him and his brothers about £40 apiece, but that David's name was not men-

---

* This was a fact.

DIFFERENTIAL DIAGNOSIS. 347

tioned in the will. The bulk of the property, he said, had been left to the poor relations of the Earl of Camperdown. This took Jobson to England, and among his papers is the copy of a memorial which he sent to the Lord Chancellor in 1859, announcing his return from America to claim a heritage of 15,000 acres of land that had been in his family for centuries, and also £26,000 that had been left by his uncle chiefly to Sir James Ferguson, nephew of Lord Camperdown, who had married his (Jobson's) niece. He complained, also, that he had been placed in a lunatic asylum by Sir James Ferguson's machinations.

"Having been unsuccessful in his application, Jobson returned to America. His career here is well known. He wrote casually for newspapers, received casual remittances of £1 at a time from Mr. Ira D. Jobson, and obtained credit where he could for clothing. In the winter of 1874-75, he became ragged and almost bare-footed, and in March last he was taken to the asylum at Ward's Island, more out of pity than on account of any danger occurring to the public from his harmless insanity." Subsequently the writer was informed by ex-Judge Joseph Koch, that while the latter acted as police magistrate Jobson frequently annoyed him by bringing complaints against the British Government, and certain city officials who had dealt with him unjustly in reference to a public candidacy. Be it known that this patient had had himself nominated for the comptrollership of this city by an ephemeral organization, and had actually received a number of votes in the ensuing election. It was also evident from his own account that he had been engaged in a riotous movement in Australia, and been imprisoned a year there in consequence.

While during the earlier periods of his asylum sojourn, he enunciated his systematized delusions of having been defrauded of the comptrollership, and wrote the average kind of newspaper poetry, complaining of the machinations of the British Government against himself, and, on the other hand, alleging the existence of conspiracies against that very government, these elaborate notions were less distinctly announced, and most of them even forgotten, some months before his death, when, with marked emotional tremor and imperfect tongue, he stuttered forth the wildest and most inconsistent ideas of aggrandizement, mingled with complaints about the vermin alluded to, and which evidently depended on the hyperæsthesia and paræsthesia which constituted marked features of the physical part of his symptoms. The autopsy revealed marked cystic degeneration of the cortex of the paracentral lobule and of the inner face of the frontal lobe; adhesion of the pia over the central gyri and superior parietal lobule; extensive atheromatous degeneration of the arteries, both of the great trunks and the primary and secondary branches; atrophy; fatty, granular, and sclerotic degeneration of the cortical nerve cells; as well as similar and more intense changes in the cells of the lower cranial nerve nuclei and of the anterior spinal cornua. There was no deviation from the normal convolutional type as such, but the writer has never seen the transverse type of corrugation of the gyri as pronounced in any human brain as in this one; the skull was moderately stenotic; the pons and oblongata, the latter particularly, were unusually small, but this could be referred to the diffuse sclerosis and atrophy, of which these parts were the seat.

An analogous case is that of a Doctor P——e, who was observed at the same institution, and in whose case there was, about thirty years ago, a history of erotic monomania from which he emerged with the systematized and well-defended delusion that he was King George the Fourth. During the time the writer observed him, he gradually developed all the physical signs of paralytic dementia, even to its trophic disturbances; but, though hardly able to articulate, and unable to defend his delusions, he still held to his royalty, and insisted on the honors due the latter in a vague and unintelligent way, in marked contrast with his previous plausible and linked reasoning on the same subject.

While the combinations of symptoms characterizing the special forms have been described in the second part of this work, and need not be repeated, we may here briefly recapitulate the special symptoms of insanity, and the forms in which they are likely to be found in their order of probability.*

I. DELUSIONS: A. *Systematized; a. fixed and permanent:* in paranoia (characteristic); *b. fixed for the time being and intermittent:* hysterical insanity, periodical insanity (latter rare). B. *Unsystematized; a. creative, with elements of exalta-*

---

* The complicating forms are omitted here; their association with the bodily states on which they depend serves to characterize them in the main, a special analysis of them lies outside the limits of this manual.

tion: in simple mania, maniacal exacerbations of paretic dementia in early stages, periodical mania, hysterical insanity; *b. creative, with predominating element of depression:* melancholia, hypochondriacal and other depressive phases of paretic dementia, periodical melancholia, chronic alcoholic insanity, katatonia; *c. monotonous and confused, with elements of exaltation:* in secondary confusional insanity, paretic dementia in latter stages, imbecility; *d. monotonous and confused, with elements of depression:* in secondary confusional insanity, paretic dementia in later stages, chronic alcoholic insanity; *e. monotonous and confused, with elements of depression and exaltation mingled:* in acute confusional insanity, senile dementia, epileptic insanity, katatonia, insanity of pubescence.

II. HALLUCINATIONS AND ILLUSIONS: *a. visional:* in paranoia, epileptic insanity, hysterical insanity; *b. of single or a few objects related to the patient's ideas:* in simple melancholia, simple mania, primary confusional insanity, periodical insanity, paretic dementia ; *c. of multifarious, usually disagreeable, objects:* in alcoholic insanity, epileptic insanity, paretic dementia.

III. IMPERATIVE CONCEPTIONS AND IMPULSES: *a. continuous:* in paranoia, imbecility ; *b. periodical:* in periodical insanity; *c. episodical:* in simple melancholia, hysterical insanity, paretic dementia, paranoia, imbecility.

IV. ABULIA: in simple melancholia, prodromal period of mania and paretic dementia, alcoholic insanity, periodical melancholia, forms ending in general mental enfeeblement, paranoia with overwhelming hallucinations and delusions.

V. HYPERBULIA: in maniacal phases of simple mania, paretic dementia, periodical insanity, expansive paranoia.

VI. MARKED EMOTIONAL DISTURBANCE : .A. *Without intellectual motive*; *a.* angry (1) *simply:* in maniacal furor, paretic furor; (2) *angry and treacherous:* in epileptic, alcoholic, and paretic furor ; (3) *angry and anxious :* in melancholic frenzy, transitory frenzy, katatonia ; *b. expansive, goodhumored, or pleasurable:* in simple mania, paretic dementia, periodical mania ; *c. depressed, sad, or anxious:* in simple melancholia, periodical melancholia, alcoholic insanity, epileptic insanity, paretic dementia in early stages, katatonia, insanity of pubescence.* B. *With intellectual motive; a.* an-

---

* This is also the order of the depth of the emotional disturbance.

gry and expansive: in episodical delirium of paranoia ; *b. depressed:* in prodromal period of mania, paranoia with depression, primary deterioration.

VII. IMPAIRED CONSCIOUSNESS, OF A MARKED DEGREE AND DEMONSTRABLE KIND : in epileptic insanity, transitory frenzy, stuporous insanity, melancholic frenzy, alcoholic frenzy, delirium grave, maniacal and paretic frenzy, cataleptic phases of katatonia.

VIII. MENTAL WEAKNESS PROMINENTLY DEVELOPED : *a. involving the mental faculties generally:* idiocy, imbecility, primary deterioration, dementia—whether terminal, epileptic, alcoholic, or from organic disease—delirium grave; *b. with "focal" lacunæ:* paretic dementia, syphilitic dementia, chronic alcoholic insanity, secondary confusional insanity, primary confusional insanity, insanity of pubescence.

IX. MENTAL WEAKNESS EXTENDING IN LIMITED DIRECTION: paranoia, moral imbecility.

X. MARKED AMNESIA, ASIDE FROM UNCONSCIOUSNESS : in epileptic insanity, delirium grave, paretic dementia, syphilitic dementia, senile dementia, dementia from organic disease, chronic alcoholic insanity, terminal dementia, chronic confusional insanity.

XI. SOMATIC STIGMATA: in idiocy, cretinism, imbecility, paranoia, epileptic insanity, periodical insanity, hysterical insanity, exceptionally and then non-essential in all other forms.

XII. ACTIVE DISTURBANCE OF THE BODILY FUNCTIONS: in delirium grave, melancholia, stuporous insanity, mania, katatonia, frenzy, initial and terminal periods of paretic dementia, senile dementia, paranoia with hypochondriacal or persecutory delirium.

XIII. SPECIAL TROPHIC DISTURBANCES: in delirium grave, paretic dementia, syphilitic dementia, dementia from organic disease, epileptic dementia, melancholia, terminal dementia.

XIV. POSITIVE DISTURBANCES OF LOCOMOTION : in paretic dementia, syphilitic dementia, delirium grave, dementia from organic diseases, epileptic insanity, alcoholic insanity.

XV. SPEECH DISTURBANCES: *a. acquired:* in paretic dementia, syphilitic dementia, dementia from organic disease, alcoholic insanity ; *b. congenital:* idiocy, imbecility, paranoia.

XVI. TREMOR:* *a.* senile dementia, alcoholic insanity, dementia from organic disease (multiple sclerosis), paretic dementia, syphilitic dementia, epileptic insanity; *b. true emotional tremor*, any form with high neural excitement such as mania, frenzy, paranoia with episodical deliria.

XVII. ODDITIES OF SPEECH: *a. echolalia:*† in imbecility, insanity of pubescence, dementia, imbecility; *b. verbigeration:* in katatonia, epileptic insanity, hysterical insanity, chronic confusional insanity, insanity of pubescence; *c. rhyming:* in katatonia, insanity of pubescence, epileptic mental states, sometimes in any episodical excitement.

XVIII. CONVULSIONS: in epileptic insanity, paretic dementia, syphilitic dementia, dementia from organic disease, katatonia, an accidental accompaniment of other forms.

While but the main disturbances are detailed in this schedule, the latter will serve as a guide to a provisional diagnosis at least, following the plan detailed in the two hypothetical cases detailed above (p. 342). Suppose that a history of a convulsion or convulsive movement is given in connection with mental disturbance; it may according to the schedule be an evidence of four mental states. The most frequent form with convulsions is epileptic insanity; if now a confused delirium of a partly aggressive and depressive character (I), hallucinations of multifarious and disagreeable objects (II), angry and treacherous excitement (VI), impaired consciousness (VII), positive disturbances of locomotion (XIV), and verbigeration (XVII) are found, the physician may positively pronounce the case one of epileptic insanity. And this he may do if only a majority of the signs enumerated are present, if, as is sometimes the case, the hallucinations are agreeable, if there is no motor disturbance and no verbigeration ; for the symptom combinations

---

* We are not yet able to differentiate by clinical signs all the varieties of tremor encountered among the insane; the possibility of confounding the tremor due to excessive smoking with that of alcoholism should be therefore borne in mind. There are certainly some smokers whose tremor cannot be distinguished from that of paretic dementia; indeed there are reasons for supposing that nicotine may affect the central nervous apparatus in a similar direction. Paretic dementia has become very frequent among Austrian army officers, owing to their habit of consuming large quantities of what they term "Virginia segars." These are perforated by a reed, so that the smoke is not, as in the ordinary rolled weed, deprived of most of its deleterious ingredients.

† Echolalia is the thoughtless repetition of words and phrases spoken by others, the subject not associating any mental conception with them.

of the other forms associated with convulsions are altogether different, and do not coincide except in unimportant details with those of epileptic insanity.

Among the evidences aiding in the differential diagnosis of the various psychoses are the age, sex, heredity, and vocation of the patient. A man at forty cannot be a pubescent lunatic (though he may have begun as one), nor a senile dement. A female is less likely to suffer from paretic dementia than a male, and a male less likely to be a periodical lunatic than a female. A patient with insane ancestors is more likely to suffer from the psychoses associated with a neurotic taint, than one whose ancestry is free from insanity. A Wall-street speculator is more likely to be a paretic dement than a farmer. But these facts are of relatively slight value, and merely collateral to those enumerated.

Where the diagnosis of a special form of insanity cannot be made, the alienist is compelled to limit himself to the question of the existence of insanity in general. To deal with so obscure a case, all his diagnostic acumen must be employed. It is then neither hallucinations, delusions, motor disturbances, nor amnesia that lie on the surface and indicate the line of examination to be followed. He must test for pyschical weakness in the abstract, for abnormal irritability, logical perverseness, abulia, hyperbulia, lack of reaction, and abnormal emotional states, whose characters language cannot portray, so that they can be appreciated in the living subject only.

CHAPTER III.

THE RECOGNITION OF SIMULATION.

The psychological diagnostician has less frequently to deal with the feigning of insanity by the sane than with the dissimulation or concealment of insanity by the insane. Persons who have once been inmates in an asylum, those who have sufficient mind to know the meaning of a medical examination, and particularly those who have an occasional glimpse of the fact that they are considered insane, or in part recognize their insanity themselves, are frequently very difficult to examine. For however much the alienist may become satisfied from the expressions, the manner and histories of such patients, that they are insane, they obstinately conceal those symptoms which it is desirable to discover for the purpose of satisfying the legal, or it may be the medico-legal, demands of a commitment or an examination undertaken for forensic purposes.

Simulation, although far from uncommon, is not as often resorted to as some writers would like the public to believe in order to facilitate the reception of testimony "arranged " to suit the demands of public prejudice and of medico-legal conspiracies. Historical instances of simulation are cited in the various treatises on insanity, and it seems that the history of feigned mental disease is almost coeval with the authentic history of the human species. It was known to Homer, who describes Odysseus as feigning insanity to achieve a special purpose. Solon shammed insanity in order to stimulate the Athenians before Salamis; and David is described in the Bible as feigning dementia, and resorting to the same artifices which are employed to-day by those simulating that condition. As might be anticipated, insanity has been feigned for special and usually selfish purposes. Such nobler objects as that of Brutus, who escaped persecution and threw sand in the eyes of the Tarquins, in behalf of Rome by this means, cannot be carried out by the aid of simulation in the present state of society. To-day it is usually resorted to by criminals who have no other hope of escape from punishment than the "insanity dodge"; by persons desirous of annulling contracts which they regret having made, and which they hope to have set aside by proving mental incompetency at the time of making them; and by sensation-seeking and enterprising newspaper reporters who desire to enter asylums and to "investigate" their management.* The possibility of prolonged simulation, as a step to the contemplated commission of a crime, by a calculating criminal, must also be borne in mind.

---

* Kiernan detected a newspaper reporter who had had himself committed to the workhouse as a pauper, and there shammed insanity and secured his transfer to the City Asylum for the purpose of publishing the abuses which there was reason to believe were enacted there. One reporter remained an inmate for nearly half a year at the Bloomingdale Asylum without detection, and accomplished his purpose to the fullest extent.

The subject of the simulation of insanity offers for the alienist's consideration two very distinct branches: the first is the simulation of insanity by ignorant persons; the second the simulation of insanity by persons who have had opportunities for studying or observing insanity. In the former case, detection is easy; in the latter it is more or less difficult; and there are instances on record where the best alienists have been puzzled or deceived by simulators whose skill in feigning reached the degree of the most consummate acting, and must have been based on skilful observation.* The truly wonderful power of endurance manifested by some simulators renders their exposure a far more difficult task than it is commonly supposed to be. Thus Vingtrinier† relates the case of one Picard who had been guilty of fraudulent bankruptcy and then shammed insanity for five years. The same person had previously simulated incontinence of urine for an entire year, in order to exempt himself from military service, and persisted in this, although his comrades in the barracks resorted to various and even cruel devices to check his disagreeable habit. The publicity of trials, and the full reports of expert and pseudo-expert testimony given to the public in the daily papers, are adding not a little to the difficulties of the subject. Our skill in the detection of simulation is increasing from year to year, but the skill of the simulator is also increasing. Shortly after the Gosling and Prouse Cooper trials in New York City, a noted criminal lawyer of the lowest possible *morale* instructed a defrauding lawyer, who had been formerly a medical student, to feign paretic dementia, and so far from overdoing matters—the fault of most simulators—the latter limited himself strictly to acting the symptoms of the prodromal period of that disorder to the best of his ability.

It was at one time, and is still with some, a commonly received test that the simulator does not repudiate the idea of being insane, which the truly insane person does. Aside from the fact that the insane do sometimes recognize their insanity and exceptionally admit it,‡ which

---

* Ollivier, Jacquemin, Ferrus and Marc were thus completely deceived by the simulating murderer Gilbert, and even Esquirol at one time suspected him to be insane. To-day, however, the writer believes, the shamming of such a person would not have proven as successful.

† Ann. d'hygiéne publique et de médécine légale, 1853.

‡ Seventeen patients, suffering from well-marked forms of insanity, in-

alone should have forbidden the adoption of so faulty a criterion, it is to-day valueless because many simulators know or believe it is considered such a test, and affect to disclaim the existence of the malady which they wish to have imputed to them.* That they overdo this, as well as other manufactured symptoms of derangement, is but consistent with the general character of the simulator. Derozier, whose interesting case is cited from Morel by Laurent—after such simulator-like answers as "245 francs, 35 centimes, 124 carriages to carry it," in response to the question, what his age was—being asked "Has your head been long out of order?" replied, "Cats, always cats . . . . . I am not insane, the insane don't turn around"; he then arose and turned around three or four times, as if to give his own assertion the lie.

Usually the physician's attention is directed to the possibility of the existence of simulation by some inconsistency in the clinical picture, exhibited or feigned by the subject examined. Such clearly marked affections as those detailed in the second part of this work are very difficult to feign correctly in every feature, and it is a task requiring consummate art for a simulator to remain within the true pathological boundary-line, and not to break through it in the direction of caricature. But there are some obscure and mixed groups, insufficiently studied, and for that reason not well recognized, in which a simulator may succeed in finding a place, he imposing his symptoms on the physician as signs of a mixed or impure form of insanity.

An important step in the determination of the existence or non-existence of simulation is the investigation of the previous character of the subject and the existence of a motive for simulation. A cunning knave, whose history is a repetition of crimes, is more apt to have had the idea of

---

cluding paretic dements, periodical lunatics, suicidal melancholiacs, one hallucinatory monomaniac, and one patient suffering from traumatic insanity with multitudinous hallucinations and violent impulses, consulted the writer at his office for their insanity. This enumeration does not include subjects suffering from incipient signs of paretic dementia, primary deterioration, alcoholic insanity, melancholia and *folie du doute*, whose number is far greater.

* This was the case with a business man who had been ruined by operations in Wall Street, had written numerous insane letters of a threatening character to a leading operator in the same line of business, and was subsequently discovered to be the author and indicted for attempt at blackmailing.

simulation suggested to him, and to have received instruction in the art of simulation, than a straightforward person of previously honorable character, or who has committed his first offence. A murderer, ravisher or abductor is more likely to sham insanity than a thief, because the former's risks of suffering the death penalty or of being long confined in jail are much more serious than the prospect of an asylum sojourn of at most a few years, while in the latter case the comparison of a sojourn in jail with the remaining in an asylum, involving as the latter does the necessity of continuing simulation day and night, is very apt to result in a choice of the prison as the lesser evil. In countries where the discharge from asylums is easily obtained and prison discipline is rigorous, prisoners sometimes feign insanity with the object of securing a change of quarters. The idea that the mere fact that a prisoner presenting signs of mental derangement is more likely to be a simulator than a real lunatic, and that simulation is frequent in jails, is, however, an erroneous one. The French and German statistics conclusively prove that simulation of insanity is rare among prisoners, and not at all frequent among criminals in general. On the contrary, real insanity is of comparatively frequent occurrence in jails, and much more common in prisoners than in the ordinary population, for there are special moral causes, remorse, isolation, vexation and despair, which, added to the physical stagnation and deprivation of prisoners, combine to break down their mental health.

Prisoners of war and recruits resort to simulation more frequently than any other classes. The melancholy case is related of two French prisoners of war who feigned insanity for a long period, with the intent of escaping by this means, and with such success that both ultimately became really insane. Most simulators who are convicted of simulation admit the distressing effects of the constant strain and effort on their nervous functions, and the warning that feigned insanity may become a real and incurable disorder should be conspicuously written in every prison corridor.

It is to be borne in mind, that simulation among female prisoners may be due to hysterical conditions. They resort to deception, to obtain a transfer, excite sympathy or postpone judicial measures. Sometimes the physical indications of the neurosis are marked, and the physician is in the unpleasant position of being compelled to admit the existence of real disease in a person guilty of fraud.

THE RECOGNITION OF SIMULATION. 357

It is known of other feigned disorders, such as epilepsy and traumatic tremor, that they may, if persisted in, develop into the real affections, and the analogous causation of actual insanity by simulation is no more problematical and just as plausible as that of the nervous disorders mentioned.

Very strange motives are occasionally observed to underlie simulation. Laurent speaks of former asylum patients who, after their discharge as recovered, shammed insanity to get back to their old quarters. A still more remarkable case is that of a young girl who feigned insanity in order to keep her sister—who was actually insane—company.

The simulation of insanity by the insane sometimes furnishes more troublesome problems to the diagnostician than that of the sane. This remarkable combination of real and feigned disease is by no means rare. The writer has not seen a single insane criminal who was not aware to some extent of the immunity to punishment which the insane enjoy, and who might not—as was the case with some—have feigned mental disturbance.* The murderer Dubourque, who was undoubtedly insane, feigned amnesia of his crime, and was convicted of the feint. An

---

* Guiteau has been erroneously supposed to have been a simulator. From the time he fired the fatal shot on the President to the moment when the drop of the gallows fell, there was not a moment in his career, not a word said or a deed done by him that supported this idea. If ever a more consistent record of the insane manner, insane behavior, and insane language has been made anywhere else in the history of forensic psychology than in the Guiteau trial, the writer does not know of it. Guiteau put in the plea of transitory mental disturbance, claiming Abrahamic inspiration. This is no more surprising than that an insane lawyer, whose practice had always been in devious channels, should, with the idea, under which the prisoner labored, of being his own counsel, use every means to escape an impending fate and carry out his "mission." Guiteau was unaware of the existence of his real insanity, repudiated it consistently, felt deeply insulted by that true opinion which wounded his self-love, and did everything in his power to fix the noose around his neck by combating it, showing off his apparent "smartness," and insulting his counsel. At no time did he make the slightest pretence of being or of having been *really insane*, or give himself the appearance of insanity, but, consistently with his egotism, he placed himself side by side with Abraham acting under an inspiration, as he claimed to be the silent partner in the firm of "Jesus Christ & Co." It is one of the most remarkable facts in the annals of medical jurisprudence, and one for which the general body of American alienists has been unjustly held responsible, that in the case of Guiteau these and other of the strongest evidences of insanity were marshalled into line as evidences of sanity and simulation.

imbecile and epileptic pickpocket with marked somatic signs of constitutional defect, feigned religious derangement, and succeeded in obtaining a change of locality from the penitentiary to the asylum. Whenever the physicians came into the ward he dropped down on his knees in an attitude of profound religious meditation, but at no other time. An imbecile murderer who presented the type of Kalmuck idiocy when arraigned for trial, knowing that his defence was to be insanity, tied a cloth around his head and buried the latter in his hands, associating the vague idea of assisting his counsel and the medical witnesses appearing in his behalf with the supposed necessity of giving the appearance of having a headache. Nichols, of Bloomingdale, according to Kiernan,* observed a case of simulation of dementia under the advice of lawyers, by a delusional lunatic, who had committed murder in obedience to the command of the Virgin Mary, appearing to him in the flame of a candle. Both the feigned and the real insanity were detected, and the latter was unmistakable throughout the patient's asylum sojourn. While a number of observations of similar complicated cases have been collected in Europe by Laehr, Stark, Delasiauve, Ingels, and Pelman—the last mentioned being led to express the extreme erroneous view that all simulators are mentally abnormal, Hughes was the first on this side of the Atlantic to direct attention to this subject. This authority says: "The insane appear at times, when they have an object to accomplish, more crazy than, and different from what they really are; this is the sense in which we use the term simulation, and this condition is akin to that of feigning by the sane. Simulation, while it presupposes a degree of sanity,† does not require that the patient should be wholly sound in mind, and it might be attempted by a convalescent patient not thoroughly recovered, or desirous of remaining longer in the hospital, or for some other cause." That this does occur is supported by several cases observed by the author cited, and by another referred to in the first part of this volume (page 34). There is not a single case on record in which a lunatic who simulated insanity recognized his real disorder.

---

* "Simulation of Insanity by the Insane." *Alienist and Neurologist*, April, 1882.
† It would be more correct to say "intelligence" than "sanity."

The popular idea of insanity, which is responsible for its frequent non-recognition by juries, judges, and some physicians, is also the cause for the ignorant simulator's failure. His belief is that the insane are either stark raving mad and incoherent at all times and on all points, or that they must be in a condition of fatuity. If he has read novels, he will model his insanity after that of some romancer who possibly has never seen a lunatic, and of course make as melancholy a failure as in any other event. There are five conditions in which gross incoherence is found combined with excitement—furor, frenzy, transitory insanity, febrile delirium and acute confusional insanity. We know that furor and frenzy, whether in the maniac, melancholiac or paretic, must have been preceded by a history of other mental signs, which it is difficult and in fact impossible to imitate; we know that transitory insanity is rare, of brief duration, and coupled with amnesia, and that the patient pays no rational regard to his surroundings during the attack.* Febrile delirium is associated with somatic phenomena which a simulator could not even approach imitating. The feigning of acute confusional insanity alone presents any chances of success, on account of the unessential character of the physical phenomena, the greater ease of imitating the incoherence of this disorder, and the absence of that deep emotional condition which the simulator of mania and melancholia usually fails to take into account. But even here, as elsewhere, the delirium of the simulator has specific characters. Real raving maniacs, if utterly incoherent, show the expression and somatic signs of their condition; in milder raving there is some connection of the thought with the surroundings; in simulated raving there is usually none. If a confusional or mildly maniacal patient stops to answer a question, which he usually does, he answers it with some degree of responsiveness. If asked his age, he will answer reasonably or err a little but never absurdly. He may assign a very great age, or a very much lower age than his real one, according as his tendencies are in the direction of megalomania or of micromania, but he will never say, as Derozier did: "245 francs, 35 centimes, 124 carriages, to carry them away." These incoherent phrases might have come

---

* Schwartzer, "Die transitorische Tobsucht" relates a case of simulated transitory frenzy.

from a confusional lunatic, but not as answers to such a question: there is too much method in madness to permit of such absurdities!

As a rule, the simulator in those quiet periods of his artificial excitement which are the expression of inability to keep up the exacting effort of simulation, does not recognize his friends, his surroundings, or recollect anything that occurred about that period of time which he has a motive to make people believe he was irresponsible in. The true maniac, however, happens to be lucid in those very periods, recollects his family and his friends perfectly well, and if he has committed a crime, while he may be acute enough to desire to conceal his recollection of it, he will not, if the examination is led up to the period of its commission, gradually claim to forget real circumstances occurring before and after it, as the simulator, who is always on his guard, does. It is only in epileptic mania and in paretic dementia that such amnesia really occurs; but here the physical signs or the history, or both, present us with unmistakable signs of these affections if they really exist.

The simulator also errs generally in allowing his feigned disorder to explode as well as to recede too rapidly. As a rule the psychoses develop gradually: those that do not, have certain characteristic features unknown to simulators. Thus transitory frenzy is characterized by a noticeable impairment of consciousness, and epileptic mania by the peculiar physical appearance and condition as well as the history of the patient. Outbreaks of furor in paretic dementia may occur quite suddenly in a remission or in the prodromal period when the patient is not supposed to have been insane by the laity; but the distinguishing marks of this furor and the accompanying physical signs are too numerous and the residual state too characteristic to be ever confounded with simulation by the expert. Sudden recovery is also suspicious, but not as much so as a sudden incubation of the malady. Mania has been known to disappear by a crisis, and menstrual insanity in exceptional instances gives way almost like a flash while the disorder is apparently at its height.

Another characteristic feature with many simulators is the intensification of their symptoms when under examination. A simulator will become more incoherent or demented and excited or obtrusive with all his symptoms when the physician approaches, than at any other times, for he has a motive in bringing his symptoms to notice. Sometimes the approach of the physician or of any one else will irritate real lunatics, but not beyond the limits of a diseased condition which is as recognizable in the *interim* as in the explosion. Then, too, if the mental capacity of the simulator is questioned in his presence he will do all in his power to strengthen the physician in what he supposes to be the latter's belief, while the real lunatic, unless sunken in abject dementia, will try to show a mental capacity which he has not, and to appear better than he is; this attempt, as so old an author as Hoffbauer knew, only makes the real lunatic appear the more deprived of reason.* That author refers to the case of a melancholiac who having a relapse of his melancholia, did the most extravagant things not natural to his illness in order that the relapse should not be expected to have occurred. It is in the nature of the case that the real lunatic suffering from those forms of insanity from which the simulator is most likely to select his model when he tries to appear sane, not being able to appreciate and to assume the sane character should show his mental infirmity only the more prominently. The simulator, on the other hand, not being able to appreciate or to assume the insane character, reveals the sham character of his malady the more he rants and acts the madman.

The simulator also labors under the mistaken impression † that the insane do not reason. It so happens that too great a degree of incoherence in a delusion justifies the alienist in suspecting its genuineness. To that kind of delusions which the simulator undertakes to imitate, the statement that there is method in madness preëminently applies. The simulator makes the common mistake of believing that insanity is a chaos of symptoms, although even as a morbid condition it has laws of its own.

Persons who feign a quiet form of insanity usually attempt to imitate dementia. A good case of the failure of this form of simulation is one detailed by Snell, of a widow who tried to have a contract set aside, and induced her children to claim that she had been and was insane. Being asked how many fingers she had on her hand, she said four. Being asked to count them, she skipped one finger, and said, "One, two, four, six." She further said that two and

---

* Applies to the forms of insanity which are most likely to be feigned, and not to monomania, hypomania and *melancholia sine delirio*.
† An error frequently announced in courts.

two equalled six, that she had nine children instead of seven, that her husband was dead ten years instead of five, that he died after an illness of over a week, when in reality he had died suddenly from an accident; gave the wrong name to a child, did not know the number of the year, nor where she lived, though previously she had admitted owning the very house in which the examination occurred; and when asked the ten commandments, said in reply to questions as to which were the first four: "1st. I am the Lord thy God. 2d. I am the Lord thy God. 3d. I don't know. 4th. Thou shalt not honor thy father and thy mother." The tendency to absurd contradiction, the feint of forgetting so important an occurrence as the mode of death of her husband, and of simple things, in their combination as above shown, settled the fact that the woman simulated. Even imbeciles have some ideas within their limited range, and adhere to them with a certain degree of consistency which the simulator rarely shows, and as far as their basis goes they reason with a show of logic. If a dement, not absolutely in a state of fatuity, reasons badly, there is always found confusion of words, while occasional glimpses of a clearer ideation struggle through ; the incoherence seems to be due to a digression from subject to subject, the main one losing its grasp on the enfeebled attention and memory. The simulator of imbecility or dementia, however, either talks more confusedly than harmonizes with the thread of reasoning he unwarily exhibits, or he talks less confusedly than he should in the utter absence of a connecting bond in his thoughts ; in short, he does not balance the defects in ideation and in their expression properly.

The idea has gained ground that the insane who have amnesia in fit-like spells, as in epilepsy, alcoholic insanity, and paretic dementia, never admit its occurrence, while the simulator is very ready with the words, "I don't remember." This is true in the majority of cases, but does not apply to the early phases of paretic dementia and alcoholic insanity. On the other hand, it must be recollected that maniacs and melancholiacs will in their convalescence often attempt to cut short inquiries as to their reminiscences, by the claim of amnesia. However, there is no likelihood of confounding these conditions with simulation of amnesia by the mentally healthy, for when amnesia is honestly claimed, and the mental condition immediately preceding and im-

THE RECOGNITION OF SIMULATION.

mediately following the alleged amnesia carefully examined, something distinctively pathological will always be found.

The absence of insomnia and impaired digestion in the acute psychosis is exceptional in real insanity, and is, to that extent, a ground for suspicion. These and other disturbances of the bodily functions are not, however, characteristic or essential features of chronic insanity,* although it supports the idea that a subject is really insane when the skin is in a bad condition, dry and yellow, or moist and clammy, when there is an effluvium, when the appetite is poor, the tongue coated, and the bowels are constipated. But the physician who needs these signs to convince him of the existence of insanity, and who would elevate them to the dignity of proofs of that condition, may take his place side by side with those who attempt to elevate the ophthalmoscope into a test of insanity for medico-legal purposes; he has not advanced much beyond the position of Rush, who thought that he could distinguish real from feigned insanity by means of the pulse. This claim could well be made in the earlier part of this century; only a novice would rely on or propose such tests to-day.

The simulator's task is rendered difficult whenever he is kept under continuous observation. The best actor may fail to adhere to his assumed character for days and nights in succession, and the necessity of being continually on his guard gives the performance the appearance of being labored. There are, however, instances recorded where the first period of simulation having been passed, it became a sort of second nature, and assumed an appearance of genuineness which has, as stated, imposed on the foremost authorities in psychiatry. The transfer of such subjects to an asylum is usually followed by a cure of the insanity, suspicious on account of its rapidity and its taking place in what was made to imitate an incurable form of mental disorder. It is in such cases that real insanity sometimes rewards the

---

* The condition of the skin was the chief criterion on the strength of which an "expert" witness for the prosecution in the Guiteau trial pronounced him a simulator. Unfortunately, the skin of Guiteau, though it revealed no form of insanity, was in a far worse condition than that of two-thirds of the insane in most institutions for the insane, many of whom have a perfectly normal complexion and are in good general bodily health, as far as can be discovered. Psychiatry is not destined to become a branch of Dermatology.

simulator's efforts.* A very important point to discover in the antecedent history of a suspected simulator is whether he has ever had an opportunity of observing the insane, or has read treatises on the subject.

The devices for exposing simulation are numerous. Zacchias, in consonance with the spirit of his age, recommended flagellation; and Campagne, the douche. Both were as wrong in believing that the confession of having played a part or of having attempted deceit under these circumstances has any value, as the mediæval jurists were wrong in believing that the truth could be discovered by means of the *peine fort et dur*. The insane may be made to recant their delusions, to conceal them, or, as Leuret claimed, even to lose them under powerful motives, and it would lead to gross mistakes to adopt any vigorous measures with them. The torture of being continually watched, and of having to keep up an unnatural effort under surveillance, are far more effective weapons for use against simulation, and the clinical observation of the really insane furnishes a countless number of devices by which the pretender can be exposed, without involving the risk of being inhuman to a genuine lunatic suspected of simulation.

Among the special signs which justify the suspicion of simulation are the following: 1st. The subject on the physician's entry may avoid looking at him and glance up at the wall, and on the physician's changing his place will look elsewhere, demonstratively avoiding looking him in the face. This is never the case in stuporous insanity, melancholia, katatonia, nor in apathetic dementia, the forms which a subject presenting these signs attempts to imitate. 2d. The simulator will give extravagantly absurd answers to simple questions, after the fashion of children in play when attempting to excel in saying impossible things; this is in agreement with the popular idea of insanity. Derozier being asked in June what month it was, said January, then looking out of the window, said, " Stop, one would say that it is warm." This alone sufficed to expose the sham. 3d. The simulator may take a long time to answer questions,

---

* Jacobi (the alienist of Germany), in conjunction with Richarz, Hertz, Böcker, and Snell, pronounced a subject to be a simulator, and these eminent authorities were undoubtedly right when they did so. The first mentioned, however, on receiving this person some years after as a real lunatic in his asylum, suspected that they had all been wrong in their first opinion. This does not at all follow.

THE RECOGNITION OF SIMULATION. 365

and hesitate in his answers. Delay in answering and drawling are found in depressed states, but here the appearance and expression harmonize with the exhibition of thought and speech, while the simulator's expression betrays an intelligence which his words are intended to mask. 4th. The simulator when he supposes himself unwatched will make furtive glances to see whether any one approaches who necessitates his being on guard. 5th. A person feigning epileptic and somnambulistic states may recollect perfectly his feigned acts and expressions, and carry them into his *quasi* lucid period. This never occurs in the real affections. 6th. Rhythmical movements are made by some simulators which have no analogy in insanity, or are out of harmony with the form of mental disturbance assumed. 7th. Simulators complain much more about odd and painful sensations in the head than the insane usually do. 8th. A clumsy simulator may say: "I have the delusion that I am lost, that the devil is after me," or, "I have hallucinations of faces and voices at night." Such a person can be readily exposed to be a deceiver on other grounds, but the feature here mentioned alone suggests simulation. A true lunatic may admit that he has hallucinations and delusions, using those words, especially when examined for the purpose of being committed to an asylum, but when he does so he affects to admit that he "imagined those things," but never does a real lunatic at the time he has these symptoms give them names which show that he recognizes their abnormal nature. He *is* lost, he *is* pursued by the devil, he *hears* voices and he *sees* faces. 9th. It is suspicions if insanity appears immediately after a crime, or after the arrest or sentencing of a criminal, while its previous existence can be disproved.

The likelihood of simulation being combined with sanity is very much diminished in case the person has already been insane, and particularly if there are somatic signs of heredity or a history of insanity in the blood relations of the suspected simulator. If such a person feigns, the possibility of real insanity underlying the feint, must not be forgotten.

The devices which may be legitimately resorted to, to expose simulation are the following: 1st. When examining the patient, let the interlocutor remark in an undertone to a bystander, that if such and such a sign were present he would know in which ward to put him, or under which form of insanity to classify the subject. This is far safer than the suggestion adopted from the French writers by Ray, and copied from him by some recent pamphleteers, of saying that if such and such a sign were present, the interlocutor would believe the man to be *insane*. This would put a cunning simulator on his guard. The writer had to deal with such a one in the case of a child abductor who had feigned insanity in a jail once before. Suspecting that the recommendation of the older writers would have failed, the writer turned to a bystander and said: " This is a most interesting case, and I have frequently remarked that these patients do not remember what city they are from." The criminal had previously assigned Baltimore as his home, and this was, according to the legal papers in the case, correct; but on being interrogated again, he said in a hesitating and whining voice, altogether unnatural to a person suffering from monomania with sexual perversion (the form claimed to exist), " Concord, Cincinnati." 2d. While being examined as to his general sensibility, the simulator may believe that anæsthesia is a desirable part of the clinical picture; he will wince when probed with a pin unexpectedly, but remain immobile when pricked after being warned. This is, however, a sign which is not constant nor of great value under any circumstances, though it may serve as a good basis for an accusation of shamming, made to test the moral effect of the charge on the simulator. 3d. When a simulator is accused of shamming he may either turn away from the examiner, or suddenly lapse into stupor, or undergo some other unnatural change of his symptoms. A real lunatic will either act like a sane person under these circumstances, or, as in apathetic states, show no change whatever.* 4th. A simulator if transferred from

---

* A writer whose inspiration may be found in Blandford's chapter on feigned insanity, the ideas of the latter being closely followed with no other change than one of language, and whose article opens with the statement that " moral " and "feigned insanity" are convertible terms, says that " if you will accuse the simulator of shamming, he will scarcely fail to change his countenance." Now if that writer had ever tried the experiment on his own patients, he would have found that the few who had some relics of their old pride left would certainly undergo a change of countenance if accused of humbugging. Such a proposed test of simulation reveals the lack of any searching and fair study of insanity in some of our asylums. With the approach of insanity, particularly of certain forms, a person does not lose all the feelings of the normal human being, and equally with a sane person might resent what he may subjectively

one ward to another of an asylum, will imitate the different forms of insanity he sees there. He may appear melancholy or demented one week, hilarious and destructive the next, and cases have been observed where simulators on being placed in the filthy ward of an asylum, with the idea that its disgusting and frightful scenes would induce them to abandon simulation—which sometimes is the case—devoured their own excrement, acting to the best of their ability as the other inmates did. Imitation may occur in real insanity, but it is limited to delusive conceptions which are accepted by weak-minded lunatics from more intelligent ones, in what the French call *folie communiqué* and *folie à deux*. A simulator whose signs indicate, say, dementia, monomania, melancholia, or mania, on being placed in a ward with paretic dements or epileptics, allowing him to overhear the statement that he must be either a paretic dement or an epileptic, and that he cannot possibly belong to any other form of insanity, will have delusions of grandeur and paralysis in the former and convulsions in the latter ward. Numerous suggestions may be made of symptoms out of harmony with the assumed mental disorder, and their adoption serves to expose the fraud. Thus the writer suggested ptyalism in a simulated monomania, and oscillatory movements of the head in simulated suicidal monomania, with this result. It may be feasible to relate cases in the hearing of well-posted simulators, where the insane had miraculous beliefs, or spouted poetry all day, and thus to prompt the adoption of inconsistent and convicting symptoms.

Of the various tests thus far enumerated, the device of charging simulation point blank should not be made until all other means have been exhausted. The simulator should not know that he is suspected until the last moment.

Various medicinal tests have been suggested for the purpose of exposing simulation, but they are of no value; perhaps the best is ether, but comparative lucidity may occur after its use in the really insane, and both in sane and in insane persons false assertions, self accusations, and accusations against others are made in ether and chloro-

---

regard as a deliberate insult. This ignorance is heir to the same feeling which fifty years ago treated the insane as wild beasts, and to-day treats them like paupers and jail-birds.

form narcosis. The application of the faradic wire brush may expose a simulator, and this test is a legitimate one, because it is one of the therapeutical appliances indicated in the treatment of those stuporous and atonic states which are most likely to be imitated by those simulators with whom the device may prove successful.

In analyzing simulation, as in studying insanity, the individual as a whole, his surroundings, his crime, and his present mental state must be taken into account. He who really has the acquired forms of insanity must at some time have undergone a change of character; he who suffers from a congenital or inherited form must have exhibited mental defect or disturbance long before, present somatic stigmata, or have an hereditary history or a neurotic taint. The crime, its motive, manner of commission, and the behavior after the crime, are important elements in the diagnosis between simulation and insanity in criminals. There are some crimes which alone suggest insanity, in others the motive and the manner of commission demonstrate its existence.*

It is incorrect to conclude that because the commission of a crime involves deliberation, premeditation, and skill, that it cannot be the deed of a lunatic. The insane, as has been repeatedly urged by the highest authorities, and as explained elsewhere in these pages, may reason very elaborately from false premises.† The simulator in cases where his crime was performed with skill and careful preparation, betrays his feint by claiming complete amnesia, or saying that he must have lost his head, by showing a desire to appear feeble-minded and of weak memory, by representing his family to have been insane, or by forcing spurious insane documents on the attention of the observer. In short, the false picture of insanity is usually a caricature, and violates the laws of insanity at almost every step taken by the simulator, and in more than one direction, in the majority of instances.

---

* The popular idea that the lunatic always slays openly is, however, grossly erroneous; it has been adopted as a test of real insanity as distinguished from simulation, and by a curious coincidence, by the very persons who pronounced Guiteau a simulator.

† The possibility of a lunatic's committing a crime from the ordinary criminal motives cannot be denied, for, contrary to statements made on the occasion of several celebrated trials, insanity does not improve the morals.

CHAPTER IV.

THE SOMATIC ETIOLOGY OF INSANITY.

Having examined his patient and made his diagnosis, the physician's duty, before proposing remedial or other measures in case of insanity, is to inquire into its causation, as the proper therapeusis is often guided by a correct etiological assignment. Naturally his attention is first directed to possible somatic causes, as when remediable they are far more readily and rapidly remediable than the mental causes.

Nearly all the known exciting causes of insanity are in the nature of somatic, emotional or intellectual accidents, to which the sane population is almost as much liable as the insane. The reason why insanity results in one case and not in another, must therefore with certain exceptions be sought for in some vice of the constitution—in other words, in a predisposition to insanity. That this predisposition may be acquired through traumatism, syphilis, alcoholism and other narcotic abuses we have already learned; but the most important predisposing cause of insanity is undoubtedly that hereditary transmission of structural and physiological defects of the central nervous apparatus discussed in the first part of this work (page 81).

As far as the treatment of the hereditary transmission of the defects, on which as a basis insanity may develop, is concerned, it can only be prophylactic. And it is left for a higher civilization than ours, one in which State Medicine will no longer limit itself to the quarantining of those diseases which produce popular panics, but take cognizance also of those which are more insidious and equally if not more destructive or damaging to the race, to deal with the great problem of adopting rational principles of natural and sexual selection in the propagation of our species.

To-day the practical alienist while regarding heredity as the most prominent subject of inquiry in regard to the etiology of insanity, is compelled to limit himself to the study of the acquired predisposition and exciting causes as the factors to be considered in prophylaxis and treatment. As far as the hereditary predisposition is concerned his advice will be rarely sought; and when sought, his advice will in the majority of cases be limited to the recommendation of educational methods adapted to the case, and calculated to divert the predisposed mind into channels which shall conduct it further and further away from its threatening goal. There can be little doubt that whether an inherited disposition exists or not, that faulty educational systems, particularly when associated with the hot-house growing plan, may be responsible for serious injuries to the nervous system which may in turn pave the way for the development of insanity.*

Among the physical causes of insanity, head injuries, insolation, meningitis and gross organic disease of the brain deserve the first consideration, because their influence is directly and often tangibly applied to the organ of the mind or its protective capsule.

Injuries of the skull affect the mind in a number of ways, and while those complicated by fracture with depression, are more likely to lead to serious mental results than simple concussion, yet even simple concussion may produce chronic incurable insanity or the disposition to it, as a number of well-observed cases attest.

Sometimes insanity is produced directly after an injury of this kind. Of this character are the delirium, the hallucinations and excitement often found intercurrent with the sopor and coma following shock, and whose prognosis is comparatively favorable. Sometimes serious lacunæ of the memory are noted, and the patient may lose the memory of a long period of his life altogether, either to regain it, or to pass into a condition very similar to primary mental deterioration.

The most serious psychoses resulting from traumatism are developed months and years after the injury. Sometimes they assume the character of paretic dementia (p. 201), but as a rule this is not of the pure type, and is apt in the prodromal period to be marked by the furious outbreaks and murderous impulses characteristic of what might be called the traumatic neurosis. There is a condition which might be properly called *traumatic insanity*, because it does

---

* The same applies to the feeding of the mind on morbid fiction, which, though not a distinguishing characteristic of the present age by any means, is to-day cultivated at an earlier period of life, and consequently does proportionately greater damage.

not accurately correspond to the ordinary psychoses, has distinct clinical characters, and is always when found, referrable to traumatism or to analogous causes; it develops on the basis of the "traumatic neurosis," just as alcoholic and epileptic insanity develop on the basis of the alcoholic and epileptic neuroses. The subjects of this disorder are noted to undergo a change of character, to exhibit a tendency to alcoholic excesses, to become morally perverse, suspicious, brutal and quarrelsome, and to manifest murderous or other violent impulses, occasionally associated with fits of maniacal self-exaltation or furor, usually of short duration. This condition is remarkable for its long duration and its frequent and sudden changes, the occasional lucidity of the patients being accompanied at the time by hypochondriasis. As a rule progressive deterioration sets in, and dementia terminates the history of the case. The diagnosis of this condition is facilitated by the presence of certain physical signs. Tinnitus aurium, photopsia scintillation before the eyes, headache of a pulsatory or grinding character, vertigo, paresis of various muscular groups, particularly of the eye-ball, without fibrillary tremor, anæsthesias and hyperæsthesias, as well as insomnia, are frequent accompaniments, and some of these enumerated signs are present in every case.

Insolation and the influence of radiant heat produce a form of insanity very much like that due to traumatism, but in the writer's experience these causes lead far more frequently to paretic dementia than the latter. Firemen on transatlantic steamers and waiters in hotels detailed to duty in the "plate-warming" room, furnish a comparatively large quota to that part of the asylum population suffering from paretic dementia in New York.

The influence of meningitis on mental life is well illustrated in the psychical disturbances, such as delirium, hallucinations, depression, stupor, and destructive impulses sometimes observed in the course of tuberculous and simple meningitis.* Meynert believes that abortive or

---

* In a patient dying with symptoms of paretic dementia of a stupidly delirious type, whose earlier history was unknown—a fact that may be appreciated when it is known that he was entered in the asylum records as John Doe—chronic leptomeningitis traceable to a suppurative process in the tympanic cavity was found post-mortem by the writer. The motor signs had in this case been well marked and characteristic of the disorder diagnosticated.

self-limiting meningitis in childhood may leave behind a weakness of the mental organ, which may manifest itself in imbecility with hallucinations and delusions usually accompanied by epileptiform symptoms. According to that writer, a prolongation of the posterior cornu of the lateral ventricle beyond the normal length is found in insane subjects who have suffered from slight hydrocephalic troubles with or without convulsions in infancy, and the white substance in the neighborhood of the ventricle often contains sclerotic patches in that event.

It is scarcely necessary in a work of this character to detail the various cerebral diseases which are occasionally the causes of insanity. The general statement may be made that genuine derangement of the mental faculties is more likely to occur with multilocular or diffuse than with unilocular or circumscribed lesions, with bilateral than with unilateral disease, with large than with small foci, with rapidly developed than with slowly developed disturbances, with hemispheral than with axial affections, and with morbid changes established at an early period of life than with those affecting the brain after the maturation of the mental mechanism.

Next to the organic affections of the brain and its envelopes it is the neuroses which are the expression of an impalpable brain disturbance that hold an important place in the causation of insanity. This influence has been discussed in connection with the clinical description of several forms of insanity in the second part of this volume (Chapters XVII. and XVIII.). It remains for us to speak here of the influence of chorea in the causation of insanity. In mild cases of chorea the mind is no more seriously affected than in any other affection annoying to children, and associated with insomnia. Even in severe cases the mental faculties may be found to be quite intact,* and such disturbance as is found in the majority of cases is the result of the motor disturbance and of the ensuing restlessness, irritability and peevishness of the child. In protracted cases of chorea, the mind suffers in the direction of actual insanity; in that case maniacal outbreaks, confused delirium, enfeeblement of the memory, rapid emotional change, and in ex-

---

* The sensational claim was made at a discussion of the subject at the New York Neurological Society, that all choreic children are morally imbecile!

THE SOMATIC ETIOLOGY OF INSANITY. 373

treme cases dementia may ensue. It is a psychosis with these symptoms which is designated CHOREIC INSANITY. This disorder must not be confounded with another whose title has a similar sound, namely, *choreomania*. The latter term was given to the epidemic impulse to dance which spread so extensively in middle Europe on several occasions in connection with religious movements, and which according to Yandell has occurred on the occasion of a revival movement in Kentucky early in this century.

Fevers exert an important influence in the production of insanity. The term POST-FEBRILE INSANITY is given to disorders which complicate the crisis, or what would ordinarily be the convalescent period, of certain acute febrile processes, such as scarlatina, small-pox, typhus, typhoid, pneumonia, and erysipelas. The insanity noted with the secondary fever of syphilis appears to the writer to belong to this group also. The post-febrile pyschoses are presumably associated with two different pathological states, one of asthenia and anæmic of the nerve-centres, the other anatomically marked by the filling of the periganglionic and subadventitial spaces with formed elements of the blood. As a rule, illusions and hallucinations, delusions of identity and anxious deliria open the scene; later there may be pleasurable deliria, or ideas of grandeur. A notable feature is the comparative lucidity of the patients during the day; they are then able to reason more clearly, but, inasmuch as they reason on the basis of their delusive conceptions, they are all the more dangerous for this lucidity. It is under these circumstances that patients recovering from febrile disorders commit suicide, usually by jumping out of the window. Episodical attacks of violent frenzy may vary the picture. Most of the patients suffering from post-febrile insanity recover very rapidly, the psychoses terminating with a critical sleep or by gradual defervescence, after a course of at most a few weeks. In some cases, however, particularly after rheumatic fever, scarlatina, typhus and typhoid, a more chronic course is observed. The patient's condition oscillates between maniacal and melancholic states, and is characterized by great stupidity and confusion of ideas throughout. Even here the prognosis is usually favorable. In some, progressive deterioration sets in, and dementia ensues; in others, fixed and subsequently systematized delusions remain behind, constituting the case one of delusional monomania; morbid impulses have been observed in others; and in several cases, as was well illustrated in the instance of a post-scarlatinal psychosis, the subject of which the writer exhibited before the N. Y. Neurological Society, profound moral imbecility remains after the more furibund symptoms disappear. Malarial fever is sometimes accompanied by mental disturbances which may present a perfect imitation of cyclical insanity, with lucid intervals corresponding to the period between the attacks. A chronic mental disorder, similar to that above referred to as following other fevers, is also noted as a phenomenon of the paludal dyscrasia.*

RHEUMATIC INSANITY, which is generally considered a distinct form, has many characters resembling those found in the post-febrile group; in cases where it runs a chronic course, it may terminate in paretic dementia of a kind whose prognosis is somewhat more favorable than it ordinarily is in this disorder. Other forms of insanity sometimes occur with rheumatism. But like the analogous type of GOUTY INSANITY, they are exceedingly rare. (Appendix VI.)

Anæmia is rarely the sole factor in producing insanity; usually other causes are added to it. It may, however, when suddenly produced, as after hemorrhages, be the single and direct cause of stuporous insanity, and is, in its chronic form, undoubtedly the most common cause of this variety of insanity in young subjects. As a rule the anæmia of other forms of insanity is a result and not a cause of the mental disorder, particularly in simple melancholia and the chronic forms (p. 69). In young girls it is frequently observed that, on the basis of an anæmia, there develops a state of mental anenergy. Such subjects are prone to become melancholic, and in two cases the writer has found a genuine stuporous state following the melancholia, so that there was here a complication of two different psychoses.

The metallic poisons produce a mental derangement which, owing to its frequent combination with motor disturbances, may be confounded with paretic dementia. It is, however, in its typical form characterized by exacerbating deliria of sudden development, and by comatous spells of equally sudden occurrence, which are not found in a

* In a case of inherited malarial fever, under the writer's observation, the subject at the age of from three to five years had a pleasurable, good-natured delirium, with a surprisingly brilliant flight of ideas, accompanied by a rise of temperature (103° F.), which on several occasions vicariated for the ordinary febrile attacks.

similar association in paretic dementia. The specific character of the metallic tremors will usually serve to distinguish the psychoses due to hydrargyrism and plumbism from the latter disease.

Pulmonary affections, particularly phthisis in its last stages, are sometimes marked by mental disorder, usually in the way of alternating depression, emotional mobility, petulance, an intensification of the egotism common to invalids, and accusatory delirium. Occasionally delusions of grandeur are found at the height of this disorder, which is designated PHTHISICAL INSANITY. (Appendix VII.)

Valvular disease of the heart is considered by some writers, particularly Witkowski and Leidesdorf, to be a frequent cause of depressed emotional and vague impulsive conditions. The writer has seen no confirmatory example of this, and believes that the view expressed by several of the Germans, that hypertrophy of the left side of the heart with aortic valvular lesion is more apt to be associated with maniacal states, and hypertrophy of the right ventricle with mitral valvular lesion with melancholic states, is supported by too limited a number of cases to merit acceptation. Those recorded are devoid of value, as the blood-pressure, which is the intrinsic factor, was not duly registered. The heart has important and direct relations to the brain, and it is very likely that just as disturbances of the vagus innervation are responsible for raptus melancholicus—in other words, just as a disordered state of the brain reacts on itself through the medium of the functional cardiac disturbance it provokes—so a valvular lesion may directly influence the emotional states without pre-existing brain trouble. When, however, we remember the large number of persons whose hearts are in the most extreme conditions of organic failure, and who die in consequence, but without having manifested any special psychical disorder, we will, when we discover a fixed delusion of persecution in a subject, with aortic obstruction, look for some other cause, such as an insane predisposition or mental overstrain, as the primary determining element, while the cardiac disorder may be admitted to act as an exciting cause, or, more accurately speaking, to determine the anxious or suspicious character of a delusion. It is a fact that patients suffering from cardiac lesions are more likely to develop anxious and suspicious delusions than those of an opposite nature.

Emminghaus* states that in two cases of Basedow's disease (exophthalmic goitre) he found pronounced mental disturbance in the shape of melancholia and periodical mania. The occasional occurrence of this disorder in members of families afflicted with a morbid heredity† would seem to indicate that the physical disease and the insanity are simply collaterals, and that both are the expressions of the same fundamental neurotic vice. It is an interesting problem for the future to solve why enlargement of the thyroid gland should in two disorders such as exophthalmic goitre and cretinism be associated with mental disorder or defect.

Disordered states of the uterus and ovaries, especially those manifesting themselves in disturbances of menstruation, have been supposed to play an important part in the causation of insanity. It is known, however, that the grossest lesions of the female generative organs are not usually complicated by such mental disturbance as justifies calling it alienation. Those pretty cases in which a delusional insanity is instantaneously cured by restoring a retroflected or retroverted uterus to a normal position, do not seem to occur nowadays, and the gynæcological epoch of psychiatry seems to have passed by, taking its adieu with the sacrifice at the Blackwell's Island Asylum of Mary Ann Mullen, a sufferer from unrecognized katatonia, on the altar of oophorectomy.‡ It would have been as reasonable to extirpate the bed-sore of a sufferer from paretic dementia, and to cut off the hæmatomatous ear of a terminal dement, with the hope of curing his insanity thereby.

Sudden stoppages of the menstrual flow are occasionally found to be the direct causes of a maniacal attack in persons not predisposed to insanity, and the mechanism of the psychosis is to be sought for in the thus far physiologically obscure connection existing between the uterus and ovaries on the one hand, and the encephalic vaso-motor system on the other. More frequently, persons affected by an hereditary taint suffer from a periodical form of insanity whose exacerbations are determined by menstruation, and

---

* Allgemeine Psychopathologie, p. 371.
† A cousin of Guiteau, now residing in St. Louis, was proven at the trial of the insane assassin to be afflicted with exophthalmic goitre. There were in three generations of the family, among the members whose history is known, over a dozen insane and defective individuals.
‡ The ovaries were perfectly healthy. (Appendix VIII.)

in persons who have what was described in a previous chapter as the monomaniacal character, the delusions are often greatly modified by the pelvic disorder. Treatment of the pelvic difficulty is imperatively demanded under these circumstances, and while the pelvic trouble is not the fundamental cause of the insanity in all cases in which it coexists, its disappearance is sometimes followed by great mental improvement.

The puerperal state, in the wider sense in which Ripping* uses the term, has more important relations to the causation of insanity than the other physiological periods, not excluding those of the two climacterics and of senile involution. During pregnancy itself, peculiar mental states are observed, such as morbid appetites, varying from the ordinary *pica* to anthropophagous desires; and melancholia is comparatively frequent. The greater frequency of the latter condition in the mothers of illegitimate children, in those who suffer from want, and in those who have a hereditary predisposition to or a taint of insanity, is a confirmation of the view to be announced that the physiological accidents to which the human frame is liable are not likely to produce insanity unless its production is facilitated by additional causes.

In the writer's experience, melancholia is more likely to ensue during the period of lactation, and is then a psychosis of exhaustion, while mania is more frequent with the puerperal period proper. The melancholia due to excessive lactation, or late weaning of the child, and the mania of the puerperal state, which is very often precipitated by suppression of the lochia and of the milk secretion, do not differ in any respect from ordinary mania and melancholia. Hence, the writer does not use the terms puerperal mania, or melancholia of lactation, but mania *in puerpero* and melancholia *ex lactatio*, to show that no clinical but only an etiological distinction is aimed at in the terms employed. Sometimes, especially in older subjects, and when there has been much loss of blood, dystocia, or some emotional depression, melancholia instead of mania develops in the puerperal period. In short, sthenic states favor mania, and asthenic ones melancholia or stupor.

Occasionally, transitory frenzy is observed, either in de-

---

* Die Geistesstörungen der Schwangeren, Wöchnerinnen und Sängenden.

pendence on the extreme agony of child-birth, or as a manifestation of the delirium of the parturient state. In this condition, infanticide, or suicide, or both, are sometimes committed. The view has been expressed that when albuminuria co-exists with the maniacal and frenzy-like explosions of the puerperal state, the uræmia is a collateral etiological factor. No substantial grounds exist for endorsing this view, and the writer has been able to satisfy himself of the absence of any constant relation of renal and mental disorder in the puerperal state. (Appendix IX.)

The development of monomania has sometimes been observed to date from a confinement, but as far as the writer's experience goes, only in predisposed subjects.

The relation between organic disorders of the male genital apparatus and insanity is far less constant and important than that existing between the female organs and mental disorder. That a connection between the development of the mind and the male genitals exists, is indisputable. Even if we assume that the defective development of the genital system found in brain monstrosities, idiots, imbeciles, original monomaniacs, and the periodically insane, is an accidental accompaniment of the neural mal-development, we must admit the convincing fact that the early extirpation of the testicles, as in eunuchs and castrated animals, exerts an influence on the mental complexion and development. The frequent delusion of mysterious inimical influences exerted against the sexual power, of sexual mutilation, and of marital infidelity, so characteristic of alcoholic insanity, are believed by Krafft-Ebing to have some connection with the fatty degeneration of the epithelia in the seminal tubuli which occurs in old alcoholic subjects.

The functional abuse of the male sexual apparatus is of more general importance to the alienist than its organic affections. Excessive venery and masturbation have from time immemorial been supposed to be the direct causes of insanity. Unquestionably they exert a deleterious influence on the nervous system, and may provoke insanity partly through their direct influence on the nervous centres, partly through their weakening effect on the general nutrition. That there is a close connection between pathological nervous states and the sexual function is exemplified in the satyriasis of mania and the early stages of paretic dementia as well as in the sexual delusion of monomania and the abnormal genital sensations of that condition. In the former case the sexual exaltation is a result, in the latter the genital sensations are collateral phenomena of the psychosis, but there are certain cases in which while an original predisposition may have existed, masturbation is the factor responsible for the production of insanity. While there is no special form of insanity attributable to masturbation, yet those psychoses due to or accompanied and modified by this vice seem to have certain characters in common. Melancholia, stuporous insanity, katatonia, and insanity of pubescence are the forms most frequently found in masturbators, and the essential characters of these psychoses are always recognizable under these circumstances. The ordinary characteristics of the masturbator are, however, found in addition. Thus such lunatics are usually retired, sly, suspicious, hypochondriacal, indolent, mean, and cowardly. They are capital simulators, and develop an art in concealing and in practising their vice which is in remarkable contrast with their stupidity, apathy, and feeble-mindedness in other respects. The prognosis of the psychoses associated with masturbation in males is bad.* A variety of primary deterioration marked by moral perversion is observed in young victims of the habit, which yields to treatment if the habit is abolished. If unchecked it culminates in complete fatuity; this has been observed by the writer in subjects between the eleventh and twenty-third year, and is one of the numerous conditions which passes under the designation of "primary dementia ;" it is the only one to which the term insanity of masturbation can be properly applied.†

---

* Genuine melancholia, usually *sine delirio*, occurs in female masturbators, and has a very good prognosis, probably because the effects of the vice are far less severe in the female than in the male sex, and because it is but very rarely practised with persistency and for long periods by the former.

† Gloomy as the prospect of the confirmed disorder is as a rule, yet occasionally very unexpected and happy terminations are seen. Thus, a young man of bad hereditary antecedents, who for days had not quitted his bed, and who exhibited feeble-mindedness and moral perversion, as a result of this habit, was about to be sent to an institution by the writer. The following day, he, suspicious as these subjects are, made a search and found the commitment papers. After perusing them, he immediately turned over a new leaf, went into his father's store, resumed his work, abandoned his bad habits, and to this day, that is, during a period of nearly two years, has filled his position in life with average ability, being remarkable only on account of his taciturnity.

In some cases, prolonged and excessive masturbation is observed to result in the formation of a neurotic state, which subsequently serves as the soil for the development of monomania, characterized by hypochondriacal or religious delusions. In the female sex nymphomania is observed to be associated with a similar form of insanity; it has not yet been determined to have that distinct causal relation to monomania which masturbation has in the male.

The view held by Maudsley and others that sexual excesses are the all-important factors in the etiology of paretic dementia, is not sustained by the writer's observations. Frequently such excesses are committed by paretic dements in the earlier exacerbations of their malady, as well as in the earlier period of simple mania; but while they undoubtedly precipitate the progress of these disorders, they must be regarded as phenomena and not as causes.

The influence of neuralgia and pain in the production of insanity is limited to the occasional development of transitory delirium or frenzy, analogous to that observed in the puerperal state. Schuele has promulgated an utterly fanciful view as to the existence of a "Dysthymia neuralgica," under which head this author believes that most of the psychoses may be ranged. It is safe to say that this theory will not be accepted by alienists until it is supported by more convincing testimony than that thus far adduced in its favor.

While a vast host of other somatic ills might be enumerated, which have all been shown to have an influence in the production of mental derangement, yet, multitudinous as these causes are, composing as they do a twenty-fold longer list than the psychical causes, it is after all but a small percentage of the insane who owe their trouble to their influence alone. Blandford,[*] with approximate accuracy and little elegance of diction, says: "Men and women become insane because it is their nature and constitution to develop insanity, and when we hear that this or that has caused their insanity, it is often their restless and half-crazed brain that has made mountains out of molehills, and given an objective existence to troubles and vexations which exist in their minds subjectively, and have no outward reality whatever." It is true, as stated in this extract, as also at the

---

[*] "Insanity and its Treatment," p. 153.

opening of this chapter, that the inherited and acquired insane constitution is the fundamental factor in most cases of insanity. This conclusion, and the assumptions based on it, do not, however, justify us in ignoring the physical diseases immediately preceding or associated with insanity, for there is more satisfaction to the practical alienist in remedying one case of mental disease by removing its physical cause, than in diagnosticating a hereditary predisposition in ninety-nine incurable ones. As to Blandford's concluding allegation that the brain of the insane "gives an outward existence to troubles and vexations which exist in their minds subjectively, and have no reality whatever," it applies to the fully developed disorder and not to the prodromal period of mental derangement.

CHAPTER V.

The Psychical Causes of Insanity.

There are a number of cases on record in which a sudden emotion, like anger, fright, or excessive joy, has led to the immediate development of insanity, either uncomplicated or associated with epilepsy. Transitory frenzy has been in several instances noted to follow angry excitement, while stuporous insanity and katatonia can in a comparatively large number of cases be traced to emotional shock of some kind. The manner in which these causes operate is still obscure. Some are inclined to attribute insanity resulting from them to the vaso-motor disturbance induced by emotional episodes. The writer believes that in the case of stuporous insanity the production of the functional suspension of all the mental faculties is comparable to that anæsthesia of the retina which results after sudden exposure to a very dazzling light. An external impression if it exceeds the physiological receiving power of a nerve centre, provokes a functional blunting of that centre for impressions of lesser intensity, and this applies to emotional influences as well as to more coarsely material ones.

Ordinarily the psychical causes of insanity do not act in as direct a manner as in the case just cited. It is usually only after a succession of assaults continued through a number of years that the mental organ breaks down. Worry and disappointment, hopes long deferred, and the attendant conflicts of the inner man, the continued resulting over-strain of an organ whose physiological state is one of equilibrium, constitute already a pathological state, a functional abuse of the brain. The intimate relation between the mind and the body is shown by the somatic disturbances which ensue. A not improperly so-called nervous dyspepsia, constipation, and functional disturbance of the heart and kidneys are common sequelæ of prolonged emotional over-strain, and all of them react on the organ whose functional disturbance is in the first place responsible for their existence. Headache, sleeplessness, *malaise*, a tendency to empty speculation either in the way of hypochondriasis, suspicion of others, or distrust of self, so common features of the prodromal period of insanity, mark this reaction, and are in part due to the continuance of the original emotional causes, and in part to the somatic state they have provoked.

This preliminary period of insanity, as it may not improperly be termed, may last for years without leading to serious developments, and it undoubtedly disappears or becomes latent, with or without treatment, in a far larger number of patients than those who become actually insane. It is those who have an hereditary predisposition, or who resort to stimulants and narcotics, lending a spurious vigor to the exhausted nervous apparatus, who furnish the largest contingent to our insane population. In case there is an hereditary predisposition, monomania, periodical mania, or melancholia are likely to develop. In case alcoholic or sexual excesses are superadded, paralytic dementia may result. In uncomplicated cases the insane explosion is usually in the form of a mania or melancholia, the development of one or the other being probably dependent on the constitutional tendency of the patient, whether this be in the direction of sthenic or asthenic reaction to pathological causes.

Intellectual labor is but very rarely a factor in the causation of insanity. It is only where the mental organ is weakened by physical disease and thrown off its equipoise by emotional crises, that mental labor exerts an injurious influence. If persisted in under these circumstances, primary deterioration is likely to appear.

Aside from this case, mental labor of a proper kind is, so

far from being a cause of insanity, one of the most efficient prophylactics against mental disorder. More than one member of an insane family has been prevented from joining his relations in the asylum by some fortunate accident which threw a routine occupation in his way. It is with the mind as well as with the body, a proper degree of exercise is essential to its health, and philosophers and scientists who have been free from worry and vexation, and have pursued the even tenor of investigation and reasoning, have been and are noted for reaching advanced years without manifesting any mental decay, or much less than other persons at the same period of life. Humboldt, Darwin, Cuvier, not to mention a host of others, are examples of this fact. On the other hand, poets, musicians, and artists rarely reach advanced years without manifesting deterioration, and contribute not a little to the insane population, if, indeed, many do not join the ranks of these professions because they have a taint of that insanity which is supposed to be allied to genius in them. While the greatest poets and artists have been persons of the highest mental integrity, it is best to discourage persons who have a predisposition to insanity from cultivating the higher arts. Lenau, Hölderlin, Cowper, Byron, Poe, and a number of others, are illustrations of the association of this tendency with the hereditary taint, and, in part, of its unfavorable influence. Our daily experience shows that even in the cultivation of the mechanical arts, and of the strictly scientific branches, there is room enough for the play of insane project-making and delusions. Dilettantic aspiration is the foe of the insanely disposed in every branch of the arts and sciences; the insane inventors, political, socialistic, and scientific would-be reformers crowding the quiet wards of some institutions strikingly demonstrate this.

The influence of mental and emotional over-strain in the production of mental derangement of certain types, has been discussed in Part Second (Chapters XI and XV). It remains for us to refer to the influence of education on the development of the mind in its relations to insanity. The earlier an injury affects the nervous centres the more profound are its results. This is illustrated, as to organic affections, by such examples as the porencephaly of Heschl. When this lesion involves the brain of a child, imbecility results, but when it is produced in the brain of the adult, the mind may remain unaffected. It is the same with the functional abuse of the organ. The earlier emotional over-strain, harsh treatment, sensational reading, and ambitious rivalry occur in the history of mental development, the more likely are they to awaken the slumbering predisposition to insanity where it exists, or to develop it where it does not exist. The most important task of the alienist of the future will be a thorough revision of our educational methods. There can be little doubt that competitive examinations, mechanical grinding of the "spelling match" variety and the "Gradgrind" principle will be among the things that were, after that revision is made.

CHAPTER VI.

The Medicinal and Dietetic Treatment of Insanity.

It may be accepted as a dogma of psychiatry that the leading morbid phenomena constituting insanity can be influenced by drugs in only very exceptional instances. As a rule, drugs act, when they act at all, indirectly, although a few of them seem to influence the fundamental pathological state of alienation, especially when located in the vaso-motor system, most happily.

Among the general objects of medicinal and dietetic treatment are the improvement of the general nutrition and the remedying of insomnia. The chief field for this branch of therapeutics is consequently among the acute forms of derangement.

In disorders of the sthenic type like mania, when the chest organs are in a sound condition and the general nutrition is good, the medicinal treatment is mainly limited to the control of motor excitement and the relief of insomnia. The best, most reliable, and safest drug for the former purpose is *conium*, the only reliable preparation obtainable being Squibb's fluid extract.* As a rule, twenty minims will suffice as a first dose, while from ten to fifteen minims may be subsequently given every half hour or hour until the excitement is subdued. In patients whose tolerance of the

---

* It is necessary to test every new sample, as the strength of the preparation varies.

drug has been tested, much larger doses may be safely administered. In one case the writer has known death to ensue in a debilitated patient who took a drachm of the drug, through the negligence of an attendant, while the same dose was repeatedly found to fall within the physiological limits in its effects in others. The physiological action of conium is still the subject of discussion. The writer, from his observations, is inclined to believe with Harley, Davidson, and Dyce Brown that it acts on the cerebral centres, and not alone on the peripheral nerves as Kölliker and Guttmann claim. In maniacal patients it is truly remarkable to find how rapidly with the progressing abolition of muscular overaction the mental processes become clearer and the flight of ideas less rapid just prior to the patient's dropping off into what usually proves to be a refreshing slumber. No drug in the whole range of those used in insanity is so certain in its action and leaves so few ill effects behind as conium. Patients who stagger around under its influence, and are compelled to take to their beds and become tranquil in consequence, awake some hours thereafter in a condition of comparative improvement. It is advisable not to push the drug after the first indications of motor relaxation are observed, but to watch it most carefully then.

A combination of equal parts of bromide of potassium and chloral * is the best medicinal remedy for the insomnia of maniacal patients. The necessary dosage varies so greatly that it is impossible to prescribe any rule. With a patient of good physique, and whose organs are in a sound condition, it will be simply useless to give the amounts ordinarily given for insomnia in private practice; the double, nay the treble, must be given; and better to produce one good night's or half night's rest with a single large dose than to fail repeatedly with a succession of small doses, whose aggregate amount and whose permanently injurious influence is much greater, while they fail to produce the good effect of a single large one. Neither of these drugs should be used night after night, except in emergencies or in epileptic insanity, for it is not so much the object of the alienist to crowd down a psychosis as to establish a series of relatively lucid periods, and thus to tip

---

* Not of two or three parts of the former and one of the latter, as is frequently recommended.

the scale sufficiently on the side of struggling nature to overcome the pathological influence. He should bear in mind that mania disappears, not suddenly, as a rule. but by a series of oscillations between the healthy and the diseased state, which finally merge into a healthy equilibrium, and that in the absence of a specific remedy it is wisest to follow physiological lines of treatment. No mania was ever choked down, but, at most, prolonged or diverted into the channel of deterioration by the excessive use of hypnotic and calmative drugs. It is even desirable to permit a maniacal patient to remain excited at some periods and within certain limits, as only under these circumstances can that mental influence which is often more important than the drug be tested and exerted so as to permanently benefit the patient. In paretic dementia it should be particularly carefully watched, and in the later stages of that affection is altogether contraindicated, owing to the fact that it may intensify the angio-paralytic phenomena of that disease, or produce cardiac paralysis directly.

Among the procedures which are recommended for the treatment of sthenic delirious conditions, and of such which, like *delirium grave*, are associated with incipient and acutely inflammatory states, venesection at the mastoid process, the use of ice-bags, the cold pack, baths, and of hydragogue cathartics are the most efficient. Venesection is very rarely applicable, because the period when it might be useful is usually past when the patient comes under the alienist's cognizance; its use should be limited to cases of suppressed menstruation and grave delirium. In insanity with excitement we have usually to deal with a condition rather of under nutrition than of over-nutrition, and it is in obedience to a rational meditation on this fact that phlebotomy and the leech are almost banished from asylums for the insane.*

The use of baths, originally recommended by de Boismont in recent cases of mania and melancholia with excitement, is most beneficial. Baths of nearly the temperature

---

* In a number of recorded cases a sudden shock, either mental or physical, has been followed by rapid improvement. Thus melancholias who have wounded themselves with razors, appeared to be relieved by the unsuccessful attempt at suicide, and thenceforth recovered. Of this nature probably was the result published by Fordyce Barker, where venesection in melancholia was followed by almost instantaneous recovery. This procedure would probably not have the same happy effect in that large proportion of cases where an anæmic condition is a marked feature. Severe burns have also been followed by amelioration.

of the body, and which may, in well-nourished maniacs, be prolonged for twelve hours and over, exert an excellent calmative and sometimes a better soporific effect than the medicinal calmatives. The cold pack is also useful in mania and agitated melancholia in a double way: first, because of its effect on the vaso-motor system; and second, because of its effect on the metabolic processes, and consequent curative effect on anæmia, if present. Neither the bath nor the cold pack should be used in patients whose temperature is below 98°, and on the whole they are less frequently applicable in melancholias than in manias. It will be found an excellent plan to alternate in the use of baths and hypnotics (paraldehyde), in order to prevent the patient from becoming accustomed to the influence of either. (Appendix X.)

In conditions like the exacerbations of paretic dementia, of which vaso-motor paralysis is a feature, ergot is very useful. The elaborate researches of Kiernan, undertaken with this drug at the suggestion of the writer, prove that it has a marked influence both on the *status*-like epileptiform and on the maniacal seizures of this disease. Amyl-nitrite is indicated in the opposite condition;* and as this drug, which on account of its evanescent action must be repeatedly given, effectually remedies vaso-motor spasm, it is of great service and sometimes directly curative in stuporous insanity and melancholia, particularly of that kind in which the "frozen attitude" occurs. Quite magical effects have been seen by the writer in cases of katatonia. In the first patient selected for trial, who was in the cataleptic phase, complete lucidity occurred immediately after the first inhalation; the cataleptic state recurred in an hour, and as quickly yielded; and the persistent use of the drug materially hastened the patient's ultimate recovery. In apathetic stuporous and cataleptic patients it is necessary to resort to the device of closing their nostrils and mouth for a few moments before presenting the amyl-nitrite to be inhaled, in order to compel the patient to take a deep breath. One or two deep inspirations mediate a more thorough action of the drug than twenty superficial ones, and it is for the reason that they neglect this precaution that physi-

---

* The writer is aware that this drug is recommended by some English writers for the epileptiform seizures of paretic dementia. Although its use is sometimes followed by a cessation of the fit, it is difficult to see how, in view of the angio-paralytic action of the drug, any other but a deleterious influence can be exercised by it in this condition.

cians so frequently fail in its use. The patient's face must flush and his pulse become full and expansive, or the drug has not been properly given. Happily, the very conditions in which it is of service are the ones in which the greatest tolerance is shown to it. Only in elderly subjects and those having arterial disease should its use be proscribed.

Opium is the most generally useful of all drugs in insanity. It has a direct influence on the mind, antithetical to the painful emotional state of melancholia and to the persecutory delirium of monomania. While it is itself a vascular stimulant, yet its influence on the heart is such as to overcome the wiry pulse of extreme melancholia and other conditions in which the blood current in the brain may be assumed to be diminished. It is because of the peculiar union of a vaso-constrictor influence with its well-known effect on the heart, that it of all generally stimulating narcotics is applicable to paretic dementia, and that we have the, to a superficial view, paradoxical fact, that while opium and morphia are counterindicated in manial furor, they are strongly indicated in the furious exacerbations as also in the quiet intervals of paretic dementia. Morphia is of excellent service in the treatment of periodical insanity whose exacerbations it may entirely check. As a rule it is best to give opium and its preparations by the mouth, for patients with persecutory or hypochondriacal ideas are very apt to interpret a hypodermic injection as an assault, an impregnation with poison or in some other delusional way, and it is a daily experience with the insane that the starting of a suspicion or a new train of delusions will undo the best therapeutical measures. Whether it is for this reason or some other, opium does not act well in passive melancholia. The deodorized tincture of opium and the bimeconate of morphia are the best preparations, and while opium produces constipation, and it is necessary to give a gentle cathartic with it at first, yet when its prolonged use is necessary it will be found that the intestinal canal soon resumes its functions, or, at the worst that these can be readily regulated. In the maniacal periods of paretic dementia very large doses are borne. In melancholia and anxious states generally it is best to begin with twenty minims, and continue giving from ten to fifteen minims every two hours, till an effect on the pupil and pulse is obtained.

In nearly all the conditions in which opium is admissible digitalis or strophanthus can be advantageously given. It may be accepted as a rule in insanity that the administration of both these drugs should be guided by the condition of the heart. Opium should not be given, or at least not given without digitalis or strophanthus, in cases where serious valvular lesion and dilatation exist, and digitalis should not be given when there is high arterial pressure, nor in one condition which is very rare in the asylum ward—cardiac hypertrophy.

The most recent fashion in psychiatrical therapeutics is the use of hyoscyamia. The large doses of this drug recommended by the English, and the still larger ones, employed by a few American physicians, are calculated to stagger one. Schüle expresses a natural surprise at the indiscriminate abuse of this drug, and sounds a well-timed warning as to the toxic effects, such as aphonia and ataxia, occasionally observed to follow very small doses of hyoscyamia. Until a more careful study of its effects shall have been made, the writer would hesitate to recommend it, as long as we have so many tried reliable and safer remedies at our disposal. The tincture of hyoscyamus has been long given by Kiernan in combination with chloral hydrate and bromide of sodium as a calmative of excited patients; and the writer is unaware of any good effects obtained from hyoscyamia which are not obtainable from this much safer combination or from that with conium.

Billod claims to have obtained excellent remedial effects in the treatment of hallucinatory conditions by means of stramonium. The writer has no experience with this drug, but its advocate's position entitles the drug to a more extensive trial than it has thus far received.

Cannabis indica, in large doses given at intervals of two or three days, sometimes has an excellent effect on depressed states, and its influence in some cases is rapid and strikingly manifest. In a case of *folie du doute* its influence seemed to mark the turning point in the favorably terminating history of the case. Unfortunately the unreliability of the preparations obtainable in America is a bar to its use, and even the English extracts so highly praised on both sides of the Atlantic are often inert.

Strychnia is one of those drugs which, while they exert no specific influence on any special morbid factor of insanity, rank high as general tonics in this disorder. Its excellent influence on the tone, both of the voluntary and involuntary muscles, is shown in all conditions of motor anenergy, and visceral torpor; hence its use is to be recommended in melancholia, stuporous insanity and other states of neural depression, as well as in all conditions in which the sphincters are relaxed. It has also a favorable effect on states of vaso-motor paresis, such as paretic dementia and the later periods of grave delirium; it is hence indicated wherever ergot, with which it may be advantageously combined, is applicable. In conditions with high vascular tension its usefulness is problematical, and in melancholia with the "frozen attitude" which is associated with a spasmodic state of the entire muscular system, and regarding the influence of strychnine on which the writer has no observations at his disposal, it will be well to employ it very carefully. It would be hasty, however, to assume that a spasmodic state due to a mechanism altogether the opposite of a central state of functional over-activity must necessarily be influenced unfavorably by a drug which produces spasm through such a mechanism. Strychnia is the most general neural stimulant at our disposal; its tonic influence is exerted on the entire central nervous apparatus, and is not at all chiefly localized in the cord, as some have believed.* It stimulates the central and peripheral sensory and the central motor and the vaso-motor systems, and directly affects the cerebral functions in a favorable way. Nowhere is this better manifest than in paretic dementia. The effect of strychnia is somewhat lessened by a prolonged administration; in addition, while the appetite is at first very happily influenced by it, it is afterwards unfavorably affected; for all these reasons, it is best to give nux vomica and strychnia on an "interrupted plan." The drug should not be suddenly administered in large doses, but in rising ones, till the desired physiological effects are obtained; then they should be gradually decreased, each cycle of administration lasting about ten, with intervals of five days.

As a general rule, the combination of several drugs in a mixture, intended to be constantly used even in the treatment of one and the same patient, is to be deprecated. Under these circumstances it is impossible to regulate the administration of remedies which, like those used to quiet

---

* The writer in the course of a large number of experiments performed on animals of all classes ("The Anatomical and Physiological Effects of Strychnia, on the Brain, Spinal Cord, and Nerves;" Prize Essay of the American Neurological Association) found that the characteristic spasms of strychnia poisoning are in part at least of a cerebral origin,

insane excitement, must more or less closely approach in their effects the boundary between life and death. To complicate the confessedly difficult task of approaching this limit sufficiently near to obtain the useful effects of a drug on the one hand, and not to pass beyond it on the other, by combining it with remedies which may or may not mask or neutralize its effects, is exceedingly unwise. What is useful in the way of combination in psychiatry has been tested in general practice, as for example the uniting with opium of a slight amount of belladonna. As to the harmonious combination of chloral hydrate and bromide of sodium,* the general caution must be made, that the nearer the patient approaches confusion or *moria*, the less bromide must be given to him; and in such cases it may be well to give the chloral alone. Unless there is an emergency demanding the instant calming of a boisterous or destructively maniacal patient, it is best to regulate the functions of the gastro-enteric tract before proceeding to administer any drugs, and when administering them to bear their possible influence on the digestion continually in mind, and to regulate their administration accordingly. In the majority of cases, the neurotic medicines are better absorbed and better borne in an alkalinized stomach than in one containing either an acid gastric juice or the acid products of fermentation found in some forms of gastritis. This is particularly the case with chloral hydrate and the bromides. Both the hypnotic purpose and the "gastric" indications are met better if a glass of hot milk is given immediately after the chloral and before retiring.

Alcoholic stimulants are of great service in all states of depression and restlessness, excluding those of a maniacal character. Their use is therefore contraindicated in the active phases of paretic dementia, and they should be sparingly used, if at all, in the quiet intervals and remissions of that disorder. They should never be given in periodical lunatics, without bearing the danger of the formation of a dipsomaniac tendency by them in mind. Malt liquors in moderate quantities, not exceeding, say, a half a pint of beer, ale, or porter daily, will be found useful both as nutrients and as calmatives in badly nourished subjects, suffering from insomnia. They should not be administered unless

---

* Which takes the place of bromide of potassium in every respect, and disturbs the stomach less.

the patients have out-door exercise at the same time, and the slightest sign of gastric catarrh is a contraindication to their use.

Phosphorus has been recommended on theoretical grounds in all forms of insanity. The belief in its virtues was of course strongest in those days when it was believed that the "phosphates" are continually drained from the brain through the urine, and that the chief functional substance in the brain is a phosphorized oil or fat. But the dictum that "without phosphorus no thought," correct as it is, does not justify the conclusion that phosphorus is to the brain in insanity what iron is to the blood in anæmia, as if delusions, hallucinations, abulia, and amnesia were comparable to strikes in a match factory. Undoubtedly the restorative nutrition of the brain is an important task for the psychiatrist, but he may crowd phosphorus into the stomach, without materially influencing abnormal mental processes. It is true that the symptoms of mania and melancholia are probably the expressions of disturbed biochemical states of the brain tissue, but we are not able to put our finger on any one component element, and say that this one is too rapidly wasted, and that one not rapidly enough removed. Even if we could do so, we might be powerless to control the vaso-motor disturbance so intimately connected with the biochemical anomalies. As it is, the chemistry of the brain has taught us nothing regarding the particular rôle which phosphorus, in the combination of a distearyl-glycerin-phosphoric acid combined with neurin as a base, plays in acute insanity. We are limited in our indications to empirical observations, and these teach us that in the deteriorating mental states, as well as in those associated with general nervous exhaustion, phosphorus is of considerable benefit. The problem of furnishing a reliable and easily assimilable preparation of phosphorus has not been satisfactorily solved. The trade-mark preparations are some of them very well borne by the stomach, but inconstant in strength, and, in two prominent instances at least, not honestly kept up to their original standard.

Iron is indicated in that large class of the insane in whom anæmia is present. No other rules are required for guidance in its administration, or for the administration of other restorative remedies, than those followed in general practice.

THE MEDICINAL TREATMENT OF INSANITY. 393

Quite extravagant hopes have been based on the alleged curative effect of electricity in insanity. Superficial theorizers have even undertaken to indicate the special kinds of current, and directions of such, to be applied to the head in various forms of insanity, and it is to be presumed that the more modern imposition of static electricity will come into vogue, and after a brief sway over the minds of the credulous, and an occasional success with a simulating or hysterical patient, share the fate of other epidemics of charlatanism. Electricity can have from the very nature of the case no specific effect on insanity.* Its applicability is limited to those forms in which there is simple atony, as in stuporous insanity, and to those which are associated with organic and functional disease of the nervous axis; in the latter case the ordinary rules of electro-therapy apply. In stuporous insanity its effect is to stir up the patient; but we should be very sure of our diagnosis before applying it, and not confound atonic melancholia with stuporous insanity, for in a melancholic patient electrical manipulations would probably provoke additional delusions of persecution to those he already entertains.

If we were acquainted with the molecular condition of the brain in health and disease, and if we understood better the exact influence of electricity on the molecular and dynamic states of that organ, we would be better able than we now are to formulate the indications for the use of this potent neurotic agent.

The diet of insane patients should be nutritious and easily assimilable. In view of the frequent coexistence of gastro-intestinal disturbance, milk and raw meat † should

---

* Unless we are to assume the correctness of such views as the one contained in the following citation from the testimony of the prosecuting witnesses in the Guiteau trial (p. 1363, Dr. H. P. Stearns testifying): *Question.* What is your theory of a person becoming suddenly insane through the excitement of fear in its operation on the brain ? *Answer.* I suppose that to be injury of the tissue of the brain from the effect produced upon it as communicated to it. *Question.* By the blood rushing to the brain or withdrawing from the brain ? *Answer.* It is very difficult to say what precisely does produce the effect. *It may be from a change in the electrical currents* that we know pass through the brain.

† The tender parts of beef, excluding the fatty and tendinous intersections, are scraped with a blunt knife in such a way as to retain the juice; the mass is then seasoned with an abundance of salt and a little pepper. Of course it should not be attempted to give this palatable and nutritious dish to patients who have, the delusions that they are compelled to eat human flesh, or meat seasoned with the blood of their friends.

be the chief food of depressed patients. Most vegetables and the prepared meats are commendable only in patients who at least occasionally take out-door exercise. Very good results have been obtained by the writer in epilepsy with and without insanity, by compelling a strictly vegetarian diet, as recommended by Browne, and this doubtless will be found applicable in the case of asylum patients. More attention than is ordinarily paid to this branch of the subject should be devoted to the examination of insane patients with reference to the existence of any disorders associated with defective or perverse assimilation; the proper treatment of lithæmia and allied conditions will often prove radically remedial.

Melancholiacs, paretic dements and monomaniacs laboring under delusions of persecution will frequently refuse food, fearing that it will injure them (*sitophobia*), while patients in atonic and stuporous states are simply unable to eat. Sometimes maniacal patients refuse to eat, but as a rule only for a short time, and their refusal rarely becomes a problem for the alienist to seriously consider. When a melancholiac refuses food, the rule is to compel him to take it forthwith. With maniacs, monomaniacs and paretic dements, temporizing is advisable, because these patients usually resume eating voluntarily, and for the additional reason that their anorexia is often due to a gastric disorder which can only be favorably affected by rest of the organ affected. With many of the sitophobic patients it suffces to lead them to the table, to put the utensils for feeding in their hands, or to feed them with a spoon. With a large number of melancholiacs and monomaniacs, with inspirational delusions of commands to abstain from eating, it is necessary to resort to artificial feeding. Temporizing, admissible under other circumstances, would here be injurious. Particularly in melancholia the nutritive disturbance is so great that not a day should be lost in waiting—frequently in vain—for a resumption of the physiological habits. The most simple and most readily extemporized apparatus for artificial feeding consists in a funnel, to whose lower end an œsophageal tube is attached. The tube, well oiled, may be passed through a nostril, or through the mouth with the aid of an oral speculum. The latter should be of strong make. In passing the tube backward the index finger of the disengaged hand should be used as a guide. In passing it through the nostril, the possibility of its entering the respiratory passages should be borne in mind. This accident provokes coughing, strangling, and cyanosis; the two former warnings may, however, be absent in paretic dements with laryngeal anæsthesia. Coughing and strangling may ensue through irritation of the pharynx even when the sound is passed properly, but in this case there are never any *inspiratory* noises produced in the tube; expiratory noises may occur when the tube is properly passed by escape of gas in the stomach. The food used in forced feeding should always be strained before being poured into the funnel, and should never be used at a temperature above or much below that of the body itself. Milk, yolk of eggs, dry wines, milk-punch, egg-nog, beef-juice, and hydroleine are among the substances which can be conveniently administered in this way. Medicines can also be given mixed with the food, and as this is done without the patient's knowledge, is often a great advantage with a suspicious lunatic.

The advantage of the funnel over the stomach-pump, which is frequently used, is that less pressure is employed, that the apparatus does not appear as formidable to a suspicious patient, and that the attendants, if feeding should be entrusted to them, will not be tempted to omit the important precaution of straining the food. The stomach pump permits the operation to be done more quickly, and with it, obstruction caused by a blocking up of the tube may be overcome by force. It should be recollected, however, that nausea and vomiting are likely to ensue in case the food is introduced too rapidly. Whether fed with the funnel or the stomach-pump, the patient should sit up, and if he is very obstructive a restraining chair will save the patient much needless muscular exertion, the physician much trouble, and diminish the chances of doing an injury. In case the œsophageal tube is passed along the floor of the nasal cavity, it is apt to encounter a resistance and be deflected forward by a prominence which is sometimes very marked on the posterior pharyngeal wall, and which corresponds to the bodies of the cervical vertebræ. Dr. Tuke advises throwing the head of the patient back at the moment when the sound approaches the posterior nares, the tube having previously been bent a little so as to facilitate its downward passage; then at the moment when it is about to glide down into the œsophagus, when there is a risk of its passing into the larynx, he advises the head to be brought forwards and downwards so as to send the point against the posterior wall of the pharynx. After passing the upper end of the œsophagus, the tube is usually swallowed, as it were, and glides down without further difficulty into the stomach through the action of the constrictor muscles. At least sixteen inches of the tube should be allowed to pass down before raising the funnel or using the syringe.

Some patients are artful in resisting forced feeding, learning to use their abdominal muscles in such a way as to compress the stomach and cause the food to regurgitate. This is rarely the case in acute melancholia, more common in monomaniacs with certain delusions, and patience as well as a judicious denial of certain privileges will be necessary to prevent the patient from carrying out his project of self-starvation.

It is frequently found that after having fed a patient by force several times, the sight of the feeding paraphernalia when brought in for use will induce him to eat voluntarily. It is therefore well in every case to make the proffer of food before resorting to extreme measures, and if it is taken, it may be advantageous to leave a little solid food in the room as if by accident. In that case it will be found that the patient will eat it stealthily, when he supposes himself unwatched. It is needless to add that no food should be left in the patient's room, under these circumstances, to eat which he would require a knife, fork, or spoon, or any vessels of porcelain, glass, or tin, which he could break and open his blood-vessels with; celluloid or rubber platters will fulfil all the requirements of the case.

Forced feeding should in stuporous and atonic patients be resorted to at least three times a day, and, if it can be done, once at night. The nutritive loss is something enormous in these cases, it is startling even in the best-fed patients, and it is the duty of the physician to fight the foe for every ounce of body-weight, as it were. With other patients, feeding by force will be necessary, at most, twice a day, and rarely will they require it for long periods. It should always be done under the physician's immediate orders. The additional demand has been made, that it should always be done in his presence, or by him in person. The writer has, however, seen trained and other nurses perform the operation with all the skill and judgment which the case could possibly have required, and many members of our profession could afford to learn the practical details of the operation from them. An anatomical demonstration of the parts involved should be given to every person to whom forced feeding is intrusted; with this, and experience acquired under skilled guidance, he may be safely relied on to carry out forced feeding himself, and without any other aid than that of another and older attendant.

It is unnecessary to specify here the hygienic requirements of the insane, which are those of hospital patients generally. Patients who are anæmic, and particularly those whose temperature is subnormal, should be in warmer apartments than ordinary hospital patients, or those whose disorder is of a sthenic type. In this climate a temperature of from 65° to 68° Fahrenheit, and in some cases even higher, is necessary in the wards where demented and melancholic patients are congregated.

CHAPTER VII.

The Psychical Treatment and Management of the Insane.

The subject of the psychical treatment of insanity in its widest sense comprises the prophylactic treatment of the insane predisposition. This vast subject it is impossible to treat of in a manual, and the writer contents himself with expressing his opinion that in the future certain educational means will be recognized to be as essential for the diverting of the mind inclined to perversion into healthier channels, as the methodical training of certain muscles is essential in preventing and remedying certain of the malformations which come under the cognizance of the orthopœdist.

The physician is ordinarily called upon to treat the fully developed disease, and the question of most pressing importance which presents itself is whether the patient can be treated at home, or whether it is necessary to send him to an asylum. In the case of primary deterioration, of stuporous insanity, and of the earlier phase of syphilitic dementia, this question may be frequently decided in favor of home treatment, provided all the conveniences for psychical and physical treatment are within the reach of the family. In most other forms of insanity the physician risks very little by positively recommending asylum treatment. He has three important questions to consider: 1st. The safety of society; 2d. The physical and financial safety of the family; 3d. The interests of the patient as an individual. Ordinarily the duty of the physician is in the first place towards the individual patient; in the case of insanity, however, there are many other interests than those of science, and of abstract humanity to the patient, involved. Where we have to choose between the endangering of the security, health and happiness of healthy and useful members of society on the one hand, and the compliance with sentimental considerations advanced in favor of decrepit, dangerous, or possibly useless ones, we need not hesitate long in our choice—no longer than the obstetrician when called upon to decide whether he shall perform an operation which would prove the certain death of a mother and the possible salvation of her unborn child, and another which would result in the death of the child but certain salvation of the mother, will hesitate to adopt the latter alternative. Fortunately, the alienist is in the position, when complying with the first two demands enumerated, of acting at the same time in the best interests of the patient himself.

The patients suffering from the ordinary psychoses are dangerous to society in the following ways: 1st. They may commit homicide, either under the influence of hallucinatory terror of imaginary pursuers, insane hatred of rivals in their affection, or of alleged seducers of their partners in life, an insane desire for notoriety, because of disappointments in insane aspirations, or under supposed inspiration from on high. 2d. They may commit arson or incendiarism either from similar motives as those just enumerated, in the thoughtlessness and carelessness of dementia, as the result of morbid projects—for example, when a paretic dement burns down his house to build a palace in its place—or in obedience to the pyromaniac morbid impulse. 3d. They may make delusional charges, or false charges from malicious motives, against others, and procure the punishment of innocent persons. 4th. They may make indecent assaults, either on account of satyriasis, or sexual perversion, and scandalous exposures of their persons from sexual motives, or in the abstraction of dementia. 5th. They may destroy valuable property under the influence of delusions or insane antipathy. 6th. They may propagate their disorder. Lunatics are dangerous to their families because: 1st. They may in the abject gloom of melancholia, or in obedience to the morbid impulses of that condition, immolate whole families in a general massacre. 2d. They may squander their property in insane speculation, absurd purchases, or in excesses. 3d. They may develop mistrust against members of the family, commit murder and mutilation on them, or become the instruments of designing persons, and disinherit or rob those who are naturally dependent on them. They are dangerous to themselves: 1st, On account of suicidal inclinations; 2d, through the occasional tendency to self-mutilation; 3d, through the continuance in a course of conduct and excesses which are calculated to intensify and prolong their malady. All these considerations demand that the insane, who are liable to indulge in such acts, should be beyond the range of damage to themselves and to others; and in a large number of cases the experienced alienist feels relieved of a heavy sense of possible danger as soon as the patient is within the walls of a properly conducted asylum.

An asylum sojourn has in the vast majority of cases a good effect on the insane. Curable patients are never injured in their prospects as to curability in a medically well-managed institution, and incurable patients should be there for practical reasons, and are usually better off in than out of the asylum.

The advantages of asylum treatment are the following: 1. Refusal of food and medicines—the great obstacles to the treatment of the insane outside of asylums—are best dealt with by a skilful corps of physicians and attendants always on the spot, with the necessary appliances at their disposal. 2. The necessary supervision of the insane at all hours can be carried on with the least expense and greatest thoroughness in the asylum ward. 3. The excessive and damaging use of narcotics, calmatives, and restraint, necessary for the purpose of preventing scandal in the neighborhood, and noise, destructiveness, and exhaustion at home, can be dispensed with in the asylum. 4. The sojourn of a patient in an asylum, the continual reminder which the restraint of its walls is to him that he is considered insane—whether he believes himself to be so or not—is in many cases a far stronger incentive to a kind of reflection which leads to the correction of delusions, than any drug.

With many delusional monomaniacs psychical treatment is of far greater value than are food and drugs. If there is any point of attack offered by these disorders when independent of perverted sensations and visceral conditions, it is the logical apparatus. It is true that it is in the overwhelming majority of cases impossible to reason such patients out of a delusion, and that where this is possible, the insane fundament of the insane thoughts becomes the soil for other delusions. But occasionally external influences can be brought to bear upon them in such a way as to effect a rapid cure. Leuret showed this in the days when the douche and other forcible measures were more commonly used than now. He cured a patient, who had the delusion that he was a king, by having him douched whenever he expressed that delusion, or responded to salutations addressed to his imaginary majesty. Other lunatics would find the strongest confirmation in such a "persecution." Undoubtedly prolonged restraint often leads to a growing conviction that, after all, the patient's own beliefs may be as absurd as those of other patients whose insanity he is able to appreciate, because he discovers that they adhere as firmly to their beliefs as he has adhered to his. More than one instance of the happy effect of a recognition of the delusions entertained by others on the patient's mind is on record.

It cannot be our purpose here to discuss the host of questions touching the internal administration of asylums, however intimately they may be related to the important medical problem of the moral effect of the institution on the patient. It may be expected of the writer to express his opinion, as he is about to take leave of his readers, with regard to two points which the profession, and through them the general public, may require information concerning—1st. The advisability of discharging lunatics on probation during remissions of their disorder; 2d. The use of restraint.

Remissions of insanity, rarely amounting to absolute lucidity, are very common in asylums, particularly among monomaniacs, the periodically insane, and paretic dements. There are many cases recorded of periodically insane subjects who voluntarily sought the protection of the asylum whenever they felt the morbid period approaching; there are a limited number of monomaniacs whose delusions are entirely harmless, and not likely to change in this respect, who may support their families and occupy a respectable position in the community when discharged; finally, there are a few paretic dements who cannot be regarded as *non compos mentis* in the remissions of that disease. All these classes are fit subjects for discharge on furlough. It should, however, be determined in some legal way that a definite responsibility is assumed by the superintendent discharging such a patient, as well as by the relatives who assume his charge. In Germany, where a furlough system and such a responsibility exist, not a single homicide or assault has occurred by a lunatic discharged under this system, and but a single theft, during five years, the insane out on furlough thus showing a far better record than the sane population. It is scarcely necessary to add that this system can be adopted only in communities in which some other factors than the political or social influence of men who turn to the asylum career, because they have failed in the general practice of their profession, are potent in determining the selection of medical officers of asylums.

Much has been written and said about the use and abuse of mechanical restraint. That this means of controlling the insane has been pushed to an extent unwarranted by the emergencies of the case, there can be no doubt. The experience of the superintendent of the Auburn criminal lunatic asylum, who is to-day able to manage an unruly and dangerous class of patients with a minimum of restraint, while one of his predecessors not only used a maximum of restraint but also fired an occasional bullet among his charges, is a significant commentary on the correctness of the position taken some years ago by the Neurological Society with reference to this question. The concealment of restraint apparatus, particularly of that variety known as the Utica crib, when foreign alienists visit a prominent institution in this State, is a confession that the apparatus and its frequent employment are, to say the least, features of which no asylum can be proud. The following citation from a distinguished American alienist* expresses I think the opinion of most humane and thoughtful asylum physicians:

"With the banishment of manual restraint many of the objectionable features of hospital life have entirely disappeared. The temptation to rely on coercion rather than kindness is removed when the power of resorting to mechanical restraint, either by threats or its actual application, is withheld, and the physician, nurse, or attendant, as the case may be,

---

* Dr. Bryce, Report of the Alabama Hospital for the Insane, 1884.

finds himself compelled to resort to gentler and more rational methods of discipline. I do not wish to be understood as maintaining the absurd position that restraint of some kind is not necessary in the management of the insane. It is a maxim of science that all development proceeds under restraint. I am simply objecting to a system of coercion that is not based upon rational grounds, and is therefore unscientific and often brutal. Human nature, insane as well as sane, revolts at the indignity heaped upon it by manacling the hands or feet. Hand-cuffs of any kind are badges of disgrace, and no display of logic can ever disabuse the patient's mind of that idea. If this kind of treatment accomplished better results than any other, there might be some grounds for advocating it in spite of its objectionable features, but the facts as shown by experience are exactly the reverse."

On the other hand, it must be admitted that the agitation against restraint has overstepped the bounds of legitimate criticism and reform. That there are some subjects who require restraint, who are better off with than without it, there can be no doubt. The demonstrative feat of a novice superintendent, who burned all his restraint apparatus as soon as he took charge of his asylum, was followed by the accumulation of black eyes, broken noses, and other minor surgical accidents, as well as by several suicides. It should be remembered that Conolly, the very apostle of non-restraint, said:

"Without a very efficient superintendence, chiefly to be exercised by the chief medical officer, the mere absence of mechanical restraint may constitute no sufficient security against neglect or even ill-treatment of patients in a large asylum. The medical officers who consider such watchful supervision not properly comprised in their duties have formed but a very inadequate conception of them."

It is with this question as with many others relating to the internal economy of asylums: reform cannot be accomplished by watchwords and catch phrases, nor by arbitrary legislation. The proper method of improving an asylum is to develop the management and supervision of asylums in a scientific direction. Scientific zeal and integrity within asylums will prove far better guarantees of humanity to the insane than associations of dilettante and newspaper editorials. Let us hope that the scientific spirit which was breathed into American psychiatry by Rush and Ray, and which has been kept alive by their immediate followers, will gain that preponderance which it merits over an unworthy opposition.

APPENDIX.

Note I. (Page 85.)

The influence of maternal impressions on the child is not limited to the production of morphological peculiarities and defects. Two sad cases coming to the author's attention within a few days of these pages going to press, illustrate this. In the first case, a child of four years, belonging to the class of Juvenile Lunatics of Kerlin, was subject to furious outbursts of violence occurring almost daily. There was no morbid heredity, but the mother while *enceinte* had an imperative conception—of which she tried to rid herself in vain—that because she recurred to the case of an imbecile child, not a relative, which had outbreaks of destructive frenzy, her child might be "marked," as it is called. The child in question had not been seen by the mother for seven years. Although the latter and the child she bore differed as to the grade of the imbecility, the mother stated that there was an almost photographic similarity between the maniacal outbreaks of the two. In the second case, a girl of six years had an attack daily in which she was seized with an irresistible propensity to undress herself and rush out of the house naked. The mother said, and in part with truth: " There is no use taking that child to a physician. I know the child cannot help it. I had the same propensity when I was in the family way with that child."

Note II. (Page 115.)

Tuke, in his capacity as a member of a committee on classification of the Antwerp Congress of Psychiatry, submitted the question of classification of mental disorders to the British Medico-Psychological Association, whose Council suggested the following:
  I. Congenital or infantile mental deficiency (idiocy, imbecility and cretinism).
    a. With epilepsy.
    b. Without epilepsy.
  II. Epilepsy acquired (meaning epileptic insanity).
  III. General Paralysis of the Insane.
  IV. Mania (Acute, Chronic, Recurrent, A Potu, Puerperal, Senile).

V. Melancholia (Acute, Chronic, Recurrent, Puerperal, Senile).
VI. Dementia (Primary, Secondary, Senile, Organic).
VII. Delusional Insanity (monomania).
VIII. Moral Insanity.

The one very positive criticism of this otherwise equally acceptable and simple classification, relates to *mania a potu*, which, notwithstanding its misleading title, is not a true form of mania, but a periodical morbid impulse. (See page 270.)

The Austrian physicians follow Meynert in adopting the following classification:
I. Idiocy.
II. Simple Forms of Insanity.
 (*a*) Acute: Melancholia.
    Mania.
    Delirium.
 (*b*) Chronic: Paranoia.
    Periodical Insanity.
    Secondary forms.
 (*c*) Complicated: Paralytic.
    Epileptic and Hystero-epileptic.
    Ordinary organic diseases of the brain.
 (*d*) Toxic: Alcoholic.
    Due to other poison.

The Scandinavian authorities either agree with or closely follow the subjoined classification submitted by Steenberg:
1. Acute Psychoses: Melancholia.
    Stuporous Insanity.
    Mania.
2. Chronic Psychoses: Chronic Melancholia.
    Chronic Mania.
    Dementia.
3. Degenerative Psychoses: Monomania.
    Hypochondriacal Insanity.
    Hysterical Insanity.
    Circular (periodical) Insanity.
    Folie raissonante (moral Insanity).
4. Alcoholic Psychoses: Acute Alcoholic Delirium.
    Chronic Alcoholic Insanity.
    Periodical Dipsomania.
5. Paralytic Insanity.
6. Epileptic Insanity: Post-epileptic Insanity.

Epileptic equivalents.
Transitory mania.
7. Idiocy: Congenital Imbecility.
Idiocy proper.

The above classifications, together with a few less elaborate ones, were submitted to the Committee of the International Congress, which after mature consideration decided that the following groups embodied all the suggestions made, and should serve as a basis for international statistics in the form of the subjoined schedule:

Patients under treatment and admitted to the Asylum of ....from........of the year......to .... of the year......

FORM OF INSANITY.	Patients under treatment on the 1st of January.	PATIENTS ADMITTED DURING THE YEAR.				Total.
		First admission.	Readmission.	Transferred from another domestic asylum.	Transferred from asylum in another State.	
	M | F	M | F	M | F	M | F	M | F	M | F
Idiocy........................						
Dementia (Simple)...........						
Mania........................						
Melancholia.						
Delusional....................						
Insanity, Acute.............						
Insanity, Chronic............						
Moral Insanity...............						
Circular Insanity.............						
Complicated Forms:						
Paralytic..................						
Epileptic..................						
Hystero-epileptic.........						
Tumors and Focal Brain Diseases............						
Toxic forms (name the toxic agent)................						

NOTE III. (Page 148.)

Some recent statistics assign a larger, and others a smaller proportion of cures than that here given. Notably small are the percentages of recovery in Italy; Riva* stating that only 24.5 per cent. of melancholiacs and 41 per cent. of maniacs recover.

---

* Rivista spérimentale di Freniatria xi., 4.

Note IV. (Page 275.)

Dr. Kerlin's classification of Idiocy and Imbecility, adopted at the Elwyn institution, happily indicates the transition of the various forms of arrested development.

I. Idiocy, { Apathetic form. / Excitable form.
II. Idio-Imbecility.
III. Imbecility, { High grade. / Middle grade. / Lower grade.
IV. Juvenile Insanity.

The lower grade of imbeciles are susceptible of some culture of the eyes and hands, are a slight command of language and some imitative power. The higher grade are capable of being taught to write and read, to calculate problems in simple arithmetic, and "approach the lower range of common intelligence in their relations to life as found among the ignorant." The middle grade are, as the name signifies, a connecting link between these two.

Dr. Kerlin adds: "Another class of imbecile children is now very distinctly recognized as an increasing product of the excesses and abnormalities of civilization. They may be found more commonly the subjects of neglect born or drifting into our almshouses, or quite frequently in the refuges, where they are sentenced for incorrigible behavior or petty crime. Under the designation of 'moral idiots' they figure in some of the works on Insanity; called so because, in many, the perversion seems to be mainly in the display of vicious or criminal intent, without responsibility or without will-power to control the impulse. There is usually a history of family idiosyncrasy, or partial insanity, back of the child." The following letter, written by one of my patients belonging to this category of Kerlin's classification, is quite expressive of the tyranny which the immoral and erratic influences exert over such subjects. The writer was in his sixteenth year and had just served a short period of imprisonment:

"Dear Parents:

"You know how I am when a thing comes into my head.
"I got to do it and if I don't want to do it I cannot help my-
"self. After I done it I feel sorry for it and dear Father you
"know I have made you lots of trouble. You know that I
"always liked riding horse backing and jumping around so
"then let me go to Texas there is where I can't get home

"again. I tell you why did not like to go to work here, is
"because I thought everybody would laugh at me because I
"always wanted (to be) a great man in the army but in Texas
"I will have a fair chance anyhow. You said that you would
"let me go. . . . Please give me a ticket to go out West
"but first I am going to ask Buffalo Bill for a place as cow-boy
"you know I like that kind, how the cow-boys act."

On the same day that he wrote this letter he forged his father's name and wrote several spurious orders, avowing these acts to me without the slightest sentiment of shame, as indeed he was incapable of feeling any. It was observed in this case that whenever excited the boy flushed on one side of his head only, the median line accurately corresponding to the boundary between the flushed and the pale halves. In other such cases, notably in one where the same cranial configuration and in addition a cleft palate and hair lip were present, similar vasomoter idiosyncrasies were observed. It is their presence which induces some authors to regard moral imbecility of this kind as a larvated form of epilepsy.

NOTE V. (Page 374.)

A few cases have been reported, where the administration of quinine had produced transitory furor. The evidence that the drug was the sole cause of the outbreak, was in several of these cases, completed by its accidental repetition,* when the alkaloid was again used by physicians unaware of the rare idiosyncrasy as which the susceptibility thus manifesting itself must be regarded. The author has seen a single such case, and was informed that the father of the patient had the same peculiarity. In such cases the physician is naturally apt to confound the results of the treatment with those of the disease treated.

NOTE VI. (Page 374.)

The acute form of rheumatic insanity presents a very favorable prognosis as regards recovery of mental integrity, although the cases in which it occurs as a complication are of a grave character in other respects. The delusions and hallucinations are often of an anxious character, a feature attributable to the intercurrent cardiac complications. In other cases the delusions are based on illusive misinterpretation of the rheumatic pains; the patients complain that they are stabbed, burnt or tortured with red-hot needles. A remarkable reciprocal rela-

---

* Insanity from Quinine, by Kiernan, Alienist and Neurologist.

tion has been supposed to exist between gout and insanity. Those exhibiting it are often moody, irritable, indulge in fantastic and changing projects and occasionally break out in a frenzy not unlike that of a furious maniac. In a masterly paper Kiernan* of Chicago has shown that many of the inconsistencies in the career of the great Pitt resulted from attacks of mental disorder due to the medicinal suppression of his hereditary gout. When the gouty manifestations reappeared this mental trouble vanished. Some German alienists do not confirm this relation.

Note VII. (Page 375.)

Seven cases illustrating a reciprocal relation between insanity and asthmatic attacks have been recorded by Conolly Norman.† Three of the cases which appear to be pure, manifested acute maniacal or melancholic symptoms; the others, imperative conceptions and systematized delusions. In the latter instances the asthmatic seizures were probably indications of the neurotic character of the patients, equally with the insane manifestations.

Note VIII. (Page 376.)

Considerable difference of opinion still exists in regard to the value of oophorectomy and other operative procedures performed for the relief of hysteria and insanity. In a discussion held this year in the British Medico-Psychological Association,‡ Dr. Percy Smith one of the physicians of the Bethlem Royal Asylum, cited the opinion in this book in connection with a case where oophorectomy had been successfully performed without altering the patient's mental condition. Other members expressed the opinion that the view in this treatise was too strongly stated. Dr. Savage, the superintendent of Bethlem, remarked—and I think that most practical alienists will agree with him—that "if the claims of gynæcologists be true, they must see cases of insanity under circumstances altogether different from those under which alienists are able to study them." In the meantime one of the most cautious German neurologists, Remak,§ has confirmed the view here announced, and used a simile in condemning oophorectomy not

---

* Contributions to Psychiatry xii., *Journal of Nervous and Mental Diseases*, Vol. x., No. 1.
† *Journal of Mental Science*, April, 1886.
‡ *Journal of Mental Science*, p. 298, July, 1886.
§ Zeitschrift für Klinische medizin vi., 5.

unlike the one used by the author. He says that even in hysterical disorders, oophorectomy is no more justifiable than would be the extraction of a tooth is true neuralgia of the Fifth Pair. I am aware that Goodell and Flechsig claim remedial results from oophorectomy, particularly in hysterical derangement. But none of their cases seem to have been observed a sufficiently long time to justify very positive statements. Indeed Israel has furnished a singular and suggestive observation calculated to throw discredit on the claims of oophorectomists by performing a sham operation followed by radical recovery for the time being, in a case of hystero-epilepsy. Claus[*] has also made a protest against what may be termed "gynæcological psychiatry" which was provoked by the assertion of Frank that neuroses in women are naught but genital symptoms, even in cases where gynæcologists are unable to detect disease of the female organs! Bondurant[†] reports two cases, one of hysteria with periodical excitement, and one of *folie raissonante* with exacerbations at the time of the menses, in both of which oophorectomy was performed without any lasting improvement; both patients were mentally as abnormal after the expiration of a year as they had been prior to the operation.

I am aware of the fact, and it has been corroborated to me by experienced practioners, that at present the operation of oophorectomy is performed in a large number of cases, without any justifying indications. Healthy ovaries have been removed by the dozen in New York City alone, in cases where almost any surgical procedure, or potent moral impression would have been as effective. This protest against oophorectomy and the abuse of gynæcological procedures in psychiatry has, as Tuke correctly pointed out in the discussion referred to, been misunderstood, owing to the failure of one of the speakers to cite the context. As the passage read in the first edition, standing by itself, it was susceptible of being so misunderstood. The author's intention was to utilize the non-existence of insanity in some of the gravest cases of uterine and ovarian disease as an argument against an intrinsic relationship between ordinary affections of the female pelvic viscera and insanity. The grave and disastrous error which is almost daily committed by those who would cure insanity by gynæcological operations and methods, is the neglect to select proper cases. Certainly those forms of insanity which, like primary paranoia and moral

---

[*] Irrenfreund, 1882, No. 6.
[†] *American Journal of Insanity*, January, 1886.

imbecility, are due to a neurotic taint or defect should be regarded as a *noli me tangere* by the oophorectomist. But the literature of the subject shows that they have furnished a large quota to the cases operated on—and uniformly without any good result.

The author has frequently observed the predominance of imperative or hypochondriacal conceptions in patients suffering from ulceration of the cervix and dysmenorrhœa. A similar condition is also observed in rectal disease. The relief of such conditions is so self-evident an indication, as stated in the text, that it is not necessary to specially urge it. No peripheral affection in the insane should be neglected. Probably nothing in psychiatry is more remote than a systematized religious delusion and a stricture of the urethra; yet the author possesses the records of a case where the relief of a stricture in a patient afflicted with such a delusion produced such amelioration that he became fit for discharge from the asylum, though his delusion remained.

NOTE IX. (Page 378.)

Mental disorder of an acute maniacal or delirious character, either transitory, or passing into dementia which appears to be merged into a terminal uræmic coma, is occasionally observed in advanced kidney disease. It is usually associated with other uræmic symptoms. Some question exists as to the part played by the cardiac complication, but inasmuch as the delirium in the observed cases was not of that anxious cast which usually results from cardiac influences, it may be assumed to have been a uræmic phenomenon.

NOTE X. (Page 387.)

Paraldehyde, the new hypnotic, appears to have failed in the hands of most American physicians as in the author's. Possibly this, as Sommer suggests, was due to the unreliability of the earlier samples of this drug; he claims to have had very good results with forty to sixty grain doses—occasionally combined with bromide of potassium—in agitated and sleepless patients. Unpleasant vasomotor paralyses and congestions have been observed to follow its use, even in young persons; and while this very feature renders it useful in some recent cases of passive melancholia, yet it should be avoided wherever there is reason to suspect a disturbance of the vasomotor regulatory centre, or where the blood-vessels are themselves diseased.

INDEX.

ABORTIVE monomania, 311
Abstraction, 22, 61; in paretic dementia, 186
Absurd beliefs, 21
Absurdity of delusions, 31
Abulia, 64, 252, 349
Accusations by hysterical lunatics, 258; of other lunatics, 398.
Active dementia, 170; organic changes, 123, 178; phase of paretic dementia, 190.
Acts, imperative, 36
Acute hallucinatory confusion, 163; confusional insanity, 161
Acuteness of maniacs, 133, 134
Adhesion of dura to cranium, 228; of pia to cortex, 110, 219
Adventitia, changes of, 106; granular material in, 102, 107
Affect, 55
Affections in paretic dementia, 186
Age of paretic dements, 216; influence of, on hallucinations, 53
Agitated dementia, 170; melancholia, 145
Agoraphobia, 36, 65
Albuminuria, 70, 211
Albutt, Clifford, on paretic dementia, 236
Alcohol, use of, 391
Alcoholic delirium, 254; delusions, 254; excesses in periodical insanity, 270; hallucinations, 253; insanity, 251; paretic dementia, 215; tremor, 252
Alternating consciousness, 59
Amblyopia, 202
Amenomania, 288
Amnesia, 57; diagnostic relations of, 350; in melancholic frenzy, 143; in transitory frenzy, 155; in paretic dementia, 186
Amyl nitrite, use of, 387
Anæmia, 374; in melancholia, 144; of brain, 108
Anæsthesia, 68
Anatomical basis of mind, 101; peculiarities in idiocy, 285; of monomania, 301, 306
Anatomy, morbid, of insanity, 92
Aneurismal changes, 107
Angry excitement, 133
Anomalies, congenital, 77; of cranium, 86
Anorexia, 72
Anthropophagy, 39, 43
Anxious hallucinations, 255
Apathetic dementia, 168; phase of paretic dementia, 189
Ape-fissure, 286
Aphasia, 193
Apoplectiform attacks, 193, 203, 206; albuminuria after, 211; recovery from, 216
Apoplexy, genuine in paretic dementia, 206
Arachnoid, blood cysts of, 109; calcareous plates in, 229; hemorrhage in, 231
Arcus senilis, 174
Argyll-Robertson pupil, 208
Arrested development, states of, 275
Artefacta, 94
Artificial feeding, 394
Ascending type of paretic dementia, 185
Asylum treatment, 397
Asymmetry of skull, 87
Atavism, 278

Ataxia, 73; in paretic dementia, 193, 211
Athetoid movements, 206
Atony in melancholia, 145; in initial period of mania, 137; in stuporous insanity, 158; in katatonia, 152; of intestinal tract, 72
Atrophy of brain, 103; of optic nerve, 236
Attitude of insane, 74; of alcoholic subjects, 253; *orgueilleuse*, 299; frozen, 72
Aura-like prodromata, 269
Austin on the pupil, 207
Automatic acts in insane, 58, 75
Axis cylinders in paretic dementia, 223

BAILLARGER on cortical changes, 219; epileptic character, 259; on *folie à double forme*, 272; on monomania, 288
Basis of mental co-ordinations, 101
Baths, use of, 386
Bedsore, malignant, 213
Beliefs, absurd, 21
Bell's typhomania, 249
Benedict on hallucinations, 68
Billod on stramonium, 389
Bladder disturbances, 211
Blanching of hair, 78
Blandford on causation, 380; on monomania, 291
Blind, insanity in the, 49
Blood in insanity, 69
Bloodcysts, 109, 230
Blood-stasis, 224
Blood supply in relation to hallucinations, 47, 53
Blood-vessels, changes of, 106, 224, 227, 232
Bones, changes of, 78
Bony plates in arachnoid, 228
Brain in insanity, 100; defects of, 90, 285; cysticerci in, 112; hyperæmia and anæmia of, 108; membranes of, 109; wasting, 164
Brierre de Boismont on extravagances of paretic dements, 187, 192
Bromides, use of, 385
Bromism, 339
Brown on brain wasting, 164
Brutality of paretic dements, 197

Bucknill on beliefs of imbeciles, 277; and Tuke on monomania, 290
Bulimia, 72, 140; in paretic dementia, 202

CALCAREOUS plates of arachnoid, 229
Calcification of nerve-cells, 104
Calmeil on epileptic character, 259
Campagne on reasoning mania, 136
Cannabalistic tendencies, 42
Cannabis indica, uses of, 389
Cardiac disorder, 375
Cataleptic periods of katatonia, 152
Cathartics, uses of, 386
Causes, somatic, of insanity, 369; psychical, of insanity, 381; of idiocy, 276; of monomania, 301; of transitory frenzy, 157; of stuporous insanity, 159; of delirium grave, 247; of primary deterioration, 164
Cavities of the neuroglia, 106
Cerebral tissues, changes in, 101
Change, sudden, of delusions in paretic dementia, 199
Character, epileptic, 260; in circular insanity, 274; in periodical insanity, 269; in hysterical insanity, 257; change of, 178; in paretic dementia, 185; maniacal, 136
Chloral, uses of, 385
Choreic insanity, 373
Choreomania, 373
Chronic alcoholic insanity, 251; confusional insanity, 170; delusional insanity, 283; hysterical insanity, 256; mania, 170
Circular insanity, 271
Civilization, influence of, on paretic dementia, 182
Clarke, Lockhart, on cystic degeneration, 219
Classification of insanity, 113
Claustrophobia, 36
Climacteric insanity, 122
Clivus, anomalies of in pyromania, 87
Complicating insanities, 121, 124, 129
Complication of one form of insanity by another, 128; of monomania, 319, 345

Concealed delusions, 323; insanity, 352
Conceptions, imperative, 35
Concussion, 370
Confession of imaginary crimes, 141
Confusion of ideas, 162; acute hallucinatory, 163
Confusional insanity, primary, 161; secondary, 170; morbid changes in, 102
Congenital anomalies, 77; monomania, 308
Congestive attacks of paretic dementia, 207
Conium, uses of, 384
Connecting links of degenerative series, 88
Connection between lesions and symptoms, 113
Consciousness, disturbances of, 57; alternating, 59; double, 59; of impending loss of reason, 61
Convalescence in mania, 139; in Katatonia, 153
Convulsions, 351
Convolutions, atrophy of, 103; in paretic dementia, 218
Coprostasis, effects of, 112
Corpus callosum in imbeciles, 286
Corrugation of brows in paretic dementia, 210; in monomania, 339
Cortex, changes in, 101; in paretic dementia, 218; adhesion of pia to, 219, 231; cystic degeneration of, 219; in imbecility, 285
Crampi in alcoholism, 252
Crania progenia, 87, 282
Cranial diploë in paretic dementia, 228
Cranium, changes of, in insanity, 86
Cretenism, 282
Crib of Utica, 401
Criminal acts of insane, 398; of paretic dements, 187
Cyclothymia, 271
Cystic degeneration of cortex, 219
Cysticerci, 112
Cystitis in paretic dementia, 211

DAGONET on morbid anatomy, 102
Dangers of insanity, 398
Death in melancholia, 148; in paretic dementia, 193

Decubitus, malignant, 78, 213
Defects in idiotic brain, 90
Definition of circular insanity, 271; of delusion, 24, 29; of dipsomania, 271; of hallucination, 43; of illusion, 44; of insanity, 17, 19; of katatonia, 149; of mania, 131; of melancholia, 141; of monomania, 301; of paretic dementia, 241; of periodical insanity, 267; of primary confusional insanity, 161; of primary deterioration, 163; of pubescent insanity, 175; of senile dementia, 171; of stuporous insanity, 158
Degeneration in monomania, 293
Degenerative changes in brain, 102; series of forms of insanity, 88
Deglutition, impairment of, 211
Deliberation of the insane, 64
Délire des actes, 270
Delirium, 341; acutum, 247; of epileptic insanity, 264; grave, 80, 247; of mania, 133; of melancholic frenzy, 143; of monomania, 296; of paretic dementia, 192, 203; tremens, 251; hysterical, 217
Delusion, definition of, 24
Delusional insanity, 288; monomania, 312
Delusions of alcoholic insanity, 254; of confusional insanity, 162; diagnostic bearing of, 340, 348; in delirium grave, 248; detection of, 327; depressive, 28, 141; erotic, 27, 300; expansive, 27, 298; genuine, 25; of grandeur, 32, 248; hypochondriacal, 26; of mania, 138; of marital infidelity, 33; mechanism of, 28; of melancholia, 141; modifying influence of external circumstances on, 34; of visceral impressions on, 33; of paretic dementia, 195, 199; persecutory, 26; religious, 27, 304; rudimentary, 36; of senile dementia, 173; spurious, 26; systematized, 25, 288; unsystematized, 25, 31,
Dementia agitated, 170; brain in, 108; epileptic, 259; frequency of, 168; nerve cells in, 104; neuroglia in, 106; paretic, 178; para-

lytic, 179; passive, 168; senile, 171; syphilitic, 243; terminal, 166; trophic changes in, 171; varieties of, 119, 120
Depressed states, differential diagnosis of, 330
Depressive delusions, 141; period of mania, 137
Descending type of paretic dementia, 214
Destruction of nerve-cells, 105, 223
Detection of simulation, 364
Deterioration, primary, 163; secondary, 166; of nerve cell protoplasm, 222
Development, states of, arrested, 275
Diagnosis of insanity, 320, 330; of circular insanity, 273
Diarrhœa, 72
Diet of insane, 393
Digitalis, uses of, 388
Differential diagnosis of the various forms, 330
Diminutives, use of, in katatonia, 152
Diploe in paretic dementia, 228
Diplopia, 240
Dipsomania, 37, 271
Discoloration of cortex, 219
Disseminated sclerosis, 223
Diurnal change of symptoms in melancholia, 149
Double consciousness, 59, 163
Down, on ethnic types in idiocy, 279
Dreams causing delusions, 33; in primary deterioration, 164
Drunkenness, influence of, on progeny, 83
Dura mater, inflammation of, 109, 228
Duration of mania, 138; of melancholia, 147; of paretic dementia, 216
Dysuria spastica, 72

ECSTATIC states, 258
Effluvium in mania, 72
Ego, 60
Egotism in senile dementia, 172; in monomania, 309
Electricity, uses of, 393
Electro-muscular reactions, 67

Emminghaus on exophthalmic goître, 376
Emotional disturbance, 56; insanity, 55; disturbance in mania, 132; in melancholia, 140; tremor, 209; diagnostic importance of, 349
Emotions in imbeciles, 281
Endyma, changes of, 220
Engelhorn on transitory frenzy, 156
Ependyma layer in imbecility, 285
Epilepsy, relation of, to periodical insanity, 91; in imbeciles, 282;
Epileptic insanity, 258: dementia, 259; morbid anatomy of, 99, 102, 107; chronic, 261; intervallary, 261; handwriting in, 266
Epileptiform attacks in paretic dementia, 193, 203, 205; pathology of, 227
Episodical attacks of paretic dementia, 203; pathology of, 237; delirium in monomania, 296
Epithelial granulations of pia, 109
Equivalent, psychical epileptic, 259
Erotic delusions, 27; monomania, 317
Erotomania, 27, 300
Erlenmeyer on syphilitic hypochondriasis, 243
Errors of pathologists, 93; of Utica school, 95; of sane as differing from delusions, 28
Esquirol on epileptic dementia, 259; on monomania, 287; on morbid anatomy, 96
Etat criblé, 107
Ethnic types in idiots, 278
Etiology of insanity, 369, 381; of paretic dementia, 240
Etiological forms, 115, 370-379
Examination of the insane, 320
Excesses of paretic dements, 190
Exhaustion, maniacal, 139
Exhileration of paretic dements, 192
Expenditures, extravagant, 192
Explosion of paretic dementia, 190
Exposure, indecent, 188
Expression of the insane, 74, 322
Extravagant projects, 198

FACIAL appearance of alcoholic subjects, 253

Factors determining delusions, 31
Faithful memory of insane, 58
Falret on epileptic insanity, 260; in folie circulaire, 271
Fatty changes of nerve-cells, 105
Feeble-mindedness, 275
Females, paretic dementia in, 196, 217; periodical insanity in, 376
Fever, secondary, of syphilis; insanity in, 244
Fibrillary change of neuroglia, 105, 106; tremor, 337
Flemming on recovery from paretic dementia, 216
Flight of ideas in mania, 132
Fluxionary states in insanity, 107
Folie à double forme, 272; *à deux*, 367; *circulaire*, 271; *communiquée*, 367; *du doute*, 308, 311; *raissonante*, 65
Fournier on pseudoparalyses, 246
Foville on cranial deformity, 87; on transitory frenzy, 155
Free nuclear bodies, proliferation of, 225
Frenzy, melancholic, 142; transitory, 154; melancholic, 143
Frequency of mania, 138; of melancholia, 148; of senile dementia, 175; of pubescent insanity, 177; of paretic dementia, 180
Fright, precordial, in melancholia, 143
Frozen attitude, 72
Furloughs for insane, 401
Furor, 135; maniacorum, 135; of paretic dementia, 192, 204; transitory, 55
Fürstner on cortical lesions in paretic dementia, 243
Fury, pathological, 55,

GAIT in paretic dementia, 210
Galloping paretic dementia, 203
Ganglionic bodies of cortex, changes in, 104; in paretic dementia, 221
Gangrene, pulmonary, 80, 213
Gauster on recovery in paretic dementia, 216
Generosity of paretic dements, 194
Goitre exophthalmic, 376
Gouty insanity, 374
Grand mal intellectuel, 260
Grandeur, delusions of, 336

Granular matter in adventitia, 102, 107; wasting, 222
Granulations, epithelial, of pia, 109; of ventricular endyma, 220
Gray, L. C., on temperature, 71
Griesinger on *Grübelsucht*, 36, 58, 311; on primary and secondary forms, 289
Ground-glass appearance, 221
Grössenwahnsinn, 295
Grübelsucht, 36, 65, 311
Gruyère cheese appearance, 219
Güntz on senile dementia, 174
Grave delirium, 247
Gyri of idiot's brain, 278

HÆMATOMA of ear, 79; of dura, 230
Hæmaturia in paretic dementia, 211
Hallucinations, 43; definition of, 43; visual, 50; auditory, 51; gustatory and olfactory, 53; unilateral, 52; in paretic dementia, 200; in katatonia, 152; in melancholia, 142; in mania, 134; in monomania, 313, 316, 318; in alcoholic insanity, 253; differential diagnostic relations of, 349; in the sane, 21
Hallucinatory mania, 134; confusion, 163
Hammond, on mysophobia, 36; on extravagances in paretic dementia, 192, 203; on Kalmuck idiocy, 278
Handwriting of insane, 76; in paretic dementia, 210; in monomania, 76; in epileptic insanity, 266
Headache in katatonia, 152; in paretic dementia, 338
Head injuries, 370
Head sensations in melancholia, 144; in confusional insanity, 163
Heart disease in insanity, 375
Hebephrenia, 176
Hecker on hebephrenia, 176
Hæmorrhage of brain in paretic dementia, 206
Hæmorrhagic pachymeningitis, 229
Hereditary transmission, 275; modus of, 83; transformations in course of, 91
Herpetic eruptions, 78
His, perivascular space of, 106, 227

Homicidal impulse, 37; in melancholia, 146; monomania, 36
Howard on temperature of insane, 69, 71
Hughes on simulation, 358
Huguenin on durhæmatoma, 229
Hyaline thrombi, 527
Hyoscyamia, uses of, 389
Hyperæmia of optic disc, 67, 236, of brain, 108; of cortex, 239; in delirium grave, 250
Hyperæthesia, 68; in alcoholic insanity, 212
Hyperalgesia, 68
Hyperbulia, 64, 349
Hypochondriacal delusions in melancholia, 141; in monomania, 316
Hypochondriasis, syphilitic, 243
Hypocrisy in hebephrenia, 176
Hypomania, 136; like condition in paretic dementia, 203
Hysterical insanity, 256

IDENTITY, changed sense of, 60; in confusional insanity, 162; in mania, 134
Idiocy, 275; brain defects in, 90; gyri, 278; varieties of, 276; atavisms in, 278; ethnic types of, 278; instincts in, 279
Illusions, 43; definition of, 44; of sight, 50; of hearing, 51; of smell and taste, 53; of identity, 50; in paretic dementia, 200; dependence of on tinnitus aurium, 53
Imbecility, 275, 280; moral, 281; partial, 281; anatomical defects in, 90, 285
Imitative tendencies of idiots, 280
Imperative conceptions and acts, 35, 36, 349; movements, 75
Impression, maternal, influence of on progeny, 84
Impulse, homicidal, 37; morbid, 36, 270
Inconsistency of unsystematized delusions, 33; in hebephrenia, 176; in paretic dementia, 195
Increased frequency of paretic dementia, 183
Indecent exposure, 187
Inebriety, 252

Inequality of pupils, 207, 336
Inflammatory process in paretic dementia, 218, 238
Influence of insanity on disease, 80
Inhibition, logical, 29; in mania, 132; in melancholia, 141
Initial stage of mania, 137; of melancholia, 147; of katatonia, 149; of paretic dementia 184
Injuries of head, 369
Insane attitude, 74; expression, 74, 321; manner, 74, 321
Insanity, definition of, 17, 19; circular, 271; delusional, 288; dissimulated, 353; intellectual, 293; morbid anatomy of, 92; moral, 281; primary, 289; secondary, 289; simulated, 352; somatic signs of, 81; trophic disturbances of, 77
Insanity of pubescence, 175; propensities of, 176; frequency of, 177
Insolation, 369, 371
Insomnia in primary deterioration, 164; treatment of, 385
Instincts in idiots, 279
Intellectual labor as a cause of insanity, 382.
Intellectual insanity, 293
Intensification of neurotic vices in transmission, 85
Intervals of paretic dementia, 193
Intervallary epileptic insanity, 261
Intestinal tract in insanity, 71
Interstitial encephalitis, 221
Ireland on varieties of idiocy, 276
Iron, uses of, 392
Irritability in paretic dementia, 188, 189
Isthmus affections in paretic dementia, 241

JACOBI of Germany, somatic theory of, 71
Jessen on delirium grave, 249
Johnson, case of, 75
Jolly on hallucinations, 68

KAHLBAUM on katatonia, 149
Kalmuck idiocy, 278, 358
Katatonia, 149; hallucinations in, 152; prognosis and frequency

of, 153; morbid anatomy of, 98, 102, 230
Kiernan on othæmatoma, 79; on katatonia, 149; on transitory frenzy, 155, 156, 157; on athetoid movements in paretic dementia, 206; on trophic changes, 212, 232; on simulation, 358; on use of ergot, 387
Kinking of blood-vessels, 107, 226
Kirchhoff on original monomania, 307
Kirn on pupillary spasm, 75
Kleptomania, 37, 270
Koster on periodical insanity, 268
Krafft-Ebing on stigmata, 89; on degenerative forms, 116; on morbid anatomy, 97; on classification, 117; on hysterical insanity, 257

LACUNÆ of memory in paretic dementia, 192; in syphilitic dementia, 244
Lamination of cortex, obliteration of, 101
Lasègue and Garel, on crania of epileptics, 87
Le Grand du Saulle on stigmata, 89; on *folie du doute*, 311
Legal aims of a definition of insanity, 19
Legal insanity, 23
Leptomeningitis, 231
Lesbian love, 42
Letterwriting tendency, 210
Leucin precipitates, manufactured at Utica, 95
Leuret, illusion of, 53
Lymphatic flow, retardation of, 102, 106, 108
Lypemania, 287
Locomotion disturbances, 350
Locomotor ataxia in paretic dementia, 210; its analysis, 240
Logical inhibition, 29
Lubimoff on paretic dementia, 223
Lucid intervals in paretic dementia, 193; in mania, 139

MAGNETIC illusions, 54; in paretic dementia, 200
Malarial insanity, 374

Maldevelopment after organic affections, 82
Malignant decubitus, 213
Mania, definition of, 131; delusions in, 134; depressive stage of, 137; duration of, 138; errabunda, 181; exhaustion in, 139; chronic, 288; frequency of, 138; in puerpero, 133, 377; identity, change of sense of, 134; hallucinations in, 134; hallucinatory, 134; gravis, 247; morbid anatomy of, 97; melancholic, 137; periodical, 269; prognosis of, 138; recurrent, 140; transitory, 55, 154; typical, 131
Maniacal excitement, 120; in katatonia, 152; fury, 135; furor in paretic dementia, 203; morbid anatomy of, 227
Manie grave, 123; *raissonante*, 65; *systematisée*, 288
Manner, insane, 74
Mannerisms of insane writings, 77
Marasmus of nerve-tissue, 172
Marcé on monomania, 288
Marital infidelity, delusions of, 33, 254
Marks of insane resembling bruises, 80
Masturbation, 378; in hebephrenia, 176, 177
Masturbatory insanity, 54, 379
Maternal impressions, 84; preponderance of influence of, 82
Maudsley on classification, 114
Mechanism of hallucinations, 44
Medicinal treatment of insanity, 384; means for detecting simulation, 368
Medico-Psychological Association, classification proposed by committee of, 114
Megalomania, 200, 295
Melancholia agitata, 144; anæmia in, 114; attonita, 145; cum stupore, 145; delusions in, 31, 141; definition, 140; duration, 147; ex lactatio, 377; from masturbation, 379; frenzy in, 142; frequency of, 147; hallucinations in, 142; head-symptoms in, 144; homicide in, 146; in puerpero, 378; mild, 146; maniacal, 142; morbid anatomy of, 98; progno-

sis of, 148, 149; periodical, 271; precordial fright in, 143; pulse in, 143; sine delirio, 69. 146, 148; suicide in, 146; self-mutilation in, 142; weight in, 144; without delusions, 146

Memory, disturbances of, 57; in melancholia and mania, 140; in imbecility, 281; in dementia, 165; in paretic dementia, 192

Mendacity in paretic dementia, 198; 215

Mendel on mania, 97; on hypomania, 136; on hallucinatory mania, 134; on paretic dementia, 231; on hæmatoma, 230

Meningitis, 370; from over-study, 248.

Meningo-myelitis, 232

Menstrual insanity, 267, 376

Metallic poisons, 374

Meynert, on hallucinations, 46; on melancholia, 98; on brain-weight in insanity, 109; on paretic dementia, 223

Mickle on paretic dementia, 197, 199, 201

Microgyria, 286

Micromania, 200

Mild melancholia, 146

Miliary aneurisms, 107; sclerosis, 95

Milky opacity of leptomeninges, 109; 231

Mind, anatomical basis of, 101

Miserly inclinations in paretic dementia 190; in senile dementia, 172

Monomania, abortive, 311; abuse of term, 287; brain-defects in, 90; congenital, 308; definition of, 300; delusional, 312; diagnosis of, 335; erotic, 317; handwriting in, 76; homicidal, 36; history of, 286; masturbatory 380; morbid anatomy of, 99. 110; persecutory, 299,314; prognosis of 318; querulous, 315; reasoning, 309; religious, 300; relation of, to imbecility, 88, 281; secondary, 170; sine delirio, 309

Monomanie gaie, 288; *triste*, 288; *vaniteuse*, 295

Moos on hallucinations of hearing, 53

Moral insanity, 56, 281; imbecility, 56, 281; perversion, 56; deterioration in paretic dementia, 190

Morbid anatomy of insanity, 92; of paretic dementia, 218; of delirium grave, 250; impulses, 36, 270; propensities.35, 38

Morel on stigmata of heredity,89; on classification, 115; on monomania, 288

Moria, 341; post-epileptic, 269

Motor disturbances, 73; in katatonia, 150; in paretic dementia, 191, 242, 209; in imbeciles, 282

Motor excitement in mania, 135; as a factor in differential diagnosis, 335

Mouvements en manège, 206

Movements, imperative, 75; rhythmical, 75

Multiple sclerosis, 214

Muscular coat of arteries in paretic dementia, 225

Musky odor of idiots, 73

Mutism, 255, 337

Mysophobia, 36, 65

NASSE on pupil in paretic dementia, 208

Necrobiosis of ganglionic bodies, 221

Necrophilism, 43

Negative pathology of mania, 97

Nerve-cells, atrophy of, 103, 104, 221; destruction of, 103; pigmentation of, 102; in acute delirium, 104, 250; in mania and melancholia, 104

Neumann on paretic dementia in females, 217

Neuroglia, changes in, 105; nuclei of, 105; in paretic dementia, 223

Neuroses, dependence of insanity on, 124

Neurosis, traumatic, 371; hysterical, 256; epileptic, 258; alcoholic, 251

New formation of vascular channels, 225

Nitrite of amyl, uses of, 387

Nuclear bodies of neuroglia, 105 in. imbeciles, 286; proliferation of, in adventitia, 106

INDEX. 419

Nutritive disturbances in insanity, 104
Nymphomania, 27, 39, 380

OBJECTIVE surroundings, recognition of, by the insane, 62
Oblongata, changes of, in paretic dementia, 223
Obscenity of maniacs, 132
Occipital headache in katatonia, 152; lobe in paretic dementia, 202, 243; in imbeciles, 286
Oddities of behavior in paretic dementia, 215; of speech in insane, 351
Oophorectomy, 376
Opium, uses of, 388; insanity from, 254, 255, 271
Optic papilla in paretic dementia, 235
Organic changes in alcoholic insanity, 128, 251; in senile dementia, 128
Originäre Verrücktheit, 301
Osteophytes of sella turcica, 87
Othæmatomata, 78
Outbreak of mania, 137; of delirium grave, 248; of periodical mania, 269
Over-study, effects of, 248

PACCHIONIAN bodies, 109
Pachymeningitis, 229; hemorrhagica, 229
Papilla optic, in paretic dementia, 235
Paralucid intervals of paretic dementia, 193
Paranoia, 286, 301, 341.
Parental influence, 82
Paresis, 73
Paretic dementia, 178; abstraction in, 186; affections changed in, 186; age of subjects of, 216; alcoholic form, 215; amblyopia in, 202; apathy in, 189; aphasia in, 193; apoplectiform attacks in, 193, 203, 206; ataxia in, 191, 193; bladder disturbance in, 211; brutality in, 197; bulimia, 202; causes of, 240; change in character of delusions of, 199; character, peculiarities in, 185; complicating other forms of insanity, 338;

course of, 184; decubitus in, 213; deglutition impaired in, 211; delusions of, 192, 195, 196, 197; diagnosis of, 338; duration of, 216; epileptiform attacks in, 193, 203, 205; episodial attacks of, 203; exhilaration in, 192, 194; explosion of, 190; extravagant expenditures in, 192; female cases of, 196, 217; frequency of, 180; furor in, 204; gait in, 210; galloping form of, 203; generosity in, 194; hallucinations in, 200; handwriting in, 209, 210; hemorrhage in, 206; hypochondriacal, phases of, 200; initial period of, 184; irritability in, 189; isthmus changes in, 241; letter-writing tendency in, 210; lucid intervals of, 193; maniacal attacks in, 203; mendacity in, 198; miserliness in, 190; morbid anatomy of, 99, 100, 218; moral deterioration in, 190; motor disturbance of, 209; neuroglia in, 223; oblongata, changes in, 223; optic papilla in, 235; perivascular spaces in, 107; pia in, 231; pons, changes in, 218, 223; physical signs of, 190; prognosis of, 215; prodromal period of, 184; projects in, 192, 198; ptosis in, 209; pulse in, 212; pyonephritis in, 211; relation of lesions to symptoms of, 236, 243; relation of physical and mental signs, 178, 180; remissions of, 193, 214; sclerosis in, 214; sensory disturbances in, 201, 211; septic complications of, 213; speech disordered in, 209; spider-shaped cells in, 223; spinal cord in, 231; spinal symptoms of, 240; suffocation in, 210; stages of, 184; temperature in, 212; tremor in, 209; trophic disturbances of, 193, 212; varieties of, 214; vasomotor disturbances in, 212; white substance, changes of, in, 223
Partial insanity, 90, 287; punishability, 52
Parturient state, 378
Passive dementia, 170
Pathological fury, 55; results of insane excitement, 110 111

Pathos in katatonia, 150
Pellagrous insanity, 124, 125
Pelman on pupil in paretic dementia, 208
Pemphigus in delirium grave, 250
Penuriousness in senile dementia, 172
Periganglionic spaces, 106; in paretic dementia, 220
Periodical insanity, lesions of, 107; relation of, to epilepsy, 91; definition of, 267; mania, 269; melancholia, 271; morbid impulses 270; prognosis of, 274
Peritonitis, influence of, on delusions, 33, 34
Perivascular spaces, 106, 220
Persecutory delusions, 26; monomania, 318
Perversion, moral, 56; sexual, 39
Petit mal intellectuel, 260
Pfleger on brain-weight, 109
Phosphates, loss of, in insanity, 69
Phosphorus, uses of, 69, 392
Photopsia in relation to hallucinations, 255
Phrenitis, 247
Phthisical insanity, 375
Physical signs of insanity, 65; of senile dementia, 174; of paretic dementia, 207; of alcoholic insanity,
Pia, changes of, in insanity, 109, 110; in paretic dementia, 231
Pigmentation of nerve-cells, 102, 104, 222
Plagiarism committed by the insane, 27
Planning of insane acts, 64
Plates, bony, in membranes of brain, 228
Ponicaré and Bonnet, theory of, as to paretic dementia, 235
Pons, changes in paretic dementia,
Popular idea of insanity, 359
Porencephaly, 384
Post-epileptic insanity, 260; stupor, 260; moria, 261
Post-febrile insanity, 373
Post-mortem changes in the brain, 95
Precordial fright, 143
Predisposition to insanity, 382; signs of, 81

Pre-epileptic insanity, 261
Pregnancy, morbid propensities in, 38; insanity in,
Preponderance of maternal influence in heredity, 82
Primary confusional insanity, 161; prognosis of, 163
Primary dementia, 160, 177, 379; insanity, 122, 289; mental deterioration, 163
Primordial-Delirien, 292
Principles of classification, 118, 119
Prisoners, insanity in, 49
Prognosis in alcoholic insanity, 256; in hebephrenia, 177; in hysterical insanity, 258; influence of hæmatoma on, 79; in katatonia, 153; in mania, 138; in melancholia, 148; in masturbatory cases, 379; in monomania, 318; in periodical insanity, 279; in paretic dementia, 215; in primary confusional insanity, 169; in senile dementia, 174; in stuporous insanity, 160.
Progressive Paralyse, 179
Progressive Paresis, 180
Projects of the insane, 27
Propensities, morbid, 35, 38
Pseudo-emotional states in paretic dementia, 209
Pseudo-monomania, 288
Pseudoparalysie générale, 246
Psychical degenerative states, 321; causes of insanity, 381; treatment, 397
Psychoneuroses, 121
Ptosis in paretic dementia, 209; in neurotic subjects, 339
Puerperal state, insanity of, 377
Pulse in melancholia, 143; in paretic dementia, 212
Pulsus tardus, 70
Punishability of the insane, 52
Pupil in insanity, 75; in paretic dementia; Argyll-Robertson, 208; diagnostic importance of, 336
Pure insanity, forms of, 121
Pyonephritis in paretic dementia, 211
Pyromania, 37; cranial deformities in, 87

Querulanten-Wahnsinn, 315

INDEX. 421

Querulous monomania, 315
Quiet type of paretic dementia, 214, 217

RACE, question of, in relation to paretic dementia, 181
Radiant heat as a cause of insanity, 371
Raptus melancholicus, 143
Rarefication of neuroglia, 223
Reactions, electro-muscular, 67
Reasoning of maniacs, 133; monomania, 309
Recurrent mania, 140
Reflexes in stuporous insanity, 159
Refusal of food, 394
Reich on transitory frenzy, 157
Reinhard on septic complications of paretic dementia, 213
Relation of skull-shape to insanity, 86, 87
Religious delusions, 27; monomania, 304
Remissions, 400; of paretic dementia, 214
Restraint, 401
Retardation of lymph-outflow from cortex, 102, 106, 108
Retina in insanity, 66
Retinal after-image, analogy of, to memory, 44
Reversion of idiots to apes, horses, sheep and other animals, claim of, 276, 277
Rheumatic insanity, 374; causation of paretic dementia, 215
Rhythmical movements in insane, 75; in delirium grave, 249; in katatonia, 151
Richarz on parental influence, 82
Ripping on insanity in puerpero, 377; on syphilitic psychoses, 245
Rush on moral imbecility, 56; on nomenclature, 288

SALIVA, dribbling of, 159
Samt on post epileptic insanity, 260
Sander on *originäre Verrücktheit*, 308
Sankey on vascular kinking, 226; on monomania, 291
Satyriasis, 39, 173
Schopenhaur, case of, 92
Schüle on delirium grave, 247; on

frequency of paretic dementia in females; on hyoscyamime, 389; on morbid anatomy of monomania, 99; on recovery of paretic dementia, 216
Sclerosis of cortex, 219; of oblongata, 223; of cord, 232; cerebrospinal, 216, 223, 240; miliary, 95; of optic papilla, 236
Secondary fever of syphilis, insanity with, 243; deterioration, 166; insanity, 122, 289; insanity with confusion of ideas, 170; partial insanity, 170
Self-consciousness, 60
Self-mutilation in melancholia, 145
Sella turcica in pyromania, 87
Senile dementia, 171; anatomical changes in, 106; definition of, 171; delusions in, 173; hyperæsthesia in, 174; frequency of, 175; memory in, 172, 173; insanity, 171
Sensibility, disturbances of, 68; in paretic dementia, 211
Sentiments distinct from emotions, 55
Sex, relation of, to mania, 138; to melancholia, 148; to paretic dementia, 217; to periodical insanity, 274
Sexual ideas in paretic dementia, 196; mutilation, delusion of, 254; perversion, 39, 308; sensations in hysterical insanity and monomania, 258, 312
Shepard's definition of insanity, 221
Shrinkage of pyramidal bodies of cortex, 221
Simon on paretic dementia, 191, 197, 202
Simulation, 352; by hebephreniacs, 176; by hysterical patients, 257; by masturbatory lunatics, 379; by the insane, 357
Sitophobia, 394
Skae's classification, 115
Skin in insanity, 72, 78, 363
Skull, condition of, in paretic dementia, 228; deformity of, 86; asymmetry of, 87, 305
Snell on syphilitic paretic dementia, 244; on *Verrücktheit*, 244
Somatic etiology of insanity, 369.

signs of insanity, 81; disturbances in alcoholism, 253
Special senses, illusions and hallucinations of, 50
Speech in insanity, 76; disturbances, 350; in paretic dementia, 191, 209; in idiots, 280
Spider-shaped cells in paretic dementia, 223
Spinal cord in paretic dementia, 231–235; lymph-spaces of, 220; type of paretic dementia, 185
Sphygmograph, revelations of, 70, 213
Stages of paretic dementia, 184
Stasis, thrombic, 229, 239
States of arrested development, 275; epileptic mental, 259
Stearns on electrical basis of mind, 393
Stigmata of heredity, 86, 89, 279, 350
Stomach pump, use of, 395
Stramonium, use of, 389
Strychnia, use of, 389
Stuporous insanity, 120; definition, 158; prognosis of, 160
Suffocation in paretic dementia, 211
Suicidal impulse, 37
Suicide in melancholia, 141, 146
Summarische Erinnerung, 59
Sutherland on the blood in insanity, 69
Symptomalogy of the various forms of insanity, 131–329
Syphilitic dementia, 244; hypochondriasis, 243
Systematized delusions, 25, 295, 314

Tædium vitæ in paretic dementia, 179
Taint, constitutional, signs of, 81
Temperature in insanity, 71; in paretic dementia, 212; in delerium grave, 249
Terminal deterioration, 166; dementia, 169
Tetanic condition of muscles in insanity, 74
Theatrical behavior in katatonia, 150, 151
Thefts committed by paretic dements, 187

Theomania, 300
Thrombic stasis, 227
Thunderstruck melancholia, 145
Thyroid gland in cretinism, 284
Touch-sense, disturbances of, in paretic dementia, 201; transformation of psychosis in hereditary transmission, 91, 275; of delusions in monomania, 316
Transition of mania to melancholia in circular insanity, 274
Transitory frenzy, 154, 377; furor, 55; mania, 55, 154
Transmission of hereditary vices, modus of, 83, 84, 275; intensification of neurotic, vice in, 85, 86
Traumatic insanity, 370
Tremor, 351; ataxic, 209; emotional, 209; paretic, 209; alcoholic, 252
Treviranus on brain in sleep, 47
Tristimania, 288
Trophic disturbances in insanity, 77, 350; in paretic dementia, 193, 212; in delirium grave, 249
Tuke, D. Hack, on classification, 115; on artificial feeding, 395
Turnbull on paretic dementia in the young, 216
Typhomania, 247

Ullrich, case of, 40
Undue influence in senile dementia, 173
Unilateral hallucinations, 52
Unsystematized delusions, 25, 31, 195, 197
Urine in insanity, 69; in paretic dementia, 212
Utica crib, 401; spurious pathology of school at, 95, 96

Van der Kolk on sympathetic insanity, 71
Varieties of idiocy, 276; of paretic dementia, 214; of mania, 136; of melancholia, 144; of monomania, 295, 314; of periodical insanity, 269
Vascular walls, changes of, 106; strain, effects of, 107, 110
Vaso motor condition in paretic dementia, 212, 237
Venesection, indications for, 386

Ventricles in paretic dementia, 220; granulations of, 220
Verbigeration in katatonia, 151
Verrücktheit primäre, 289; *originäre*, 301
Vesicular degeneration of cord, 233
Violence as a cause of othæmatoma, 78
Virchow on durhæmatoma, 229
Visceral sensations as basis of illusions, 54; lesions in insane, 111, 112
Visions, 341; in paretic dementia, 236; monomania
Vogt on idiots, 275
Voisin on classification, 116, on anosmia in paretic dementia, 190; on pulse in paretic dementia, 212
Vulgarity of paretic dements, 190

Wahnsinn, 289
Wasting of brain, 164; in paretic dementia, 218
Weakness, mental, in its differential diagnostic relations, 350

Weight of brain, in insanity, 108; of body in circular insanity, 273; in melancholia, 144
Westphal on imperative conceptions, 40; on confusional insanity, 163
White substance of brain, changes in, 103, 223
Will, disturbances of, 64; in alcoholic insanity, 252; in paretic dementia, 188
Winslow on double consciousness, 59
Wolff on pulsus tardus, 70
Word-stoppage in paretic dementia, 207
Wundt on normal temperament, 189

YANDELL on epidemic insanity, 373
Young, paretic demetia in the, 216

ZACCHIAS on recognition of insanity by the insane, 62; on detection of simulation, 364

www.ingramcontent.com/pod-product-compliance
Lightning Source LLC
Chambersburg PA
CBHW050844300426
44111CB00010B/1125